Biology and the Mechanics of
the Wave-Swept Environment

MARK W. DENNY

~~~~~~~~~~~~~~~~~~~

Biology and the Mechanics of the Wave-Swept Environment

~ ~ ~ ~

Princeton University Press

Published by Princeton University Press, 41 William Street, Princeton,
New Jersey 08540
In the United Kingdom: Princeton University Press, Guildford, Surrey

Library of Congress Cataloging-in-Publication Data

Denny, Mark W., 1951–
    Biology and the mechanics of the wave-swept environment.

    Bibliography: p.   Includes index.
    1. Seashore biology. 2. Wave mechanics. I. Title.
QH541.5.S35D46 1988 574.5'2638'01532593 87-32806
ISBN 0-691-08486-6 (alk. paper)   ISBN 0-691-08487-4 (pbk.)

This book has been composed in Aster by Syntax International

Printed in the United States of America by Princeton University Press,
Princeton, New Jersey

Designed by Laury A. Egan

To My Family

CONTENTS

~ ~ ~ ~ ~ ~ ~ ~ ~ ~ ~

PREFACE

~ ~ ~ ~ ~ ~ ~ ~ ~ ~ ~

In this book we will explore the mechanics of wave-induced water motion and how they affect the lives of organisms along the shore. Because this is a relatively new field of study, there is little biological literature upon which to draw. Therefore, the text deals more with ideas and perspectives than with established facts, and is intended more to teach the reader how to think about the subject than to present a compendium of what is known. Original journal articles are cited where they are directly pertinent and can serve as an entrance to the literature; but this book is not a review, a fact confirmed by the short bibliography. Even within the subject of wave-swept biomechanics, the treatment is less than all-encompassing. Much of the text is focused on those mechanics that are particularly applicable to rocky shores and coral reefs, and sandy beaches have largely been neglected. There are two reasons for this limitation. First, the macrofauna and flora of rocky shores and coral reefs are much richer than those of sandy beaches, and the possibility for applying the concepts of biomechanics is therefore greater on rocks and reefs. Second, I have spent much of my professional life working on rocky shores and coral reefs, and I am therefore more familiar with their characteristics.

This text is directed at several audiences: marine biologists (particularly ecologists) who would like to know more about the mechanisms by which plants and animals live on exposed shores; engineers, physical oceanographers, and fluid dynamicists who would like to see how their knowledge of mechanics can be used to study biology; and scientists in general who would appreciate an offbeat approach to the study of form and function. The presentation has been adjusted to accommodate this diverse audience, and a few of these adjustments require a brief explanation.

First, there is the issue of mathematics. A knowledge of basic physics and fluid dynamics is fundamental to all studies of mechanics in the wave-swept environment, and assimilating and using this knowledge requires that one be comfortable with calculus. In many fields this fundamental mathematical background is assumed, but we biologists are renowned for our aversion to mathematics, and many are the minds that switch off at the sight of an integral sign. As shortsighted as this attitude may be, it is a fact of life with which an author must cope. As a result, the mathematics of the wave-swept environment are presented here in a slightly unconventional fashion. First, I have geared the presentation to those I perceive to be the average biologists: those who have had an introductory course in calculus but whose grasp of the subject has become rusty through lack of use. To a reader with this background the appearance of differentials and integrals might be a nuisance at first, but it should not be totally baffling. Second, I have made a concerted effort to explain the math in a fashion that is easily digested: the important underlying assumptions are stated, many of the basic equations are derived informally, and intermediate steps are included in calculations when it would otherwise not be obvious how I moved from the initial equation to the final result. In large part the imposing number of equations is due to this attempt to present the math in coherent, small steps. I hope that this approach will coax the average biologist into flexing his or her mental machinery. Perhaps this will knock some of the rust loose, lower the activation energy required for the study of biomechanics, and give the reader some appreciation for the tremendous utility of a little mathematics; on the other hand, I hope that those to whom math is second nature will bear with the protracted explanations. Finally, I have assumed that the average biologist has not had a course in ordinary or partial differential equations. Therefore, with few exceptions, the concepts requiring a knowledge of these techniques are presented without derivation.

Next, there is the problem of presenting the organisms. I assume that the reader has a basic knowledge of invertebrate zoology and phycology: if you know what a limpet and a kelp look like, you are in good shape, but if you do not know a sea cucumber from a zucchini, you may want to consult an introductory text on nearshore biology. Carefoot (1977) or Ricketts et al. (1968) are informative and readable.

The diversity of topics covered in this text leads to problems with nomenclature. There is a limited number of Roman and Greek letters, and the same symbol is used with different meanings in different fields. With some effort it would be possible to give each variable and coefficient its own symbol by making unconventional assignments. But this approach would require the reader to transpose symbols when making any reference to the literature—an arduous task. Because this text is intended to serve as an entry into the literature, I have chosen where possible to use the symbols that are conventional for the particular field being discussed. As a result, several symbols are used in multiple ways, but each symbol is carefully redefined when it is given a new meaning. The appendix at the back of the book provides a list of symbols and the equations where they are first used.

ACKNOWLEDGMENTS

~ ~ ~ ~ ~ ~ ~ ~ ~ ~ ~

This book was written backwards. I had always assumed that one wrote a book after teaching a subject for many years, gradually building up a set of notes that could easily be transcribed into a book. But when the urge to teach a course on the mechanics and biology of the wave-swept environment took control of me, I was faced with the lack of an appropriate text. It seemed reasonable at the time to write the text as preparation for the course. Little did I know how large a task it is to write an interdisciplinary book, and how complex are the intertwinings of various fields. I have done my best to ensure that, however casual the explanation, the facts and ideas presented are correct. In this endeavor I enlisted the help of James Gere, Michael Isaacson, Arthur Nowell, Tom Daniel, Robert Guza, John Gosline, Michael LaBarbera, Steve Gaines, Mimi Koehl, Celia Smith, Chuck Baxter, Warren Via, Judith Connor, and Joan Oltman-Shay. They did their best to correct my errors and keep a tight rein on my analogies, for which I am grateful. I am responsible for any remaining errors.

In addition, Steve Vogel helped to streamline the text, and an anonymous reviewer suggested several ways in which the flow of ideas could benefit from reorganization. My understanding of nearshore flows was aided by discussions with Ed Thornton, who provided valuable insight on a number of topics (and, unwittingly, the wire for a number of the experiments).

Thanks are also due to the many others who helped in the preparation of this text: the class that served as guinea pigs for this foray into intertidal biomechanics—Francis Putz, Missy Holbrook, Susanne Lawrenz-Miller, Suzanne Ullensvang, Susan Keen, Steve Katz, Lani West, Michael Judge, Brad Gallien, Petros Levounis, Martha Krump, Val Connor, and Mark Shibata—whose enthusiasm and creativity were a joy to behold; Sharon Nugent, who patiently retyped the text; Emily Carrington, who organized the bibliography; Judith May of Princeton University Press, who stood by the project through thick and thin; Alice Calaprice, the "flexible purist" who polished my prose; and Freya Sommer, who lent her artistic talents and diving experience to the figures in Chapter 2.

Many of the ideas and some of the data presented here were gathered while my research was supported by the Office of Naval Research (Contract No. N00014-79-C-0611), and by the National Science Foundation (Grant No. OCE-83-14591). The wizards of Silicon Valley made this publication project feasible by inventing the word processor.

And finally, I would like to acknowledge the debt I owe to my mentors: my father, Floyd Denny; and Steven Wainwright, John Gosline, and Robert Paine. They helped shape the way I think about science. My wife Susan drew the figures, endured my preoccupation, and taught me how strong a family can be.

Biology and the Mechanics of
the Wave-Swept Environment

CHAPTER 1

~ ~ ~ ~ ~ ~ ~ ~ ~ ~

Introduction: The Need
for Proper Tools

"You can buy trick wrenches which make [the] task easier."
—*Chilton's Repair and Tune-up Guide* (1982)

As a graduate student in biology, I had somehow thought I was prepared. After all, I had dissected my share of pickled invertebrates, and I had thumbed through *Between Pacific Tides* so many times that it became an old friend. With this kind of preparation, I asked myself, how many surprises could there be left in my first trip to a wave-swept rocky shore?

I was on my way to Tatoosh Island, a small dot on the map lying about a half mile off of Cape Flattery, Washington. It is the most northwestern point in the lower forty-eight states and as wave-beaten a piece of real estate as one is likely to find. I was accompanying some friends for a bout of intertidal experimentation and was excited at the prospect of seeing firsthand all the strange organisms I had only read about.

We arrived at high tide and, having stowed our gear, sat on the edge of the cliff for a spell of sightseeing until the tide was low enough to begin work. The Pacific swells rolled in, one after the other crashing over the rocks, and someone mentioned that every spring the foghorn had to be cleared of all the rocks thrown up by the winter surf. The foghorn on Tatoosh is more than 90 feet above the sea. Listening to the stories and watching the waves, I continually reminded myself that the books are full of intricate descriptions of the plants and animals that live down there: kelps that can hold onto the rocks in those breakers without being shredded; sea stars that walk around, hunting prey in the surf. The more I watched, the more certain I became that the books must be describing places less exposed than this. I might be a naive terrestrial biologist, but I was certain that the wave-swept maelstrom down there had to be a desert.

Come low tide the surprises began. Everything in the books was true, but now it seemed that they had deliberately understated their case. As I wandered from one tide pool to the next, I was reminded of a passage from John Steinbeck's preface to *The Log from the Sea of Cortez*: "The exposed rocks had looked rich with life under the lowering tide, but they were more than that: they were ferocious with life." Indeed. Nothing could be less of a desert than the intertidal rocks of Tatoosh Island—acres of mussels and barnacles, orange and purple sea stars, urchins, tunicates, algae of every conceivable shape. The diversity of life was overwhelming. The closer I looked, the more weird and wonderful organisms I found hidden among the kelp and anemones. But how could all these diverse forms survive in such a stressful environment?

This paradox of diversity in the midst of apparent physical adversity is the reason that wave-swept shores have long been recognized as "the most fascinating and the most complex of all the environments of life'' (Yonge 1949). Perhaps the first to take scientific notice were the British disciples of Linnaeus, who in the early nineteenth century scoured the globe for new and exotic organisms to catalog and categorize. In their search, these naturalists were particularly drawn to two habitats: tropical rain forests and wave-swept shores. Both abounded with species new to science, and both piqued the curiosity of those who studied them. But while rain forests became the model system for much of modern biological thought and a classroom for Darwin and Wallace, the seashore became a curiosity of Victorian society. The wave-swept environment was simply too stressful to encourage on-the-spot inquiry, and instead, spurred on by the romantic descriptions of P. H. Gosse, amateur naturalist and taxonomist alike raided the shore for ''specimens'' to grace their aquaria. Safely removed from their dynamic natural environment, wave-swept plants and animals became living knickknacks for genteel society and dissection fodder for evolutionists.

Only in the last thirty years, thanks to the work of J. H. Connell, R. T Paine, and others, have wave-swept shores begun to fill their potential as ''the best training ground for zoologists'' (Yonge 1949). Techniques have been developed that allow ecologists to manipulate experimentally the biological interactions of intertidal communities, and the results have been insightful and important (e.g., Connell 1961a,b; Paine 1969; Sousa 1979). With the advent of modern experimental approaches, the wave-swept environment has begun to reveal itself as an attractive research system (Paine 1977). Many of the organisms are sessile, reasonably accessible, and large enough to be easily identified and tagged without being too large to be manipulated. The rocky substratum provides a convenient, more or less two-dimensional construction surface to which cages and other experimental hardware can be readily attached. The turnover rate

in intertidal communities is sufficiently rapid to complete many experiments in a practical length of time. For instance, wave-swept rocky shores are an ideal system for studying the role of physical disturbance in the maintenance of community structure. Processes such as fire and wind damage that reoccur in temperate forests with a timescale of decades can be modeled by analogy in intertidal mussel beds where wave-induced damage reoccurs with a time scale of years.

However, despite its demonstrated utility as a model system for ecological studies, the potential importance of the wave-swept environment as a research system has not been fully realized. The problem again lies with the stressful nature of the environment. All too often, those aspects of community ecology that cannot be fit into the prevailing hypotheses concerning predation, competition, and recruitment are passed off with vague references to ''wave exposure'' and the role of ''the physical environment.'' Certainly it is valid to lay blame on the environment; nowhere on earth does fluid move as violently as it does along the shore. But there is also little *mechanistic* understanding of how wave-induced water motion affects ecology. Nearshore ecologists simply do not possess the proper tools for understanding the environment in which they work.

And therein lies the reason for this book. Many of the tools required for a quantitative study of the wave-swept environment exist, but only as scattered pieces: wave theory from the physical oceanographers, materials mechanics from the engineers, concepts of turbulent mixing from meteorologists. In this text I draw these pieces together to provide a basic toolbox of concepts and techniques for understanding the interaction between biology and the mechanics of the wave-swept environment.

The text is roughly divided into four parts. Chapter 2 is an introduction to wave-swept organisms from which biological examples are chosen in the rest of the book. Chapters 3 to 15 are an introduction to the various areas of mechanics relevant to the biological study of the wave-swept environment. I have tried to include the discussion of biological examples in this exposition of mechanics, but this has not always been possible. The study of near-shore biomechanics is a relatively recent and underexploited field, and there simply are not very many examples to cite. As a result, Chapters 3 to 15 are heavy on physics and light on biology. However, the reader's perseverance will be repaid; the mechanical tools developed in these chapters are ultimately put to use in a biological context. Chapters 16 to 18 are a synthesis wherein the knowledge gained in the previous chapters is used to examine the mechanical determinants of size, shape, and longevity in wave-swept plants and animals. Finally, Chapter 19 is a brief review of the instruments and paraphernalia available for measuring various aspects of the nearshore environment.

CHAPTER 2

~ ~ ~ ~ ~ ~ ~ ~ ~ ~ ~

The Organisms

"Biology is wet and dynamic."
—Berg (1983)

Throughout this book we approach the topic of wave-swept organisms from the viewpoint of a team of engineers assigned the task of designing a new, improved plant or animal. A potential difficulty is associated with this mechanistic view, in that it works exactly backwards from the process of evolution: we start with a set of design criteria and end with the mechanical structure(s). But the present structures of plants and animals are in fact the result of natural selection (a process of trial and error), not of a goal-directed design process. Natural selection starts with the mechanism (a plant or animal), and the design criteria are applied implicitly through the survival of the fittest. Thus, if we were to study organismal design from a biological perspective, we would start with the evolved mechanism (the functional plant or animal) and try to deduce the design principles that have been favored by natural selection. Several problems are inherent in working backwards from the finished product to the original design criteria. For instance, selection pressures on a lineage of organisms have certainly varied through evolutionary time either as the environment has changed or in response to evolved changes in the organisms themselves. In either case, structures have evolved in response to a varying set of design criteria, and it may be futile to look for *the* original criterion. Furthermore, once a certain body plan has evolved, it can limit the possibilities for future change. Unlike an engineer who can decide to scrap an

entire design, go back to the drawing board, and start anew, natural selection works with the structures that are already available. This mediates against abrupt changes in form. In addition, most structures have evolved in response to many simultaneous selection pressures. For instance, a lineage of plants could not have evolved a structure optimally designed for photosynthesis if that structure were hopelessly inappropriate for resisting the forces of wind or flowing water. Evolved structures inevitably represent some compromise among selection pressures. As a result of these confounding characteristics of natural selection, clear-cut examples of design principles are few and far between in the natural world. For this reason it is more educational to examine organismal design from the engineer's standpoint: begin with a simple, mechanistic approach, and work up to the complexity. This approach should pose no problem so long as one realizes that the teleological language used here is employed as a shorthand for an evolutionary argument.

Before we embark on this task, it is well worth surveying the existing plants and animals in an attempt to determine what design criteria Mother Nature has found important. Unfortunately, no single chapter can do justice to the flora and fauna of the wave-swept environment. A day spent poking around a tide pool or snorkeling over a coral reef will convince anyone that the plants and animals that inhabit wave-swept shores are far too diverse in their morphology, physiology, and ecology to be described with economy. Therefore, I will introduce the plants and animals of wave-swept shores in the context of a few basic principles of mechanical design. Those who want a more traditional ecological or taxonomical introduction to the plants and animals of the shore should consult the many useful texts—for example, Ricketts et al. (1968); Abbott and Hollenberg (1976); Morris et al. (1980); Carefoot (1977); and Wertheim (1985).

The design criteria I have chosen to introduce the wave-swept organisms are very basic. By no means do they form an exhaustive list, but they serve well in setting the stage for the discussion of mechanics that follows in Chapters 3 to 18. This introduction is accompanied by drawings of five typical wave-swept habitats. Most of the organisms I discuss can be found in one or more of these drawings.

A Source of Energy

The most basic criterion in the design of organisms is seen in the contrast between plants and animals (Fig. 2.1). In a mechanistic sense, this is a contrast between autotrophs that can manufacture food from inorganic compounds using sunlight as a source of energy, and heterotrophs that must rely on the chemical energy stored by autotrophs. In a few cases the dividing line between plants and animals gets a bit fuzzy—for example, corals and some anemones contain symbiotic algae that provide a fraction of the animal's metabolic needs. These exceptions aside, the choice between autotrophy and heterotrophy constitutes one of the basic design decisions in the wave-swept environment.

The requirement for light limits the depth at which a plant can grow. Although several genera have evolved efficient photosynthetic apparatuses that allow them to live at depths in excess of 100 m, most

macroalgae are found at depths of less than 30 m, presumably because below these depths the daily light levels are not high enough to result in a net carbon gain. In fact, over half the biomass of algae in the ocean is present as macroalgae (the seaweeds) in the shallow nearshore. As we will see in Chapter 4, the water flow induced by surface waves typically extends to a depth of 50 to 75 m. Thus benthic macroalgae are most abundant at the depths that are subject to wave action. Because of their requirement for light, plants also cannot effectively hide in cracks and crevices to avoid flow forces. In addition, although sunlight provides the energy for photosynthesis, the raw materials (carbon dioxide in the form of bicarbonate and inorganic nutrients) must be absorbed from seawater. The influx of these raw materials is enhanced by water flow, which again ties the macroalgae to the wave-swept environment.

In contrast, animals, which do not directly depend on light for their energy, potentially have a much wider choice in their habitats. They can (and do) live in the dark abyssal depths where wave-induced flows are not a problem. In the wave-swept environment they may hide in cracks and crevices to avoid flow forces without compromising their food intake. Although animals do not directly depend on light, they do strongly depend on plants, and this dependence can limit their habitat in the same way.

Herbivores must be close enough to plants to be able to use them conveniently as a food source. Those who prey on herbivores are similarly bound to a proximity to plants. However, this potential limitation is avoided by organisms that are suspension feeders and strain their food from the water around them, enabling them to live wherever the "soup" is sufficiently thick. Relieved of the need constantly to chase their food source, most benthic suspension feeders are sessile, relying on water movements to bring

Figure 2.1. Sources of energy: The flora and fauna of the California nearshore.

Heterotrophs

A   anaspidean, *Aplysia californica*
B   señorita fish, *Oxyjulus californica*
C   turban snail, *Tegula brunnea*
D   red abalone, *Haliotis rufescens*
E   ostrich plume hydroid, *Aglaophenia sp.*
F   bat star, *Patiria miniata*
G   compound tunicate, *Cystodytes lobatus*
H   lined chiton, *Tonicella lineata*
I   dunce-cap limpet, *Acmaea mitra*
J   compound tunicate, *Didemnum sp.*
K   giant keyhole limpet, *Megathura crenulata*
L   crab, *Cancer antennarius*
M   sea anemone, *Tealia sp.*
N   brittle star, *Ophiothrix spiculata*
O   stalked tunicate, *Styela montereyensis*
P   plumose anemone, *Metridium senile* (expanded)
Q   plumose anemone, *Metridium senile* (contracted)
R   sea cucumber, *Parastichopus californicus*
S   gumboot chiton, *Cryptochiton stelleri*
T   red sea urchin, *Strongylocentrotus franciscanus*
U   predatory starfish, *Pisaster giganteus*

V   keyhole limpet, *Diodora aspera*
W   sand dollar, *Dendraster excentricus*
X   sun star, *Pycnopodia helianthoides*
Y   striped surfperch, *Embiotoca lateralis*
Z   acorn barnacle, *Balanus nubilus*
AA nudibranch, *Anisodoris nobilis*
BB volcano sponge, *Acarnus erithacus*
CC tube anemone, *Pachycerianthus fimbriatus*
DD inarticulate brachiopod, *Terebratalia transversa*
EE blood star, *Henricia leviuscula*
FF sea cucumber, *Cucumaria miniata*

Autotrophs

1   sea lettuce, *Ulva sp.*
2   giant kelp, *Macrocystis pyrifera*
3   seaweed, *Laminaria dentigera*
4   sea grape, *Botryocladia pseudodichotoma*
5   seaweed, *Dictyoneuropsis reticulata*
6   Turkish towel, *Gigartina corymbifera*
7   crustose coralline algae, various spp.
8   erect coralline algae, *Calliarthron sp.*

food to them. In many cases these water movements are provided by the environment, allowing animals to act as passive suspension feeders. (For instance, barnacles in the wave-swept environment are primarily passive suspension feeders. Many acorn barnacles have the ability to generate feeding currents by pumping with their cirral net, but they refrain from pumping when sufficient ambient water flow is available, presumably to save energy.) Many anthozoans—for example, anemones, corals, hydroids, and gorgonians—have also chosen this tactic. Other invertebrate phyla among which passive suspension feeding is common include the bryozoans and annelids, and there are many scattered examples of passive suspension feeding among phyla that are generally known as foragers or predators. For example, in the echinoderms, brittle stars and sea cucumbers often use their tube feet as filters, and sea urchins may become facultative suspension feeders, hunkering down in a burrow and relying on their tube feet to snag drift algae.

Passive suspension feeding may appear to be an ideal way of making a living. There are, however, limiting mechanical constraints. Some structure must be held into the flow to strain the water, and if the water velocity is high, holding that structure may be difficult or impossible. Furthermore, by holding a feeding structure into the flow an animal may open itself to mechanical damage or predation. This problem can be circumvented by active suspension feeding. For instance, mussels, clams, tunicates, and sponges use ciliated or flagellated epithelia to pump water through internal filters. The cost of this strategy lies in the energy expenditure of pumping water.

It is possible to effect a compromise between active and passive suspension feeding: some subtidal sponges augment the flow through their filters by using the pressure difference created by the ambient flow (Vogel 1974, 1978). It is less clear whether this same mechanism works in intertidal sponges.

Mobility

A second design criterion differentiating plants from animals concerns the ability of the organism to move (Fig. 2.2). For reasons that are not obvious from a purely mechanical viewpoint, plants have never evolved the ability for locomotion. Still, it is interesting to muse upon the possibilities presented by a mobile plant, such as the capacity to run away from herbivores or to track sunflecks as they move through the day.

Crabs, shrimps, lobsters, and other crustacea walk using jointed legs. This type of movement can be quick and agile. Perhaps its only drawback is its lack of a mechanism for providing strong adhesion to the substratum. Most organisms that move by using jointed appendages are relatively easily dislodged and many hide from the full brunt of wave forces.

Snails, limpets, and nudibranchs crawl on a single, ventral foot. They adhere quite well to the substratum by using a specialized mucus glue; movement is powered by a series of muscular waves that pass along the foot. Many sea anemones can move slowly using what appears to be a similar form of locomotion.

Many forms of swimming animals abound in the wave-swept environment. Copepods and shrimps swim by beating their appendages with an oarlike motion. Fish swim by beating either their tail or their pectoral appendages or both. Those who use their pectoral appendages (for example, rays) often do so in a manner reminiscent of bird or insect flight: the primary force of locomotion comes from the lift on the moving appendage. Also, many species of octopus inhabit the wave-swept environment. When disturbed, these animals escape using a form of jet propul-

Figure 2.2. Mobile vs. sessile: The fauna of a tropical coral reef.

Mobile

A crinoid, *Comanthus parvicirra*
B butterfly fish, *Chaetodon punctatofaciatus*
C damsel fish, *Abudefduf xanthonotus*
D nudibranch, *Favorinus japonicus*
E sea star, *Fromia monilis*
F shrimp, *Lysmata sp.*
G brittle star, *Ophionereis sp.*
H sea urchin, *Echinometra sp.*

Sessile

1 porous coral, *Porites sp.*
2 sea fan, *Melithea squamata*
3 staghorn coral, *Acropora cervicornis*
4 tube coral, *Dendrophyllia sp.*
5 false finger coral, *Stylophora mordax*
6 Christmas-tree hydroid, *Pennaria sp.*
7 sponge, *Psammaplysella purpurea*

sion, rapidly expelling water from their mantle through a siphon. Similarly, the scallop, a bivalve mollusc, escapes using jet propulsion.

Echinoderms exhibit some of the more bizarre forms of locomotion found in nature. Starfish, sea urchins, and sea cucumbers move by using tube feet, which are hydrostatically supported hollow appendages. The tube feet of animals that live on hard substrata have mucus-coated suckers on their tips, allowing the animal to adhere to the surface over which it is walking. Sand dollars crawl by manipulating their spines, and crinoids swim by rhythmically waving their feather-like arms.

Although most animals have the ability to move about, many species have never evolved that ability or have secondarily given up the potential for locomotion. Prime examples are hermatypic corals, gorgonians, sponges, and bryozoans, whose adult forms are strictly sessile. Other examples include the bivalves (the clams and mussels) and the barnacles.

Mussels adhere to the substratum with a system of byssal threads. These threads can be released and new threads are attached nearby, allowing the animal to slowly change its position; however, this form of locomotion is very slow, and it is probably used only by juveniles to any great degree. Clams have an extensible foot that can be used for burrowing in sand and mud, and there are species (*Donax*, for instance) that inhabit the surf zone of wave-swept sandy beaches and are constantly readjusting their position in response to the flow forces they encounter. However, on the whole the bivalves are sessile. Adult barnacles have almost entirely given up the ability to move themselves about. Both stalked barnacles and acorn barnacles glue themselves permanently in place upon metamorphosis, and aside from some gradual sliding in response to crowding by other barnacles, they cannot change their position.

Many animals that are sessile as adults have mobile larval forms. More than 70% of all marine invertebrates have some form of planktonic larval stage, and these organisms maintain at least the potential for dispersal. Larval dispersal is not without its problems: besides the possibility of being eaten while floating about, larvae have the problem of getting back to the substratum in some acceptable spot. How this is accomplished is one of the larger open questions in marine science today, and we will examine one aspect of this process in Chapter 10.

The ability to move about is a requirement for certain types of feeding. Herbivores that feed on the abundant but sessile macroalgae of the nearshore or on the diatom scum that coats many marine surfaces must be able to move at least a bit to reach greener pastures. This movement may be extremely limited, such as that exhibited by the South African limpet *Patella cochlear*, which sits in one spot and slowly rotates, harvesting the algal garden it maintains in its immediate vicinity. More typical, however, are herbivores such as the littorine snails and acmaeid limpets that wander far and wide across the rock surface, scraping it clean of its algal coating. Similarly, a predatory life style is generally accompanied by the ability to move about actively. Common predators such as sea stars, crabs, and snails are among the faster-moving and more agile of the invertebrates present in the wave-swept environment.

There are definite costs associated with movement. Energy is expended during movement, and it is a general rule of thumb that a moving organism cannot adhere to the substratum as well as an organism that is stationary.

Reproduction

The physical nature of the wave-swept environment may place severe constraints on the manner in which plants and animals reproduce (Fig. 2.3). For example, many plants and animals reproduce sexually by liberating either sperm or eggs, or both, into the water. Sperm and eggs are generally not noted for their locomotory abilities, and this form of fertilization probably relies heavily on ambient water movement to bring gametes into contact. This is certainly true for the red algae, where the female gametes are retained in the gametangia on the adult plant, and the sperm are nonmotile. External fertilization is a chancy business at best (see Chapter 10), and accomplished effectively only if the males and females are closely packed. Thus, mobile animals may choose to congregate before spawning. For sessile species (both plant and animal) the requirements for effective external fertilization may determine the limits to spacing among individuals.

Many species avoid the problems of external fertilization by fertilizing their ova internally. For mobile species this poses no additional problem, since males

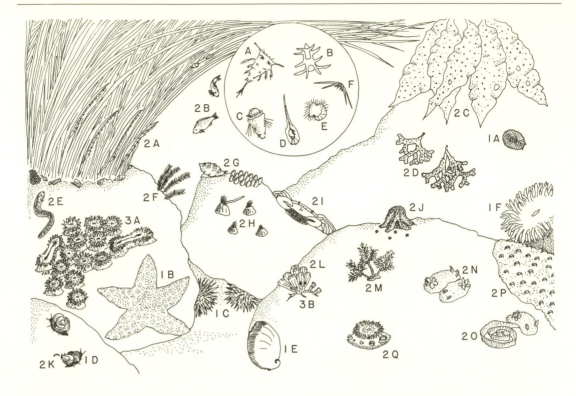

Figure 2.3. Modes of reproduction. The flora and fauna of the intertidal zone along the Monterey coast.

Sexual reproduction

1. External fertilization

A   file limpet, *Collisella limatula*
B   ochre star, *Pisaster ochraceus*
C   purple sea urchin, *Strongylocentrotus purpuratus*
D   black turban snail, *Tegula funebralis*
E   black abalone, *Haliotis cracherodii*
F   great green anemone, *Anthopleura xanthogrammica*

2. Internal fertilization

A   sea grass, *Phyllospadix coulteri*
B   reef surfperch, *Micrometrus aurora*
C   irridescent alga, *Iridaea sp.*
D   seaweed, *Mastocarpus papillata*
E   eel blenny, *Cebidichtys violaceus*
F   ostrich-plume hydroid, *Aglaophenia sp.*
G   whelk, *Thais emarginata*
H   acorn barnacles, *Balanus glandula*
I   shore crab, *Pachygrapsus crassipes*

J   six-rayed star, *Leptasterias hexactis*
K   slipper limpet, *Crepidula adunca*
L   lightbulb tunicate, *Clavelina huntsmani*
M   erect coralline alga, *Corallina sp.*
N   nudibranch, *Doriopsilla albopunctata* (mating)
O   nudibranch, *D. albopunctata* (laying eggs)
P   sponge, *Haliclona sp.*
Q   proliferating anemone, *Epiactis prolifera*

3. Asexual reproduction

A   elegant anemone, *Anthopleura elegantissima*
B   lightbulb tunicate, *Clavelina huntsmani*

Inset: larvae

A   barnacle nauplius
B   brachiolaria
C   polychaete worm
D   tunicate tadpole
E   gastropod veliger
F   ophiopluteus

and females congregate to copulate. Among sessile species, internal fertilization has led to the evolution of some amazing structures. For instance, the penis of an acorn barnacle may be several times as long as the animal's shell is wide. Even given this large penis a barnacle must settle within one penis length of a neighbor, enforcing a gregarious life style. Barnacles have also adapted to this situation by becoming hermaphroditic, thus avoiding the possible embarrassment of settling close to a group of neighbors only to discover that all are of the same sex as the settler.

It is possible that the wave-swept environment is so physically rigorous that sexual reproduction can be a losing proposition, and many wave-swept organisms have the ability to reproduce asexually. Macroalgae such as *Endocladia* and *Rhodymenia* commonly reproduce by vegatative propagation. Among animals, the most notable of the asexual reproducers are the anthozoans. Coral colonies generally grow by asexual reproduction (although sexual reproduction is possible), and several species of intertidal anemone typically reproduce by binary fission or budding, at times forming large mats of cloned individuals.

Rigidity of the Structure

Skeletal flexibility is a basic design criterion for many plants and animals, and wave-swept organisms can be cleanly divided into two categories, those that have opted for flexibility and those that have opted for rigidity (Fig. 2.4). This basic option carries with it other design constraints. As will be discussed in Chapters 11 to 14, rigid structures are often subjected to larger fluid-dynamic forces than flexible structures, and a rigid organism must be designed to cope with this duress. For example, the rigid shell of an acorn barnacle requires that the animal

produce an exemplary glue to ensure that the shell stays attached to the substratum. The decision to be rigid also imposes constraints on form and locomotion. Consider, for example, organisms that produce a calcium carbonate skeleton. The decision (in an evolutionary sense) to build a rigid exoskeleton from calcium carbonate dictates many other design constraints. Calcium carbonate is relatively hard, and in this respect can be used to build armor that is effective in discouraging predators, but it is also heavy and weak. The construction of a pure calcium carbonate skeleton has the consequence of restricting hermatypic corals to a sessile existence. Molluscs have learned how to add sufficient organic matrix to their calcium carbonate skeletons to make them relatively strong. These strengthened materials are used to form shells that can be lugged around on the backs of snails and limpets, but one supposes that there is a substantial price paid in terms of energy expenditure both in producing this rigid shell and in toting it around. Again, the presence of a rigid structure implies the presence of large wave-induced forces, and these organisms must adhere strongly to the substratum. The crustacea have refined rigidity into an art. The shells of crabs are stiffened with calcium carbonate, although the volume percentage is much lower than in either coral skeleton or mollusc shell. The remainder of a crab's shell is formed of chitin, a strong, light polysaccharide embedded in a protein matrix. The result is a rigid shell that is light enough to be carried around by a very agile animal. Interestingly, hermit crabs have forsaken these advantages, and have taken refuge in uninhabited gastropod shells. They are much less agile than their unencumbered cousins.

In general, flexible structures can afford to be much less strong than rigid structures. This has opened up a number of avenues in the evolution of biological form. Organisms from many phyla, in-

cluding the majority of macroalgae, many hydroids, gorgonians, anemones, and bryozoans, have achieved apparent success in the wave-swept environment by being able to bend under the influence of fluid-dynamic forces. As we will see in Chapter 11, bending reduces the area of the organism that is projected into the flow, effectively streamlining it and thereby reducing the fluid-dynamic force (drag) imposed. In addition, it takes less material to produce a flexible skeleton than it does to produce a rigid skeleton of the same size. Flexibility is likely to produce a savings in material expenditure.

Finally, there are those organisms that are able to control their flexibility. Crabs can cantrol the overall flexibility of their rigid exoskeleton by muscular action at the joints. Sea stars can be highly flexible when moving about, but they can become quite rigid when the situation warrants (for instance, when attempting to open a mussel). In this particular case, the mechanism controlling rigidity is not well understood.

Size

Large size is an advantage to an organism for many reasons (Fig. 2.5). The larger the organism, the more progeny it can produce, and presumably the greater its reproductive success. In the intertidal zone where heat and desiccation stresses may be important, larger organisms have the advantage of a smaller surface area per body volume and thus may be better buffered from the exigencies of the environment. In certain instances, large size can serve as a refuge from predation or herbivory. In the age-old tradition of "them that has, gets," size often confers an advantage in competition for space.

Despite the advantages of large size, most wave-swept organisms are quite small. Perhaps the largest wave-swept organisms (in terms of mass) are the hermatypic corals, where individual colonies can grow to be several meters long in their greatest dimension. However, this is a misleading example. Each individual in a coral colony is quite small, and most of

Figure 2.4. Rigid vs. Flexible: The rocky intertidal zone along the Monterey coast. There is a continuum from very rigid to very flexible, and at times the line dividing rigid from flexible is drawn somewhat arbitrarily. For instance, a gooseneck barnacle is flexible compared to an acorn barnacle, but it is rigid relative to most of the macroalgae.

Rigid

A   rough limpet, *Collisella scabra*
B   acorn barnacles, *Chthamalus sp.*
C   periwinkles, *Littorina sp.*
D   chiton, *Nuttallina californica*
E   shield limpet, *Collisella pelta*
F   owl limpet, *Lottia gigantea*
G   purple urchin, *Strongylocentrotus purpuratus*
H   black abalone, *Haliotis cracherodii*
I   gooseneck barnacle, *Pollicipes polymerus*
J   mussel, *Mytilus californianus*
K   chiton, *Katherina tunicata*
L   chiton, *Mopalia ciliata*
M   keyhole limpet, *Fissurella volcano*
N   ochre star, *Pisaster ochraceus*
O   acorn barnacle, *Tetraclita rubescens*

Flexible

1   rockweed, *Pelvetiopsis limitata*
2   sea palm, *Postelsia palmaeformis*
3   rockweed, *Fucus distichus*
4   rockweed, *Pelvetia fastigiata*
5   shore crab, *Hemigrapsus nudus*
6   sea grass, *Phyllospadix scouleri*
7   elegant sea anemone, *Anthopleura elegantissima*
8   feather boa kelp, *Egregia menziesii*
9   ochre star, *Pisaster ochraceus*
10 great green anemone, *Anthopleura xanthogrammica*
11 seaweed, *Laminaria sp.*
12 elephant-ear tunicate, *Polyclinum planum*
13 erect coralline algae, *Calliarthron sp.*
14 polychaete worm, *Nereis sp.*
15 ostrich-plume hydroid, *Aglaophenia sp.*
16 nudibranch, *Rostanga pulchra*

the mass of a large coral is formed from the dead skeletons of ancient corrallites. If we limit ourselves to a consideration of single individuals and measure size as mass of living tissue, it is clear that organisms that are flexible have a decided edge in the size to which they can grow. The largest organisms in the wave-swept environment on the west coast of North America are the kelps *Macrocystis* and *Nereocystis*. These highly flexible plants can grow to be 40 to 50 m in length, easily weighing 20 to 50 kg. Next in size are the flexible macroalgae of the nearshore such as *Egregia*, *Laminaria*, and *Pterygophora*. Among the animals, the largest found on wave-swept shores are sea stars (which, in exposed sites on Pacific shores, may grow to about 20 cm, arm tip to arm tip, and weigh 1 or 2 kg) and large gastropods such as abalones and limpets. These organisms, which have rigid skeletons, are quite small compared to the flexible macroalgae. Even though there is a clear contrast in maximum size between rigid and flexible wave-swept organisms, all wave-swept organisms are still small when compared to plants and animals of other habitats, none coming even close in size to a blue whale or redwood tree. The reasons for their small size will be examined in Chapter 17.

A wealth of very small organisms exists in the wave-swept environment. If an organism is small enough (generally less than 4–5 mm in maximum dimension), it is able to avoid the brunt of the wave-induced fluid-dynamic forces (see Chapters 9 and 11). The unique nature of the wave-swept environment is much less a factor in determining the mechanical structure of these organisms, making them only of secondary interest in this book.

The list of design criteria given here is obviously incomplete. For example, internal circulation, osmoregulation, and sense organs have not been discussed. These neglected criteria deal more with the physiological aspects of biology than they do with structural and ecological aspects, and are therefore left for others to examine.

Figure 2.5. (facing) Big vs. small: The flora and fauna of a kelp bed on the California coast.

Very big

1   bull kelp, *Nereocystis leutkeana*
2   giant kelp, *Macrocystis pyrifera*

Big

3   blue rockfish, *Sebastes mystinus*
4   seaweed, *Laminaria dentigera*
5   seaweed, *Pterygophora californica*
6   seaweed, *Costaria costata*
7   SCUBA diver, *Homo sapiens macho*
8   seaweed, *Eisenia arborea*
9   copper rockfish, *Sebastes caurinus*
10 giant keyhole limpet, *Megathura crenulata*
11 plumose anemone, *Metridium senile*
12 fish-eating anemone, *Tealia piscivora*

13 sun star, *Pycnopodia helianthoides*
14 colonial worms, *Dodecaceria fewkesi*
15 sponge, *Polymastia pachymastia*
16 lingcod, *Ophiodon elongatus*
17 wolf eel, *Anarrhichtys ocellatus*
18 sea cucumber, *Parastichopus californica*

Small (inset)

A   urn sponge, *Leucilla nuttingi*
B   bryozoans, *Heteropora, Thalamaporella sp.*
C   tube worms, *Serpula vermicularis*
D   hydrocoral, *Allopora sp.*
E   nudibranch, *Rostanga pulchra*
F   hydroid, *Abietinaria sp.*

CHAPTER 3

~ ~ ~ ~ ~ ~ ~ ~ ~ ~ ~

An Introduction to
Fluid Dynamics

In this chapter we begin to explore the physics of moving fluids. In doing so, we set forth on a long and tortuous road. The subject of fluid dynamics is extremely complex, and the accurate and complete description of even the simplest of flows involves the use of mathematics. In this introduction we will take the first few steps toward understanding moving fluids and keep the mathematics to a minimum (a more mathematical approach will come in Chapters 5 to 11).

As far as possible, the basic concepts will be explained through intuitive examples. But there is some danger in this approach. Much of the behavior of fluids is simply nonintuitive, and following one's intuition very far is likely to lead to spurious conclusions. Several concepts presented here simplistically will be reexamined in later chapters, and others that need special attention will be noted. The cautious reader should refer to standard fluid dynamics texts on these points (e.g., Fox and MacDonald 1978; Massey 1983; Sabersky et al. 1971; Streeter and Wylie 1979).

Which Way Is Up?

Before examining the physics of the wave-swept environment, we need to establish a system whereby we can precisely locate objects in space. In this book we use the coordinate system adopted by physical oceanographers (Fig. 3.1). The x and y-axes are horizontal, and the x-axis usually coincides with the direction of mean flow. The

z-axis is vertical and positive in the upward direction. As an object moves through space, its velocity may have components along each of these axes. The velocity in the x-direction is u; in the y-direction, v; and in the z-direction, w.

Figure 3.1. The oceanographic coordinate system.

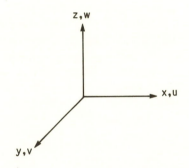

Force

In Chapter 2 terms such as "force" and "energy" were used without rigorous definition. For the descriptive manner in which the interaction of morphology and water motion was treated, this laxity caused no problem. But any quantitative exploration of the mechanism behind fluid phenomena requires that our terms be well defined. We start at the very beginning by reviewing Newton's three laws of motion:

1. A force is required to change the state of motion of an object.
2. The acceleration (the time rate of change of velocity) caused by a force is

proportional to the magnitude of the force and inversely proportional to the mass of the object.

3. If object A places a force on object B, object B places an equal but oppositely directed force on object A.

These three laws are familiar to anyone who has had an introductory physics course. In introductory physics, however, it is likely that they were applied only to situations involving solid objects—cannonball trajectories, bats hitting baseballs, and the like. It is important to realize the universality of Newton's principles: the laws apply just as well when the object in question is a fluid. A simple example serves to illustrate this point. If you have ever ridden a bicycle in the rain, you have experienced the stinging splatter of raindrops pelting your face. As each raindrop falls, it has a velocity relative to your face, and upon impact on your forehead this velocity is abruptly changed. By Newton's second law this change in velocity, an acceleration, requires a force. It is this force that causes the sting.

Newton's second law provides us with a definition of force:

force = mass · acceleration.

For example, the force in the x direction is

$$f_x = m \frac{du}{dt}, \qquad (3.1)$$

where u is the instantaneous velocity of the fluid. Force is measured in newtons (1 N = 1 kg m/s^2) and is a vector, that is, it has both a magnitude and a direction. The acceleration of a gravity g (= 9.81 m/s^2), acting on a mass of fluid, produces the force mg, which we normally think of as the fluid's weight.

Alternatively, we can express force in terms of a fluid's *momentum*. Momentum (or *inertia*) is the product of mass and velocity. For instance, momentum along the x-axis is

$$\text{momentum} = mu. \qquad (3.2)$$

Like force, momentum is a vector having both magnitude and direction. If mass is constant, the force in the x-direction is

$$f_x = m \frac{du}{dt} = \frac{d(mu)}{dt}. \qquad (3.3)$$

In other words, force is the rate of change of momentum. The two definitions (eqs. 3.1, 3.3) are equivalent, and we interchange them as is convenient.

Mechanical Energy

Mechanical energy is defined as force times the distance through which the force moves its point of application. Mechanical energy is tangible as the capacity to do work, and we use the terms "energy" and "work" interchangeably. Energy is measured in joules (J): 1 J = 1 Nm.

The mechanical work that can be done by fluids comes in three forms:

1. *Gravitational potential energy* is the work that can be done by a mass of fluid as a consequence of its vertical position. For example, a mass, m, of fluid in a bucket exerts a force mg when acted upon by the acceleration of gravity. If the mass changes its height by a value h, the change in potential energy (= force × distance) is

gravitational potential energy = mgh. (3.4)

This change in potential energy is equal to the work done on or by the fluid. For instance, consider the situation shown in Figure 3.2a. A bucket is suspended from a rope that winds around a pulley. Pulling on the rope raises the bucket by a height h, increasing its potential energy by mgh. In the process, an amount of work equal to mgh has been done on the water. Lowering the bucket back to its starting point reduces it potential energy by mgh, and work is done by the water on the pulley (by turning an electric generator, for instance). Note that potential energy can be measured only as a relative quantity; a change

Figure 3.2. (a) A suspended bucket is capable of doing work due to its gravitational potential energy (see text). (b) The calculation of kinetic energy (see text). (c) Flow work. A fluid can transmit energy along a pressure gradient (see text).

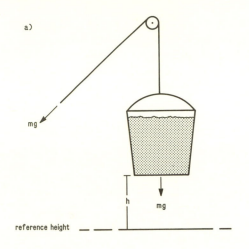

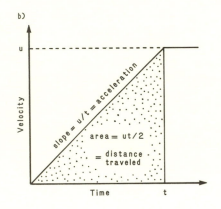

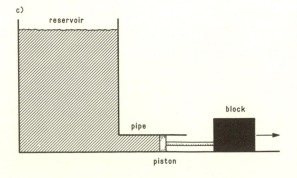

in the height used as a reference results in a change in potential energy.

2. A second type of potential energy is *pressure-volume energy*, the potential energy stored in a compressed fluid. For instance, consider a bicycle pump with a clogged nozzle. If you push down on the plunger, you apply a force over a distance and do work on the air in the pump. If no air escapes, the work done results primarily in a compression of the air. A small amount of mechanical energy is lost to heat in this procedure, but most is stored and can be recovered.[1] When you slowly release the plunger, it moves back to its original position, doing work on your hand as it does.

Pressure-volume work is available only when a fluid is compressed. If you fill a bicycle pump with water (which is very nearly incompressible), you can push on the plunger with all your might but it will not move. Because work is the product of force and the distance moved, the application of even a large force in this example results in no expenditure of mechanical energy. Because we are here primarily concerned with the dynamics of seawater and are dealing with a fluid that does not allow for the expenditure or storage of appreciable pressure-volume work, we will not explore the concept further.

3. If a net force is placed on an initially stationary mass of fluid, the fluid accelerates. As the fluid moves, the force acts through a distance, and a third type of work is done. The amount of this work is easily calculated by noting the relationship graphed in Figure 3.2b. Under the action of a constant force, a mass of fluid has constant acceleration and its velocity increases linearly with time. Over the interval, *t*, how much work has been done on the

---

[1] The degradation of mechanical energy to heat energy is a subject that will not be treated here in any detail. The inquisitive reader should consult an introductory text on thermodynamics for an explanation of the many ramifications of exchanging mechanical and heat energy.

fluid? The fluid has accelerated from rest to velocity u during this period, so its acceleration is u/t and the applied force is mu/t. The distance covered in time t is equal to the average velocity, $u/2$, multiplied by the time traveled at this velocity, t. Thus,

$$\text{work} = \left\{\frac{mu}{t}\right\} \cdot \left\{\frac{ut}{2}\right\} = \frac{mu^2}{2}. \quad (3.5)$$

This work, the work required to accelerate the mass up to speed u, is the *kinetic energy* of the mass of fluid at speed u. This kinetic energy is available to do work.

Each of these forms of mechanical energy can be transmitted from one place to another. The simplest example is the work done by a fluid when it is acted upon by a pressure. Consider the apparatus shown in Figure 3.2c. Water flows from a reservoir into a pipe and pushes on a piston. The water in the pipe has pressure p above the ambient pressure outside the pipe due to the weight of water in the reservoir. This pressure, acting over the area of the piston, applies a net force to the piston and to the block to which the piston is attached. When this force is first applied, the block accelerates, but soon the frictional force that accompanies the block's movement equals the applied force. At that point the fluid, piston, and block are in equilibrium, and subsequent movement continues at a steady speed. What work is then being done by the water in the pipe? Recall that work is the product of force and distance. Therefore force ($=$ pressure $\times$ piston area) multiplied by the distance moved by the block is the work. But the product of piston area and distance moved is the volume of fluid that enters the pipe. Thus,

$$\text{work} = (\text{pressure} \cdot \text{area}) \text{ distance}$$
$$= pV. \quad (3.6)$$

Now, the fluid in the pipe does not change height as it flows, so it is not expending gravitational potential energy in moving the block. The fluid flows at a constant velocity, so kinetic energy is not changed.

Clearly, the work done by the flowing fluid in this situation is different from that previously described, and it is known as *flow work* or *flow energy* (Streeter and Wylie 1979; Massey 1983). Note that this type of work is available only when the fluid doing the work connects areas of different pressure. If the pressure of the water in the pipe were equal to the pressure outside, no net force would act on the piston and no work could be done. In turn, the maintenance of a pressure difference requires the expenditure of energy elsewhere. Although the potential energy of the water *in the pipe* does not change as the piston moves, the potential energy of water in the reservoir decreases. In this sense, the water in the pipe plays a role equivalent to the rope in Figure 3.2a— it transmits energy from one place to another. Thus flow work is not an energy possessed by the fluid itself (in the sense that a particle of fluid possesses kinetic or gravitational potential energy); rather, it is a measure of the maximum amount of work a flowing fluid can do by physically connecting areas of different pressure.

For an incompressible fluid such as water, the sum of gravitational potential energy, kinetic energy, and flow energy is called *available energy*.

The No-Slip Condition

We now turn to an odd but fundamental property of fluids. We can show empirically (and confirm on the basis of molecular interactions) that the fluid immediately in contact with a solid surface does not slip relative to that surface. This property of all fluids is known as the *no-slip condition*. It is not difficult to find confirmation of this fact in everyday experience. For instance, dust on a hood or windshield is not disturbed even when a car is driven at high speed. Similarly, a small amount of ink injected near the wall of a water-filled beaker takes an amazingly long time to disperse when the water is stirred. The importance of the no-slip condition will become apparent as we discuss viscosity.

Viscosity

What, exactly, is a fluid? We all have an innate understanding of the difference between solids and fluids, but it is useful to state this difference in precise terms.

If you deform a solid object, the amount of force required is proportional to the amount of deformation. As long as a constant force is applied, this one particular deformation is maintained, a fact that is easily demonstrated by hanging weights from a rubber band. The implications of this characteristic of solids are discussed in detail in Chapter 12.

In contrast, if you deform a fluid, the amount of force required depends not on the *amount* of deformation, but rather on the *rate* of deformation. For instance, the force required to stir a jar of honey depends not on how far the honey is stirred, but on how fast you stir it. We can use this relationship to define a basic characteristic of fluids—viscosity.

Consider a cube of fluid sandwiched between two plates, each of area S (Fig. 3.3a).

The lower plate is held stationary, and in accordance with the no-slip condition the fluid adjacent to it must stay put. We now apply a force to the upper plate. In response to this force, the upper plate accelerates in the x-direction, carrying the adjacent fluid with it. As a result, the fluid between the plates is sheared. The upper plate accelerates until the resistance from the fluid is equal to the applied force, at which point the system is at equilibrium. We can use this equilibrium velocity of the upper plate to quantify the relationship between applied force and the rate of deformation.

First, we must account for the size of the plates. Obviously, the larger the plates the more fluid they affect and the more force is required to move them. Accordingly, we divide the applied force by the area of sample over which it is applied. It is this value of force per area, the *shear stress* τ, which is proportional to the rate of deformation.

Now, the proportional deformation at any point in the fluid cube is

$$\text{deformation} = \frac{dx}{dz},$$

so that the rate of deformation of the fluid between the plates is

Figure 3.3. (a) Fluid sandwiched between a moving plate and a stationary plate develops a velocity gradient. (b) The rate of deformation is equal to the velocity gradient.

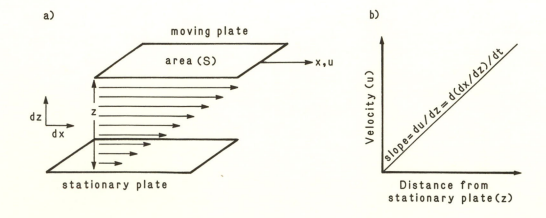

rate of deformation $= \dfrac{d(dx/dz)}{dt}$

$$= \frac{d(dx/dt)}{dz} = \frac{du}{dz}. \qquad (3.7)$$

In other words, the rate of deformation is equal to the velocity gradient in the fluid (Fig. 3.3b). Thus,

$$\tau = \mu \frac{du}{dz}, \qquad (3.8)$$

where μ, the *dynamic viscosity*, is the proportionality constant between the applied force per area and the rate of deformation. It has units Ns/m^2 or pascal seconds (1 Pa = 1 N/m^2).

The dynamic viscosity is a measure of the tendency of a fluid to stick to itself; it is a characteristic of each fluid and varies among fluids. For seawater, μ decreases with increasing temperature (Table 3.1). Seawater in the tropics (20°C) is about 23% less "sticky" than seawater in temperate oceans (10°C).

The dynamic viscosity often appears in conjunction with the fluid's density, ρ. To simplify the writing of equations, the ratio μ/ρ is called the *kinematic viscosity*, ν (Table 3.1).

Table 3.1
Viscosity of Seawater (from Vogel 1981)

°C	Dynamic Viscosity, μ Ns/m^2	Kinematic Viscosity, ν* m^2/s
10	$1.391 \ 10^{-3}$	$1.36 \ 10^{-6}$
20	$1.072 \ 10^{-3}$	$1.05 \ 10^{-6}$
30	$0.868 \ 10^{-3}$	$0.85 \ 10^{-6}$

* Kinematic viscosity is the ratio of dynamic viscosity to density

It should be evident from this derivation that viscosity and the no-slip condition work hand-in-hand in exerting forces between moving fluids and solid objects. Take a

pertinent example. Fluid moves at a constant rate past a solid surface (Fig. 3.4a). If the no-slip condition did not hold, the fluid could slide by the object without a velocity gradient being established. If there is no velocity gradient, there is no rate of deformation in the fluid, and hence no force is required to move the fluid past the object. Conversely, by Newton's third law, the object would feel no force in the direction of flow due to the movement of the fluid. However, it is our empirical experience that fluids moving past objects do indeed impose a force. For instance, a breeze blowing across a sandy beach often exerts enough force to move the sand downwind.

Figure 3.4. (a) If fluid can slip by a solid surface, no velocity gradient is established. (b) As a result of the no-slip condition, a velocity gradient is established and a shear force exerted on the substratum.

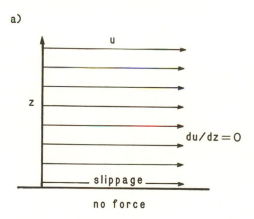

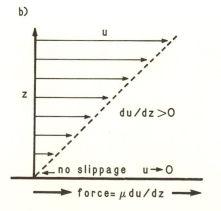

A simplistic explanation for the origin of this force can be arrived at by examining Figure 3.4b. The bulk of the fluid moves past the solid surface at a constant velocity; but right at the surface the no-slip condition is invoked, requiring the fluid to be stationary. A velocity gradient is necessarily established (drawn, for simplicity, as a straight line), and accompanying it is a shear force per area equal to the product of the velocity gradient and the dynamic viscosity. This force is called *friction drag* or *viscous drag*.

The above is an overly simplified explanation of the cause of hydrodynamic forces, and friction drag usually accounts for only a small portion of the hydrodynamic force exerted on an object in the wave-swept environment. We will return to this subject in Chapter 11.

Loss of Mechanical Energy

Because the viscous force exerted by a solid object on a fluid is imposed as the fluid flows, friction drag acts as its point of application moves through a distance and work is done. This is the work necessary to slide fluid molecules past each other, and like other types of frictional work it ultimately appears as heat. Now, in the absence of external forces, the overall energy of a moving fluid is conserved, so any energy that appears as heat must reduce the available mechanical energy of the fluid by an equal amount. This is the reason eddies and swirls in a cup of coffee gradually subside after you stop stirring; the kinetic energy of the moving coffee is converted to heat through the action of viscosity.

Conservation of Mass: The Principle of Continuity

Under everyday circumstances mass is neither created nor destroyed, a fact that is an extremely useful tool in fluid dynamics. In many cases one can understand how a fluid behaves simply by noting that its mass must be constant. For example, consider the water running from left to right through the pipe shown in Figure 3.5. The left-hand end of the pipe has area S_1 and the right-hand end has area S_2. The volume of water entering the pipe is $S_1 x_1$, where x_1 is the distance the fluid moves. The water has density ρ_1 (in kg/m³) so the mass of this volume is $\rho_1 S_1 x_1$ as it enters the pipe. If the walls of the pipe are rigid, the mass of water that enters must also leave. Thus, the mass moving out the right-hand end, $\rho_2 S_2 x_2$, must equal the mass entering the pipe:

$$\rho_1 S_1 x_1 = \rho_2 S_2 x_2. \qquad (3.9)$$

Figure 3.5. The principle of continuity: in flow through a rigid pipe $S_1 x_1 = S_2 x_2$.

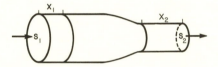

Because water is a nearly incompressible fluid, at a constant temperature its density does not change during flow. Thus, $\rho_1 = \rho_2$, and the conservation of mass implies the conservation of volume:

$$S_1 x_1 = S_2 x_2, \qquad x_2 = x_1 \frac{S_1}{S_2}. \quad (3.10)$$

As a result, the distance the fluid moves in the right-hand section of pipe is greater than in the left-hand section ($x_2 > x_1$) because the area of the right-hand section is less ($S_2 < S_1$). The same reasoning can be applied to the speed at which fluid enters and leaves the pipe:

$$u_2 = u_1 \frac{S_1}{S_2}, \qquad (3.11)$$

where u_1 and u_2 are the water velocities in the left- and right-hand sections, respectively, and density is assumed constant. Decreasing the area of the pipe increases the velocity of the fluid, as when you put your thumb over the end of a hose when you want to squirt someone.

The *principle of continuity* can hold true even in situations where physical pipes are not present. Consider the volume of water shown in Figure 3.6a. At every point in this volume we know the direction in which the fluid is moving and its speed. For the moment we assume that the direction and speed of movement at each point are constant through time, i.e., the flow is *steady*. One can, in principle, follow a path through the water so that the direction of movement at one point on the path leads directly to the next point. A line drawn along this path is a *streamline*. By definition this streamline is tangential to the direction of flow everywhere along its length[2].

We could, if we were diligent, draw enough adjacent streamlines to form a so-called *streamtube* (Fig. 3.6a). Because the wall of this tube would be made of streamlines, it is always tangential to the direction of fluid flow. By defining the streamtube in this fashion, we have specifically ruled out the possibility that any water that enters the streamtube can cross its walls. Defining streamlines and streamtubes this way provides a method whereby a region of flowing fluid can be treated in the same manner as that encompassed by a rigid pipe. The mass (and volume) of water flowing into a streamtube must be equal to the mass (and volume) coming out the other end.

The utility of this thought experiment is easily demonstrated. A biological object is placed in flowing water (Fig. 3.6b), and we would like to know what the velocity is near the object. Upstream of the object small amounts of dye are injected into the fluid

Figure 3.6. (a) Streamlines forming a streamtube. At every point the streamlines are tangential to the direction of flow. As a result, no fluid crosses the walls of the streamtube. (b) Streamlines as a tool for measuring velocity. Because streamlines are closer together above the object, the flow must be faster to satisfy continuity.

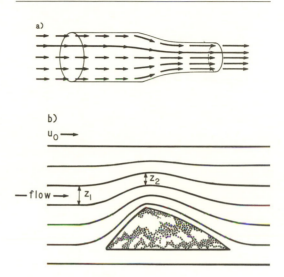

at equally spaced intervals across the flow. The dye follows the path of the fluid, and if at every point the fluid moves with a time-invariant velocity, the dye streaks trace out a series of streamlines. At a position near the origin of the dye streaks, two streamlines are separated by a distance z_1. Near the solid object the distance between streamlines has decreased to z_2. These two streamlines can be regarded as a two-dimensional slice through a streamtube, and hence (by analogy to eq. 3.11) $u_1 z_1$ must equal $u_2 z_2$. Thus u_2, the velocity we seek, is equal to $u_1(z_1/z_2)$. Knowing the velocity at one point on the two-dimensional streamtube, we can, by the principle of continuity, calculate the velocity at any other point along the tube.

This method of mapping velocity has its limitations. For example, the paths followed by streaks of dye (or any other means of marking the flow) define streamlines *only* if at every point along the dye streak the velocity of the fluid is time-invariant.

---

[2] An interesting consequence of this definition is that streamlines can never cross. If they did, the fluid at the point where they crossed would have to be traveling in two directions at once.

Figure 3.7. The misleading dye stream. Because the flow is accelerating, the dye stream does not lie in the direction of the present flow. Thus the flow's history biases present appearances.

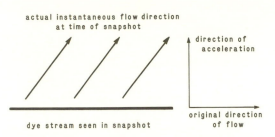

This criterion is not difficult to violate. Consider a dye streak formed while the fluid is moving from left to right at a constant velocity—a horizontal streamline (Fig. 3.7). The fluid is then subjected to rapid acceleration at right angles to the previous direction of flow. At a time soon after the acceleration commences, a snapshot is taken of the dye streak. The snapshot shows a dye path still extending directly from left to right even though the actual instantaneous velocity has a vertical component due to the applied acceleration. The dye path is not tangential to flow and is therefore not a streamline. Because the path exhibited by a dye streak may be biased by the history of the fluid motion, it is not an accurate method of analyzing velocities in time-varying or *unsteady* flow. Most flows in the wave-swept environment are time-varying, and for our purposes this method of analysis has only limited experimental use. Problems also arise in attempting to apply the principle of continuity to time-varying streamtubes. If flow is unsteady, a streamtube can change its shape and volume, and consequently the mass of a fluid entering the streamtube may not be equal to the mass leaving. In other words, the principle of continuity as we have examined it here need not hold in time-varying flows. Despite its limitations, frequent use will be made of the method outlined above for mapping velocities, and the concept will be important in explaining

the mechanism by which fluid-dynamic forces arise.

Conservation of Energy: Bernoulli's Equation

It is useful to know how the various components of a fluid's available energy change as the fluid travels along a streamline. For example, in Chapter 4 this information will allow us to calculate how quickly waves travel. However, such a description is simple only when the available energy of the fluid remains constant. This is a problem for real fluids, where viscosity can result in a loss of mechanical energy. To avoid this problem and thereby gain access to a formidable tool, we invoke a bit of fluid-dynamic legerdemain and consider the case of a hypothetical "ideal" fluid, that is, the fluid is incompressible and has no viscosity (it is said to be "inviscid").[3] When there is no viscosity, there is no viscous loss of mechanical energy. Because we used the notion of viscosity in defining the term "fluid," our approach may not seem altogether valid. This quandary is resolved by noting that the following ideas apply to real, viscid fluids in any exact sense *only* in those situations where the viscosity of the fluid is not substantially in evidence. In practice, we are then restricted to fluids sufficiently far away from a solid surface so that no appreciable gradient in fluid velocity exists perpendicular to the direction of flow—the velocity of a fluid particle may change as it flows along a streamline, but the fluid particles in neighboring streamlines must have the same velocity. Under these conditions, viscosity does not result in the exertion of any force on the fluid, no mechanical energy is lost to heat, and the fluid behaves as if it were "ideal."

The available energy of an incompressible

---

[3] To be absolutely correct, we would also have to require that the "ideal" fluid be irrotational. We will deal with the concept of irrotationality in Chapter 5, since it has no bearing on the present discussion.

fluid is equal to the sum of its kinetic energy, flow energy, and gravitational potential energy:[4]

available energy

$$= (\tfrac{1}{2})mu^2 + \underbrace{pV}_{} + \underbrace{mgh}_{}$$
$$\underbrace{\quad}_{\text{kinetic}} \quad \underbrace{\quad}_{\text{flow}} \quad \underbrace{\quad}_{\text{gravitational}}$$
energy   energy   potential energy

$$(3.12)$$

In many situations it is easier to talk about a particular volume of fluid than about its mass. Dividing eq. 3.12 by volume ($= m/\rho$), we arrive at the conclusion that

$$\frac{\text{available energy}}{\text{volume}} = \tfrac{1}{2}\rho u^2 + p + \rho gh. \quad (3.13)$$

Note that volume-specific mechanical energy has the units of pressure. The volume-specific kinetic energy $\tfrac{1}{2}\rho u^2$ is often called the *dynamic pressure*, and the volume-specific gravitational potential energy is the *static pressure head*.

In the absence of viscosity, the available energy is constant along a streamline. For example, choosing two points, 1 and 2, both on the same streamline and assuming the density to be constant,

$$\tfrac{1}{2}\rho u_1^2 + p_1 + \rho gh_1 = \tfrac{1}{2}\rho u_2^2 + p_2 + \rho gh_2$$

$$\tfrac{1}{2}\rho(u_2^2 - u_1^2) + (p_2 - p_1) + \rho g(h_2 - h_1) = 0. \quad (3.14)$$

This expression describes how velocity, pressure, and height must covary along a streamline, and is one form of *Bernoulli's equation*. The usefulness of this equation is best made evident by example. Consider the cross section through a streamtube shown in Figure 3.8. It should be clear (from the principle of continuity) that the velocity of the fluid flowing along the bottom streamline increases while going from point 1 to point 2. Points 1 and 2 are at the same

height so the term in eq. 3.14 containing an expression for the change in height becomes zero, leaving an equation that relates velocity and pressure:

$$\tfrac{1}{2}\rho(u_2^2 - u_1^2) = p_1 - p_2. \quad (3.15)$$

The velocity at 1 is less than at 2 so the left-hand side of this equation is greater than zero. This can only be true if the pressure at 2 is less than at 1. In other words, the increase in velocity from 1 to 2 is accompanied by a decrease in pressure. Only in this manner can the available energy of the fluid remain constant. The fluid velocity decreases from point 2 to point 3, reaching a velocity at 3 equal to that at 1. In doing so the pressure increases again so that $p_3 = p_1$. The kinetic energy per volume at point 2, which is high due to the high velocity, has been traded for an increase in pressure at point 3. The overall available energy per volume, of course, remains constant.

Figure 3.8. An application of Bernoulli's equation. Pressure and velocity are inversely related along a horizontal streamline (see text).

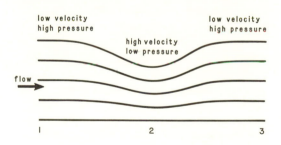

The fact that Bernoulli's equation is strictly applicable only in situations where the viscosity of the fluid is not evident (i.e., well away from solid objects) is no real hindrance to its use in the real world. We can apply Bernoulli's equation to many problems of practical interest as long as we realize that the answers we get, although qualitatively correct, will not be quantitatively exact because some mechanical energy is lost because of viscosity.

---

[4] The cautious reader should consult a standard fluid-dynamics text for a rigorous justification of this assertion. It involves the integration of Euler's equation for the flow of an inviscid, incompressible fluid.

Reynolds Numbers

The applicability of Bernoulli's equation in practical terms raises a general question. How do we know when viscosity is an important factor in the pattern of fluid flow? Because this problem will arise in a variety of disguises in this book, it is useful to provide a tool now for its examination.

In many situations the pattern of flow is determined by the relative magnitude of the force associated with the fluid's momentum, which tends to keep the fluid moving, and the force caused by the fluid's viscosity, which tends to bring things to a halt. Again, fluid in a cup serves as an example. If we stir a cup of coffee ($\rho = 1000$ kg/m^3, $\mu = 6 \cdot 10^{-4}$ Ns/m^2), swirling patterns of eddies are formed that die out in 20 or 30 s. If instead of coffee we fill our cup with glycerine ($\rho = 1261$ kg/m^3, $\mu = 0.22$ Ns/m^2), we have to stir vigorously to work up any sort of eddy, and those produced quickly die out. The increased viscosity has changed the pattern of flow. Alternatively, we could fill our cup with mercury ($\rho = 13,546$ kg/m^3, $\mu = 1.55 \cdot 10^{-4}$ Ns/m^2); a brief swish of the spoon sets numerous eddies in motion, which, due to the large momentum of the dense fluid, persist seemingly forever. Again the pattern of flow is changed. In each situation, how much the pattern changes depends on the relative magnitudes of the inertial and viscous forces involved in the flow. This ratio is called a *Reynolds number* (Re) and is the tool we are after.

The form of the Reynolds number can be derived in the following fashion. Consider a hypothetical cube-shaped volume with sides of length L (Fig. 3.9). We locate this volume at an arbitrary fixed point in the flow and allow fluid to move through it as if it were not there. We can then calculate the inertial and viscous forces acting on the faces of the cube. Recalling that force is the time rate of change of momentum (eq. 3.3), we can see that the inertial force acting on the upstream face

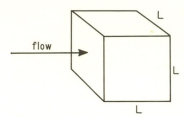

Figure 3.9. The forces exerted by fluid flowing through this cubic volume are used in the derivation of a Reynolds number (see text).

of the cube is equal to the rate at which momentum enters the hypothetical volume. Now the upstream face has area L^2, so if the fluid moves with velocity u perpendicular to this face, a volume uL^2 moves across the face every second. The mass of this fluid is ρuL^2, and, multiplying by velocity, we calculate the *momentum flux*, the rate at which momentum enters our hypothetical volume:

$$\text{momentum flux} = (\rho uL^2)u = \rho u^2 L^2. \quad (3.16)$$

This momentum flux, a force, tends to push the cube downstream. Note that as long as fluid does not change velocity as it crosses the cube, momentum is carried out at the same rate it comes in, leading to an equal but opposite momentum flux on the downstream face of the cube. This second inertial force would balance that on the upstream face, and there would be no net force on the cube. However, for our purposes here we are only concerned with the magnitude of each individual inertial force rather than their combined effect.

We now examine the viscous force acting on the bottom of the cube. From eq. 3.8 we know that the shear stress τ, the force per area acting on the bottom, is equal to the dynamic viscosity multiplied by the vertical velocity gradient across the cube. In many cases the velocity gradient will not be accurately known a priori, so we characterize the gradient in a general fashion and note that the velocity will change by u m/s in a distance $k_{Re}L$, where the constant k_{Re} depends on the particular flow situation. Thus the velocity gradient is $u/(k_{Re}L)$. Given this definition we see that

$$\tau = \text{viscous force}/L^2 = \mu(u/k_{Re}L)$$

$$\text{viscous force} = (1/k_{Re})\mu u L. \qquad (3.17)$$

The ratio between inertial force (eq. 3.16) and viscous force (eq. 3.17) is the Reynolds number:

$$Re = \rho u(k_{Re}L)/\mu. \qquad (3.18)$$

The higher the Reynolds number, the larger the relative importance of inertial forces as compared to viscous forces. Conversely, the lower the Reynolds number, the greater the relative importance of viscosity. The Reynolds number is thus the criterion we need to answer the question posed above.

In this formulation, $k_{Re}L$ is a characteristic length for the particular flow situation. If we denote $k_{Re}L$ by a separate symbol, $L_{characteristic}$, we can write eq. 3.18 in a more standard form:

$$Re = \rho u L_{characteristic}/\mu. \qquad (3.19)$$

Note that $\rho/\mu = 1/\nu$. Thus the Reynolds number may be written as

$$Re = u L_{characteristic}/\nu. \qquad (3.20)$$

Obviously, the Reynolds number we obtain in any situation depends on the length chosen to be characteristic. How do we know what length to choose? Ideally we would choose $L_{characteristic}$ by measuring the velocity gradient (or equivalently, the shear stress), but this is often technically difficult. Besides, the gradient is likely to change from place to place, so which do we choose? Instead of dealing with these problems directly, it is common practice to circumvent them by establishing standard characteristic lengths for certain situations. Thus, there are many different Reynolds numbers, each defined with a particular characteristic length. For example, when calculating a Reynolds number for flow around an object, the length of the object along the direction of flow is taken as char-

acteristic. Similarly, the Reynolds number for flow in pipes is characterized by the pipe diameter.

The use of a characteristic length in the definition of the Reynolds number means that Re is actually one step removed from value we set out to measure. Only if $L_{characteristic}$ is chosen so that the Reynolds number is scaled appropriately to the actual velocity gradient will Re be quantitatively equal to the ratio of inertial and viscous forces. In most situations this is not the case, and Re should be viewed as *proportional* to the ratio of inertial and viscous forces, and the actual value of Re below which viscous forces become significant must be determined empirically for each general type of flow and the pattern being examined.

An example of the utility of the Reynolds number will show how this works.[5] In small-diameter pipes at low flow velocities, water moves in an orderly, *laminar* fashion. Dye introduced into the flow will travel many pipe diameters as a distinct streak line. Under these conditions any disturbance in the orderly progression of the fluid is quickly damped out by viscosity. However, as the pipe diameter is increased or the flow is speeded up (or both), a critical point is reached at which the flow becomes disorderly or *turbulent*. Small disturbances in the flow are no longer damped by viscosity, so they grow and intensify. Soon the fluid moves in a chaotic fashion, and dye introduced into flow is quickly mixed throughout the pipe cross section. By independently varying velocity and pipe diameter, we can experimentally show that this transition from laminar to turbulent flow always occurs at a Reynolds number (based on pipe diameter) of about 2000. However, because we have based our Reynolds number on the dimension of the pipe rather than tying it directly to the velocity gradient, a

---

[5] This example is, appropriately, that in which Osborne Reynolds (1883) experimentally demonstrated the utility of the ratio $\rho u L_{characteristic}/\mu$. This ratio bears his name in honor of these experiments.

value of 2000 does not necessarily mean that inertial forces are *precisely* 2000 times as large as viscous forces at the transition to turbulence. Instead, the Reynolds number serves as an index to tell us, first, that inertial forces are large relative to viscous forces, and, second, that for the particular characteristic length we have chosen, the flow pattern changes drastically when our index reaches 2000. Herein lies the utility of the Reynolds number. By serving as an index of the relative importance of inertial and viscous forces, it provides a wonderfully practical tool for predicting the pattern of flow. As various Reynolds numbers are introduced in this text, their characteristic lengths will be defined, and in each situation the critical values for that particular index will be discussed.

CHAPTER 4

~ ~ ~ ~ ~ ~ ~ ~ ~ ~ ~

An Introduction to
Water Waves

Waves are commonplace. It would be difficult to imagine a world that does not resound with light waves, sound waves, radio waves, and water waves. The attempt to understand wave motion fully has occupied whole legions of physicists and mathematicians for the last three centuries, and the size of the literature on wave mechanics is staggering. Much of this research has been directed toward wave forms that are not readily tangible. For instance, an electron may be thought of as a wave, but it is difficult to pick one up and examine it. In such cases, we must resort to the dry language of mathematics to describe the wave at hand. Not so with ocean waves. The waves that toss ships at sea and the breakers that come foaming and pounding upon the shore are phenomena that are often all too tangible. It is a part of the wave-nature of the universe that we can experience first hand.

We can begin our exploration of the physics of water waves with such a personal experience. We will examine the flows and forces caused by ocean waves in some considerable detail, and along the way we will encounter nonintuitive concepts that are best treated mathematically. However, we can assimilate these concepts more easily if we first examine our tangible experience with water waves.

Terminology

The ocean outside of the surf zone is described as *offshore* (Fig. 4.1). The *near-shore* or *inshore* area extends from the point where waves begin to break up to the beach. Here the term "beach" is used in a generic sense—it is that portion of the land that meets the sea. The beach can be rocky or sandy. The portion of the beach that is covered and uncovered by each wave is the *swash zone*. As the tide rises and falls, the swash zone rises and falls. Those areas on the shore that lie in the swash zone at some time during the tidal fluctuation comprise the *intertidal zone*; below this is the *subtidal* zone. The shoreline is the line at which the still water level (SWL) intersects the shore; it varies with the tides.

Wave Viewpoints

There are two ways in which one typically experiences an ocean wave: floating on the surface of the water, or standing on the shore looking out to sea. Each point of view illuminates different aspects of wave motion, and we will consider them in turn.

Imagine yourself in a small rowboat well out on the ocean. It is a sunny day with no wind to speak of. Despite the benign weather there are large, long waves present (the technical term is *swell*), and you can easily see the waves as they travel. The pattern of waves is very regular; one arrives every 10 s or so. The hull of your boat is embedded in the water and moves as the water moves. What is your experience as you sit here on the ocean's sur-

Figure 4.1. Terms used to describe aspects of the shore (redrawn from Paul D. Komar, *Beach Processes and Sedimentation*, 1976, p. 13, by permission of Prentice-Hall, Englewood Cliffs, N.J.).

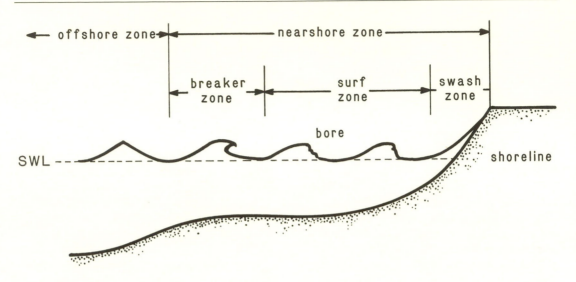

face? The first thing you probably notice is that the boat goes up and down at regular intervals, ad nauseum. If you pay close attention, you can notice a second motion associated with the swell. As you sit in the trough of one wave with the crest of the next wave approaching, your boat is drawn toward the crest so that you are moving not only upward but horizontally as well. After the wave has passed you are again drawn toward the crest, moving horizontally in the opposite direction from before. Your overall motion is backwards (relative to the direction of wave propagation) and up, then forward and up; forward and down, then backwards and down. This is analogous to the motion you experience on a Ferris wheel (Fig. 4.2). If your boat is in sufficiently deep water (we will define "sufficiently" later), the path moved by the boat during the passage of one wave is a circular orbit. The boat moves around this orbit, returning to its starting position in time for the arrival of the next wave, at which time the entire procedure is repeated.

An important lesson is learned from this examination. The motion of the *wave*

Figure 4.2. Motion in a surface wave. (a) The small, numbered arrows show the direction of movement at four points in a wave. The wave form is moving to the right. (b) The analogous motion of a Ferris wheel.

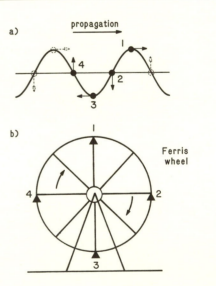

form can be separated from the motion of the *fluid*. As you sit in the boat you can clearly see the waves approach, pass under you, and continue on. The wave form, then, has a net movement in a particular direction, and for this reason these waves are called *traveling waves*. On the other hand, your boat (which moves with the water in which it floats) has no *net* movement; it returns to its starting position with the passage of each wave.

Now consider waves as viewed from a sandy beach. If you look well out to sea, the waves are recognizably the same as those you experienced sitting in your boat. However, as waves approach the shore, the nature of the wave form changes gradually but dramatically. Whereas the waves and present a low and rolling landscape, the waves near shore are more bunched together, and the steepened peaks of waves are separated by nearly flat stretches of water. Although the distance between wave crests decreases as the waves move toward shore, the period between crests is the same as offshore.

The difference between nearshore and deep-water waves extends to the water motion as well as the shape of the wave form. If we throw a tennis ball into the surf zone, it is often gradually washed ashore, demonstrating that in nearshore waves the strictly orbital motion of water changes to a situation where water is transported with the wave form. A moment's thought should convince us that there can still be no net shoreward movement of water. If there were, each wave would pile an additional mass of water onto the beach, and the ocean would quickly empty itself onto the shore. Since this local buildup of sea level does not occur (at least not to any noticeable degree), we are left to conclude that the inshore transport of fluid at the surface (the water movement that brings the ball ashore) is offset by an offshore movement of water somewhere else.

As waves approach shore, they eventually arch over and break. In the process of breaking, the regularity of form so noticeable at sea is reduced to a chaotic, tumultuous mess. The foaming water runs ashore, stirring up the sand, and eventually dies out.

The examples discussed so far assume a sunny, windless day, with no local chop and an orderly swell. This is nature at its most benign, and anyone who has spent time near the sea knows that these conditions are more the exception than the rule. To gain a feeling for the sea in its natural state we must explore a more representative example.

Consider waves as viewed from a rocky coast. It is a blustery day, the wind gusts at 30 knots from offshore, blowing the water off the tops of breakers and forming white caps. The waves raised locally by the wind add *seas* of 1 to 2 m on top of the 2–3 m swell. Despite the rough conditions you are obliged to go down to the shore to monitor a set of experiments. Having no desire to be washed away, you stand on the cliff and watch the surf roll in, trying to discern some pattern in the arrival of waves in the hope that you can correctly predict when it will be safe to scramble down to your test site. As each large wave slams into the shore, it throws spray 5–10 m into the air, and you can feel the concussion of the wave transmitted through the rock. Despite 20 minutes of observation, your attempts to discern a pattern in wave arrival are fruitless. In its present state the sea appears almost random. If you try to focus your attention on one wave crest it often disappears, and waves come from several directions at once. The whole situation is confusing but, being a diligent scientist, you decide to risk it all in the name of progress, and you clamber down the cliff to the shore. No sooner have you descended to your site than a wave that looks like the great-granddaddy of them all looms on the horizon. Muttering vile curses against all stochastic processes, you sprint for safety. But the wave is 5 m high, traveling at roughly 20 mph, and you are soon overtaken. The water knocks you down, scraping you over barnacles and mussels, and try as you might, you do not have the strength to hold onto the shore. Luckily, the wave decides to deposit you on the rocks and retires peacefully rather than sweeping you out to sea.

There are two lessons to be learned from this scenario. First, during rough weather the arrival of waves at the shore is often unpredictable. This unpredictability applies to the time and direction of arrival as well as to the height of the wave. (Much of the discussion in Chapters 6 and 16 will be devoted to finding ways of dealing with the unpredictable nature of water waves.) Second, waves can exert tremendous forces. The lore of the sea is rife with examples of the destructive power of breaking waves. In his engaging introduction to wave mechanics, Willard Bascom (1980) cites cases where storm waves have washed away entire breakwaters, broken the backs of ships, and created inlets in the Outer Banks. One of the most tangible of these stories is that of the lighthouse located at Tillamook Rock near the mouth of the Columbia River. The structure stands with its base 28 m (91 feet) above the sea, and the light reaches a height of 42 m (139 feet) above low water. During one December storm a rock weighing 61 kg (135 pounds) was torn from the base of the cliff and thrown *over* the top of the light, landing on the keeper's house and causing extensive damage. The next time you walk by a thirteen-story building, try to imagine the force necessary to throw your best friend over the structure and you can begin to get a feeling for the power of ocean waves.

Our exploration of the physics of water waves is divided into two parts. In this chapter we discuss the nature of water waves using a semiempirical approach based on observations made in a simple wave tank. For many purposes this approach is sufficient—the answers it gives are qualitatively correct and the mathematics is simple. There are, however,

many situations that require a more thorough mathematical treatment (see Chapter 5).

Laboratory Models

Waves in Deep Water

In attempting to understand a phenomenon as complex as a water wave, it is often useful to begin with a simple laboratory experiment in which all independent variables are under our control. After we play with the laboratory model sufficiently to understand its behavior, it becomes much easier for us to look at the complexity of the real world and make some sense of the apparent chaos.

Generating waves in the laboratory is quite easy, and it would serve the reader well actually to build the apparatus described here. A long tank with a rectangular cross section serves as a model ocean (Fig. 4.3). At one end of the tank is a flap hinged to the bottom. By connecting the top of the flap to a variable speed electric motor one can cause reproducible swinging of the flap. At the end of the tank opposite this wave generator is a gently sloping "beach" that functions to dissipate the energy of any waves propagated down the tank, preventing them from reflecting back toward the flap and thereby greatly simplifying the observation of wave-induced motion. The side of the tank contains a window through

Figure 4.3. A schematic drawing of a laboratory wave tank.

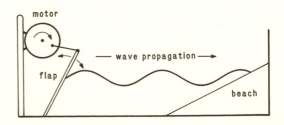

which one can see the internal motions of the wave.

When the motor is turned on, the flap moves back and forth in a sinusoidal fashion at a rate of about two cycles per second. As the flap is pushed down the length of the tank, water is pushed in front of it, piling up to create a wave crest. As the flap moves in the opposite direction it pulls water with it, creating a wave trough. The waves created in this fashion travel down the tank, with the surface of the water forming a sinusoid.

The crests of adjacent waves are separated by a distance, L, the *wave length* (Fig. 4.4). By noting the water level in the troughs and crests as they move past the window, we can measure the total fluctuation in height of the water surface. This is the *wave height*, H, and it is equal to twice the amplitude of the sinusoid, A.

Figure 4.4. Terms used to describe surface waves (see text). For clarity, the height of the wave has been exaggerated; in reality, $H \ll L$.

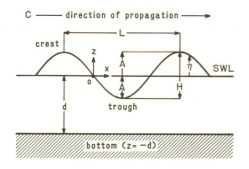

Here we assume that $H \ll L$. The crests of the waves move down the tank with a certain velocity. To avoid confusion between the velocity of the water and the velocity of the wave, the speed at which the wave crests move is called the *celerity* or *phase velocity*, C. The time it takes for a wave crest to travel one wave length is the *wave period*, T. This definition requires that the wave length, period, and celerity be interconnected:

$$C = \frac{L}{T}. \qquad (4.1)$$

To use the sinusoidal waves in our tank as a model, we must hypothesize that they are analogous to waves occurring on the ocean. We can test this hypothesis by examining in the tank the motions of individual particles of fluid. This is best done by suspending in the water small solid particles with the same density as water. These particles can be assumed to move in the same fashion as the water in which they are suspended. By watching the movement of a single such particle, we find that, so long as wave height is small relative to wave length, water particles in our laboratory waves move horizontally the same distance as they move up and down, and that they travel in very nearly circular orbits.[1] Thus waves in our tank are indeed an appropriate model for the waves of the ocean.

A simple experiment can be conducted by varying the speed of the motor without varying the amplitude through which the flap moves. By doing this we can create waves with a constant height but whose periods are different. It is immediately noticeable that waves of different periods have different wave lengths and move with different celerities. Within limits, the longer the period, the longer the wave length, and the faster the wave form moves. For example, a wave with a period of 0.25 s has a wave length of about 0.1 m and a celerity of about 0.4 m/s. For a wave with a period twice as long ($T = 0.5$ s), the wave length has increased to 0.4 m and $C = 0.8$ m/s. If we repeat this experiment but with a different amplitude through which the flap moves, we create waves with a different height. How-

---

[1] A careful examination shows a slight net transport of water particles in the direction of wave propagation, and the same is true of real ocean waves. The mechanism of this transport will be dealt with in Chapter 5. For the purposes of this introduction, the deviation from a circular orbit is negligibly small.

ever, as long as $H \ll L$ (see Fig. 4.4), the height of the waves has no apparent effect on their length and celerity. How can we account for these experimental facts?

The fluid mechanics we have discussed so far have dealt with fluids moving in a time-invariant fashion, and here we are dealing with flows that by definition are oscillatory. To apply our existing knowledge, we require some means by which the oscillating wave flow in the tank can be converted into a steady flow. This can be accomplished in either of two ways. To an observer walking alongside the tank at the speed C, the wave appears stationary and the tank moves backward at velocity $-C$. The same result could be obtained by applying a uniform counter-current to the water through which the wave travels. Thus if the wave travels to the right (in what we define as the positive direction) with a celerity C, a counter current with celerity $-C$ brings the wave form to a halt relative to a stationary observer. Although this second strategy would be difficult to implement in practice, it is no problem to do so in theory, and we employ it in the following explanation.

The application of a countercurrent brings the wave form to a halt, but it should be clear that we have not brought the water to a standstill. Water at the crest of the wave travels in the same direction as the wave itself (the positive direction), and water in the trough travels in the direction opposite the wave (Fig. 4.5a). The speed at which the water is traveling in these two situations can be calculated from the following facts: the water at the surface travels through a circular orbit with a diameter equal to the height of the wave (Fig. 4.5b), and it travels once around this path every wave period. The circumference of the orbit is thus

$$\text{circumference} = \pi H \qquad (4.2)$$

and the speed is

$$U = \frac{\pi H}{T}, \qquad (4.3)$$

where T is the wave period and $U = \sqrt{u^2 + w^2}$. Knowing this orbital speed, we can now calculate the overall velocity at the crest and trough. At the crest the water's speed is in the direction of wave propagation ($U = u$), opposite the direction of the imposed current. Adding the two velocities, we see that here the water moves at a rate

$$u - C = \frac{\pi H}{T} - C, \qquad (4.4)$$

relative to our stationary frame of reference. In the trough, the water moves opposite the direction of wave propagation (in the same direction as the imposed current), so that the velocity relative to our frame of reference is

$$-u - C = \frac{-\pi H}{T} - C \qquad (4.5)$$

Figure 4.5. Calculation of wave celerity for a sinusoidal surface wave. (a) Flow in the wave. (b) Water moves in a circular path with a diameter equal to the wave height. Thus $U = \pi H / T$, where H is the wave height and T is the wave period.

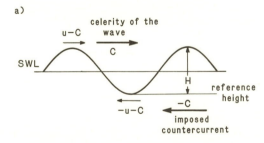

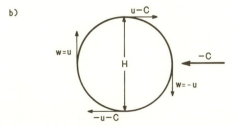

Because the countercurrent has brought the wave form to a halt, the location of the water's surface is stationary relative to our stationary frame of reference. This of course does not mean that the water itself is stationary; there is water flowing along this surface at speeds varying from $u - C$ to $-u - C$. The fact that this water moves along a time-invariant path should ring a bell in your mind: the water's surface defines a streamline. We thus have expressions for the velocity of water at two points on a streamline (the crest and the trough) and we know the difference in height between these two points (the wave height, H). Because both points are at the water's surface, and therefore exposed to atmospheric pressure, we can assume that the pressure at the two points is the same. We now have all the ingredients for a useful application of Bernoulli's equation.

Recall from Chapter 3 that Bernoulli's equation is a statement of conservation of available energy. The sum of the volume-specific kinetic, flow, and potential energies at any point on a streamline (e.g., the trough) is equal to that at any other point on the same streamline (e.g., the crest):

$$\tfrac{1}{2}\rho u^2_{\text{trough}} + p_{\text{trough}} + \rho g h_{\text{trough}}$$
$$= \tfrac{1}{2}\rho u^2_{\text{crest}} + p_{\text{crest}} + \rho g h_{\text{crest}}. \quad (3.15)$$

Because the pressure at the trough and crest are assumed to be the same, the pressure terms in this equation can be subtracted out. For simplicity, we define the height of the wave trough, h_{trough}, to be our reference height. Thus, for the wave form shown in Fig. 4.5,

$$\tfrac{1}{2}\rho \left\{ \frac{-\pi H}{T} - C \right\}^2 = \tfrac{1}{2}\rho \left\{ \frac{\pi H}{T} - C \right\}^2 + \rho g H. \quad (4.6)$$

When the squared terms are expanded and like values canceled from both sides of the equation, we find that

$$C = \frac{gT}{2\pi}. \quad (4.7)$$

This is the first of the relationships noted in the wave tank: waves with longer periods have higher celerities.

Noting from eq. 4.1 that $L = CT$, we can restate eq. 4.7 as

$$L = \frac{gT^2}{2\pi}. \quad (4.8)$$

Thus, the longer the period, the longer the wave length.

Similarly, note that $T = L/C$ (eq. 4.1). Substituting for T in eq. 4.7, we see that

$$C = \frac{gL}{2\pi C}$$

$$C = \sqrt{\frac{gL}{2\pi}}. \quad (4.9)$$

This is the last of the relationships observed in the wave tank; the longer the wave length, the faster the wave travels.

These equations can be used to calculate the wave length and celerity expected for ocean waves in deep water (Table 4.1). For example, a wave with a period of 10 s (a common value for a swell) has a wave length of 156 m and a celerity of about 15.6 m/s.

We have made several questionable assumptions in working through these calculations. For instance, the assumption

Table 4.1
Periods for Surface Waves and Their Associated Celerities and Lengths

Period, T (seconds)	Celerity, C (m/s)	(mph)	Wave Length, L (m)
1	1.6	3.5	1.6
2	3.1	7.0	6.2
3	4.7	10.5	14.0
4	6.2	14.0	25.0
5	7.8	17.5	39.0
10	15.6	34.9	156.1
15	23.4	52.4	351.2
20	31.2	69.8	624.4

that the water at the surface is subject to atmospheric pressure is true only if the waves are of sufficient length, so that nowhere is the curvature of the water very great. At short wave lengths the surface of the water is substantially curved, and the action of surface tension serves to raise the pressure under the crests and decrease the pressure under the troughs. These short waves (called *capillary waves*) obey slightly different rules. However, waves must be less than about 10 cm in length (small compared to the lengths of ocean waves) before surface tension effects become appreciable.

We have also ignored any possible effects that the solid bottom of the tank might have on the fluid motion. To examine this assumption, we must have a closer look at the movements of water under laboratory waves. We have already demonstrated that the water at the surface travels in a circular orbit with a diameter equal to the wave height. However, this motion does not extend indefinitely down into the water column. By observing particles suspended in the tank, we find that water below the surface still travels in circular orbits, but the diameter of the orbit is decreased the deeper in the tank we look (Fig. 4.6a). By

careful observation we can determine that the orbits have decreased to a negligible diameter at a depth of one half a wave length below the still water level. In other words, below a depth of $L/2$, the water does not move appreciably in response to waves moving on the surface. Thus, if the wave tank is deeper than one half wave length, the water at the bottom would be virtually stationary whether or not the bottom is there, and the presence or absence of a solid bottom makes little difference. For instance, a wave with $T = 0.5$ s has a length of 0.4 m. If the water depth in our tank is greater than 0.2 m, the surface waves will not substantially move the water at the bottom. Conversely, when the depth of the water in the tank is less than half the wave length, the bottom is very likely to affect the flow patterns, and one has no assurance that ignoring the bottom still leads to valid conclusions.

These potential effects of the bottom provide us with an operational definition for the term "deep." Waves in water depths of greater than one-half wave length behave as if the bottom were not there at all, and depths greater than $L/2$ are referred to as "deep." This means, of course, that one cannot draw on a chart

Figure 4.6. The orbital motion of water beneath waves. (a) Deep water. (b) Shallow and intermediate depths.

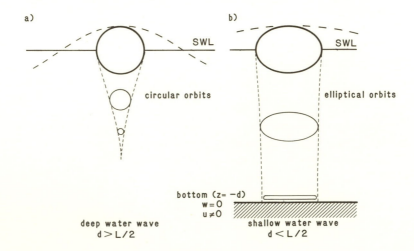

the point at which water becomes deep, since this point depends on the length of waves present. Periods of 4–15 s are common for real seas and swell, corresponding to wave lengths of 25–351 m (eq. 4.8; Table 4.1). The point at which ocean waves "feel" the bottom can thus vary from about 12 to 175 m.

Waves in Shallow Water

The information we have gleaned from wave-tank experiments applies primarily to waves in deep water. When ocean waves travel into shallow water, their period stays the same, but the water motion changes in several ways. When the water depth is less than half the wave length, the observed celerities are less than those predicted by eq. 4.7; one effect of the bottom in shallow water is to slow the wave down. Another effect becomes apparent when we look at the motion of particles suspended in the water beneath the surface of a wave in shallow water. Instead of moving in closed circular orbits, water moves along flattened, elliptical paths (Fig. 4.6b) and may exhibit a substantial net movement in the direction of wave propagation. The closer to the bottom, the more flattened the orbits, until at the bottom itself the motion is strictly horizontal. The waves become more peaked and closer together, and the troughs between waves become flattened. In very shallow water, the changes in shape and orbital motion are so pronounced that one must wonder whether we are still dealing with the same phenomenon. In fact, the wave motion in very shallow water can be modeled approximately in terms of a different sort of wave form—the *solitary wave*.

We examine solitary waves by returning to the wave tank. We start by placing the flap at the end point of its travel and leaving it stationary until all water motion stops. The flap is then moved down the tank to the opposite end of its travel, where it is again held stationary. The result is a single

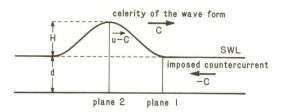

Figure 4.7. Definitions for terms used in calculating the celerity of a solitary wave (see text).

wave crest that travels down the tank. Because the flap has not been allowed to go through the second half of its sinusoidal cycle, no wave trough is formed and all the water in the wave form is above the still water level. This is a solitary wave (Fig. 4.7).

As a first approximation, we can think of a solitary wave as the upper half of a sinusoidal wave, but we must also note the differences. As with the upper half of a sinusoidal wave, the water in a solitary wave is traveling in the same direction as wave propagation. However, because the solitary wave is not accompanied by a trough, there is no corresponding movement in the opposite direction, and the passage of a solitary wave results in the net transport of water. This is appropriately analogous to the situation noted for real ocean waves. A tennis ball floating in shallow water is stationary until a wave crest approaches. It is then carried a distance shoreward, and again becomes stationary as the waves passes.

The second major difference between solitary waves and sinusoidal waves involves the definition of the wave period. Because a solitary wave consists of a wave crest with no trough, the water never returns to its original position. Since the wave period can be defined as the time it takes a fluid particle to travel around its orbit and return to its initial position, a true solitary wave has no defined period. This aspect of a solitary wave is at odds with ocean waves which, despite their change in shape, remain periodic in shallow water.

Thus the solitary wave is not an exact model of ocean waves in shallow water, and we must erect a mental warning sign regarding the extrapolation of our laboratory findings. Despite its failings as an exact model, the solitary wave is useful in explaining several aspects of shallow water wave flow.

The celerity of a solitary wave in shallow water can be calculated by a procedure similar to that used for sinusoidal waves. As before, the solitary wave form is brought to rest through the application of a uniform countercurrent with velocity $-C$, equal in magnitude but opposite in direction to the wave celerity, C. The undisturbed water well in advance of the solitary wave therefore moves with velocity $-C$ relative to a stationary observer (Fig. 4.7). Consider two planes fixed in space and perpendicular to the length of the tank. One slices through the tank well in front of the wave crest, and the other at the crest. These planes, along with the walls of the tank and the time-invariant surface of the water can be thought of as forming the boundaries of a rigid pipe through which water flows. Water flowing into the pipe through the plane in front of the wave must exit through the plane at the crest, and the equations of continuity can be applied. For the sake of simplicity we assume that the velocity does not vary with depth at the ends of the pipe.

The wave tank has a fixed width y, and in front of the wave it has a depth d. The volume of fluid flowing into the pipe is thus $-Cyd$ per second. At the crest of the wave, the total depth of the water is $d + H$. The velocity at this point is that of the imposed current $(-C)$ plus the velocity of the water moving in the direction of the wave, u:

$$(u - C)y(d + H) = Cyd$$

$$u = \frac{-Cd}{d + H} + C. \quad (4.10)$$

As before, we assume that atmospheric pressure is exerted everywhere on the water's surface. We thus know the velocity, pressure, and height at two points on the surface streamline, and can apply Bernoulli's equation. We take the still water level to be our height reference. By analogy to eq. 4.6, we write

$$\tfrac{1}{2}\rho(-C)^2 = \tfrac{1}{2}\rho(u - C)^2 + \rho gH. \quad (4.11)$$

Squaring the term in parentheses and canceling equal terms, we see that

$$C^2 = u^2 - 2uC + C^2 + 2gH. \quad (4.12)$$

This equation can be simplified if we assume that u is sufficiently small compared to C so that u^2 is negligible. Given this assumption,

$$C^2 = -2uC + C^2 + 2gH$$

$$uC = gH. \quad (4.13)$$

Inserting the expression for u from eq. 4.10 and rearranging, we get

$$C^2 = g(d + H), \qquad C = \sqrt{g(d + H)}. \quad (4.14)$$

Thus the celerity of a solitary wave in shallow water depends on the depth of the water and the wave height and is independent of wave period. This is in sharp contrast to the case of a sinusoidal wave in deep water, where the celerity is independent of the water depth and wave height and depends solely on the wave period.

If we were to take the value for the velocity at each point under a solitary wave and multiply by the time spent at that velocity as the wave passed by, we could calculate the increments of displacement for each point. If we add all these points together, we will get the distance forward a bit of water is carried during the passage of a solitary wave. According to McGowan's (1891) theory, a close approximation is

$$\text{translation distance} = 4\sqrt{\frac{dH}{3}}. \quad (4.15)$$

The higher the wave, the farther the water is carried as the wave passes.

When checked in the wave tank, we find

that the result obtained from our calculations is quite accurate. However, this correspondence between prediction and observation is somewhat fortuitous. In the above calculations, we have assumed that the velocity did not vary with depth in any vertical plane. In a real solitary wave the velocity does vary with depth, being higher than the values used here (eq. 4.10) near the surface and lower near the bottom. The actual velocity distribution in a solitary wave is shown in Figure 5.8c.

We will now examine two consequences of the fact that the celerity of waves in shallow water (as modeled by solitary waves) depends on the water depth.

Breaking Waves: A First Look

Consider an ocean wave as it travels shoreward into shallower and shallower water. Its profile becomes less symmetrical as the front of the wave steepens relative to the back. At some point the wave form becomes unstable, and the front of the wave collapses. The process leading to this collapse is most obvious in waves breaking on beaches with moderate bottom slopes. Here the breakers are of a type aptly termed *plunging* (see Fig. 7.4). These are the sort of waves one sees in surfing magazines. At the point of breaking, the water at the crest of the wave begins to travel faster than the wave form itself, forming a jet that arches over the front of the wave. The same phenomenon happens in all breaking waves, but depending on the slope of the bottom (among other things), the portion of the water that moves faster than the wave form varies. The result can be waves where the crest seems to spill down the front of the wave (*spilling waves*) or waves where the entire wave form seems to collapse (*collapsing waves*). The factors affecting breaker type are discussed in some detail in Chapter 7. For the present we concentrate on the general sequence of breaking—a steepening of the wave (the decrease of its length relative to its height), followed by breaking when the wave becomes unstable.

The steepening of periodic waves in water of decreasing depth is a direct result of the decrease of the celerity of the wave, C, in shallower water while the period, T, remains the same. Because $L = CT$ (eq. 4.1), the length of waves decreases in shallow water. Unless the wave height decreases in proportion to the wave length, the wave must get steeper. In fact, wave height increases in shallow water as waves shoal (see Chapter 7) and this steepening is augmented.[2] At some point the wave becomes unstable, much as a pile of sand is unstable if its sides are too steep. This instability occurs when water at the crest travels at the same speed as the wave form. As the wave breaks, the momentum of the water at the crest carries it ahead of the wave, forming the plunging jet.

Wave Refraction

The dependence of wave celerity on water depth also provides a ready explanation for *wave refraction*, the tendency for waves in shallow water to bend until their crests are parallel to the shore. Consider the case shown in Figure 4.8a. A wave with a straight crest approaches the shore at an angle α. To keep the situation uncluttered, we give the sea floor a simple configuration. Water near the shore has a constant depth, and at a certain distance from the shore this suddenly steps down to a greater depth, which is maintained out to sea. We follow what happens to a short segment of crest as it moves along. At time $t = 0$ the segment is traveling in deep water. All portions of the segment travel over the same depth, and therefore have the

---

[2] One might think that the increase in wave height during shoaling would offset the decrease in depth to allow for a constant celerity (eq. 4.14). Actually, the increase in H is not sufficient, and C generally decreases with decreasing depth (see Chapter 7).

Figure 4.8. Wave refraction. (a) The change in direction as a wave moves at an angle over a step decrease in water depth (see text). (b) The refraction of waves on a gently sloping beach. Waves gradually change direction, becoming more parallel to the shoreline.

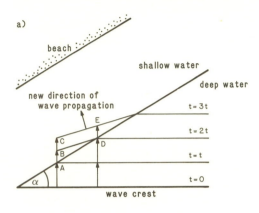

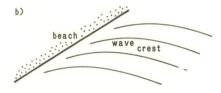

the right end of the row moves at a constant speed and each soldier to his left moves a bit slower; as a result the line gradually turns to the left.

It would be unusual to find a shore with this sort of simple bottom topography; a shore with a bottom that slopes steadily down would be much more typical. This sort of continual slope can be viewed as a series of many small step changes in depth and treated in the same fashion as outlined here. On such a shore, waves continually move into shallower water, and therefore continue to turn as they move toward shore; the crests make sweeping curves as shown in Figure 4.8b. Only when the crest line is exactly parallel to the shore do all parts of the crest move at the same speed, and at that point the crest stops turning. The practical effects of refraction are examined in greater detail when the energy relationships of shoaling waves are examined in Chapter 7.

Wave Diffraction

The direction of wave propagation can also change as waves move past an object in water of constant depth. For instance, wave crests will "wrap around" an island or the tip of a breakwater. This phenomenon is called *wave diffraction*. Diffraction may have important practical consequences for the design of harbor entrances and large offshore structures, but it is of little apparent consequence to nearshore organisms and will not be treated further here. For more information on wave diffraction, consult the U.S. Army Corps of Engineers' *Shore Protection Manual* (1984).

Wave Energy

To this point the energy associated with wave motion has been used as a means for predicting wave celerity. It is now time to

same speed (eq. 4.14). At time t our segment of wave crest has traveled forward sufficiently so that its nearshore end is just over the step decrease in depth at point A. From time t to $2t$ the nearshore end of the segment travels at a slower velocity than the offshore end, and therefore it travels a shorter distance. By the time the offshore end reaches the decrease in depth at point D, the nearshore end has only traveled to point B. A line connecting points B and D (the ends of the inshore portion of segment of crest) lies at a different angle to the shoreline than did the crest while offshore. In effect, the crest has turned so as to become more parallel to the shoreline. If we follow the wave further, from time $2t$ to time $3t$, we see that both the inshore and offshore ends of the segment travel the same distance, and the crest stops turning. The effect is much the same as watching a line of marching soldiers making a left turn. The soldier at take a closer look at the energy itself. We first examine the case of sinusoidal waves in deep water.

The mechanical energy associated with wave motion is divided between kinetic and gravitational potential energy.[3] The kinetic energy, that due to the velocity of the water's mass, is associated with the movement of water particles around their circular orbits. Under a sinusoidal wave the speed of a particle, U, remains constant as it moves around its circular path (eq. 4.3), and its kinetic energy per volume ($[\frac{1}{2}]\rho U^2$) remains constant. For a water particle at the surface, $U = \pi H/T$ (eq. 4.3). Thus,

$$\text{kinetic energy per volume} = \frac{1}{2}\rho\,\frac{\pi^2 H^2}{T^2}. \quad (4.16)$$

We know from eq. 4.1 that the period T is related to wave length and celerity, $T = L/C$. In eq. 4.9 we calculated that the celerity in deep water can be related to wave length, $C^2 = gL/(2\pi)$. Combining these facts, we see that

$$T^2 = \frac{L^2}{C^2} = \frac{L^2}{\left\{\dfrac{gL}{2\pi}\right\}} = \frac{2\pi L}{g}. \quad (4.17)$$

Substituting this expression for T^2 into eq. 4.16, we find that the kinetic energy per volume of water at the surface of a sinusoidal wave is

$$\text{kinetic energy per volume} = \frac{1}{4}\rho\pi g\,\frac{H^2}{L}. \quad (4.18)$$

Kinetic energy is thus proportional to the square of wave height and inversely proportional to wave length. However, this is energy per volume calculated for the water right at the surface. To calculate the total kinetic energy it would be necessary to repeat this exercise for all the water affected by the wave motion. Because the diameter of orbits decreases with depth below the

[3] Flow energy, the energy that can be transmitted by a fluid between areas of different pressure (Chapter 3), is not an energy possessed by the fluid itself, and thus will not be considered here.

surface, kinetic energy decreases with depth, approaching zero at a depth of one-half wave length. This dependence of orbital diameter on wave length tends to offset the effect of wave length on kinetic energy. Although short-wave-length waves have more kinetic energy per volume at the surface, the wave motion does not extend deeply into the water column and therefore affects a small volume. Conversely, long-wave-length waves have less kinetic energy per volume at the surface, but affect a larger volume. If the volume of water affected is taken to be proportional to L^3, we arrive at the relationship

$$\text{kinetic energy} \propto \rho g H^2 L^2,$$

or that the kinetic energy per area of water surface (proportional to L^2) is

$$\text{kinetic energy per area} \propto \rho g H^2.$$

From a theoretical treatment of the variation in orbital velocity with depth, this energy per area can be calculated:

$$\text{kinetic energy per area} = (\tfrac{1}{16})\rho g H^2. \quad (4.19)$$

Thus the kinetic energy varies with the square of the wave height and is constant for all portions of the wave. The kinetic energy per surface area is independent of wave length.

Wave energy is often expressed as energy per length of wave crest. This value is obtained by multiplying the kinetic energy per area by the area of a section of surface one unit wide and one wave length long:

$$\text{kinetic energy per length of crest} = (\tfrac{1}{16})\rho g H^2 L. \quad (4.20)$$

The gravitational potential energy of a wave occurs because water is displaced from the height it occupies when the water is undisturbed. The potential energy is maximal at the crest and minimal in the trough, and it is zero for those parts of the wave that are momentarily at still-water level. Thus, unlike kinetic energy that is the same for each portion of the wave, the potential energy varies from place to place.

Because the vertical displacement of water decreases with increased depth, the contribution to the overall potential energy also decreases with depth. By following a process similar to that described for kinetic energy, we can calculate the potential energy for all particles under one entire wave. Although potential energy varies from place to place, a spatially averaged potential energy can be obtained for one wave length. When this is done, the average potential energy per area is found to be

$$\text{average potential energy per area} = (\tfrac{1}{16})\rho g H^2. \tag{4.21}$$

Comparing this result to that obtained in eq. 4.19, we conclude that the total energy of the wave is split evenly between kinetic and potential energy.

To give this discussion of wave energies some tangibility, we calculate the energy per area for a typical ocean wave. The total energy of a sinusoidal wave is $(\tfrac{1}{8})\rho g H^2$. The acceleration of gravity is 9.81 m/s^2, and the density of seawater is approximately 1025 kg/m^3. Thus the energy per area for a 4-meter-high wave is

$$\begin{aligned}
&\text{energy per area} \\
&= (\tfrac{1}{8})(1025 \text{ kg/m}^3)(9.81 \text{ m/s})(4 \text{ m})^2 \\
&= 2 \cdot 10^4 \text{ J/m}^2;
\end{aligned}$$

there is enough energy in each square meter to lift a metric ton (1000 kg) two meters against the acceleration of gravity. A thousand square kilometers of such ocean surface have a total energy of $2 \cdot 10^{13}$ J—enough energy to supply all of Rhode Island with electricity for three days (U.S. Department of Commerce, 1980). This assumes, of course, that all the wave energy could somehow be extracted and converted to electrical energy, a process that has not yet proven to be practical on a large scale. One problem in extracting wave energy concerns the "transport" of this energy. The speed at which energy is transported by surface waves is not always equal to the celerity of the wave form. Rather, energy is transported at the *group velocity*, C_g. In shallow water C_g is equal to the celerity of the wave form, C, but in deep water $C_g = C/2$. Because energy is being transported at the group velocity, the rate at which energy would arrive at a deep-water wave-driven power station would be only about half of what one might expect. Unfortunately, there is no simple, intuitive explanation of group velocity, and the inquisitive reader should consult Kinsman (1965) for the mathematical derivation.

Summary

In deep water:

1. Water moves in approximately circular orbits with no appreciable net movement in the direction of wave propagation.
2. The diameter of the orbit decreases with depth.
3. The celerity of the wave depends on the wave period (and thereby the wave length) and is independent of wave height and water depth.

In shallow water:

1. There is net movement of water in the direction of wave propagation.
2. The celerity of the wave depends on the depth of the water and the wave height and is independent of wave period.

At all depths wave energy is proportional to the square of wave height.

These are useful "rules" to remember regarding wave motion, but like all generalities they should be used with caution. For instance, we have assumed in this chapter that the height of waves is small relative to their wave length. When this is not true the celerity of deep-water waves may depend slightly on their height, and in this respect they mimic shallow-water waves. This and other such complications are discussed in Chapter 5.

Wave Theories

The simple, often qualitative approach taken in Chapter 4 provides an intuitive understanding of wave action, but its utility is severely limited. For instance, what if we need to know the water's acceleration at a depth of 3 m beneath a periodic wave 2 m high? This simplified approach cannot provide an accurate answer. What are the pressure changes at the bottom as a wave passes overhead? These pressure changes are a useful way of measuring wave heights (Chapter 19), but the approach taken so far does not specify the relationship between wave height and bottom pressure. To be able to use a knowledge of wave action as a practical tool, we must have access to more accurate theories.

In this chapter we explore the simplest of these—linear wave theory. The process through which this theory is constructed is typical of many wave theories, and therefore should provide a general understanding of how such theories are devised. Linear wave theory is itself very useful in a variety of practical applications and is an excellent heuristic tool.

In our exploration of linear wave theory we will examine both deep- and shallow-water waves. The attention paid to deep-water waves may at first seem out of proportion to their biological relevance; after all, waves can interact directly with the plants and animals of the shore only when they have moved to intermediate and shallow depths. Still, there are several compelling reasons for examining deep-water waves. First, wave-induced fluid motions and the equations that describe them are simplest and most intuitive for deep-water waves. In this respect, deep water is the best classroom for introducing wave theory.

Second, waves that arrive at the shore were born in deep water, and their deep-water characteristics can affect their inshore behavior. And finally, the transition from deep to shallow water waves may itself be of biological importance. For instance, we will see that shoreward mass transport is greatest at the surface in deep water but greatest at the bottom in intermediate and shallow water—a switch which could be an important determinant of behavior in larvae attempting to reach the shore.

Linear Wave Theory

Nomenclature

When studying waves, we locate the origin of our coordinate system at the still water level (SWL) (Fig. 5.1). The x-axis lies in the direction of wave propagation, and the y-axis is parallel to the wave crests. The z-axis is vertical and positive upwards. Because one usually deals with water below SWL, z is often negative for the areas of interest. A separate variable, η, is used to denote the displacement of the surface from SWL. The depth of the water below SWL is d. Distance from SWL is measured by z, and distance from the bottom is measured by s; $s = d + z$. Wave height, H, is measured from crest to trough. Wave length, L, is the distance from one crest to the next.

In constructing any mathematical theory to describe wave motion, we must make certain assumptions and specify certain conditions. Linear wave theory operates under the following assumptions and conditions.

Figure 5.1. Definition of terms used to describe surface waves.

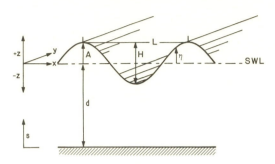

Assumptions

1. We assume that the flow is "two-dimensional." By this we mean that flow conditions are allowed to vary along the x and z axes, but for any x, z coordinate the flow does not vary as a function of y. This assumption allows us to look at one vertical slice through a wave and know that we have described the flow for every part of the wave. In practice, the assumption of two-dimensionality requires that waves be long-crested, that is, that the wave height is constant for a length of crest on the order of the wave length.

2. We assume that waves are periodic, with a constant period and height. This assumption is often violated by ocean waves, a problem we will deal with in Chapter 6.

3. We assume that the wave height is infinitesimal—in a practical sense, that $H \ll L$. Like Assumption 2, the assumption of small wave height is often violated by ocean waves. We will deal with this problem in the last section of this chapter.

4. We assume that the fluid is incompressible, and that flow will therefore exhibit *continuity*. The concept of continuity (i.e., the conservation of mass) was introduced in Chapter 3 but without any generally applicable mathematical method for describing which flows have it and which do not. In mathematical terms, incompressible fluids fulfill the criterion that

$$\frac{\partial u}{\partial x} + \frac{\partial v}{\partial y} + \frac{\partial w}{\partial z} = 0. \qquad (5.1)$$

This is the *continuity equation*, an informal derivation of which is provided in Appendix 5.1.

5. We assume that surface tension is negligible.

6. Last, we assume that the flow under waves is *irrotational*. This concept requires a bit of explanation. In Chapter 3 we noted that well away from a solid boundary water acts as if it were an inviscid fluid. This lack of apparent viscosity allowed us to propose and use Bernoulli's equation. We now explore another aspect of the behavior of inviscid fluids. Take an ice cube and float it in the middle of a bowl of water as a convenient means of marking the fluid. Put the bowl of water on the turntable of your stereo and start the turntable rotating. What happens to the ice cube? Initially nothing. As the bowl rotates, the ice cube (and the water around it) stays stationary. Eventually, of course, the no-slip condition at the walls of the bowl interacts with the viscosity of the water and the water in the bowl begins to rotate. But this is only true because the water in the bowl is everywhere close to the wall. If you were to float an ice cube in the middle of a circular swimming pool and somehow rotate the pool, it would take much longer for the water and the cube to begin to rotate. Alternatively, if the fluid in our bowl were less viscous, the tendency of the fluid to rotate would be similarly decreased. If the fluid could be made entirely inviscid, the tendency to rotate would disappear altogether. If an entirely inviscid fluid has no rotation to begin with, there is no simple way it can be made to rotate.[1] An initially stationary inviscid fluid will thus remain *irrotational*. Real fluids sufficiently far

[1] There are some ways, however, in which an initially stationary inviscid fluid can be made to rotate. For instance, if the density of the fluid varies so that denser fluid is layered on top of less dense fluid, the resulting instability can lead to rotation. For a rigorous description of irrotational flow, consult a standard text in fluid dynamics.

from a solid surface behave as if they are inviscid, and therefore do not easily change their state of rotation. *Under these circumstances* water, which at some time was not rotating, will remain irrotational and exhibit *irrotational flow*. The water motions caused by ocean waves on the whole take place quite far from any solid boundary, and these flows are approximately irrotational.

In the example used above (a bowl of water on a turntable), viscosity eventually caused the water to rotate about the axis of the turntable. In general, a water particle is not confined to rotating about one particular axis. Consequently, to ensure that flow is totally irrotational, we need to know that fluid does not rotate about either the x, y, or z axes. For the x-axis, flow is irrotational if

$$\frac{\partial v}{\partial z} - \frac{\partial w}{\partial y} = 0. \qquad (5.2)$$

Similarly, for the y-axis, flow is irrotational if

$$\frac{\partial u}{\partial z} - \frac{\partial w}{\partial x} = 0, \qquad (5.3)$$

and for the z-axis if

$$\frac{\partial u}{\partial y} - \frac{\partial v}{\partial x} = 0. \qquad (5.4)$$

The skeptical reader can find the derivation of these equations in any standard text on fluid dynamics, and an informal verification is given in Appendix 5.2.

The Velocity Potential Function

Having made these assumptions about wave motion, we must, in constructing our wave theory, next specify what is known as the *velocity potential function*. Again, this requires a bit of explanation. Consider a section of pipe with water flowing through it at a certain velocity. Why is the water flowing? This apparently simple-minded question need not have a simple answer. One obvious possibility is that the

pressure at one end of the pipe is higher than that at the other. Given such a situation, the water can reasonably flow away from the area of high pressure. This need not be the case, however. Recall the discussion of energy conservation presented in Chapter 3. Water with a high velocity and a low pressure can flow into a region of higher pressure by losing speed. For example, water in a small pipe section will flow into a section of larger diameter even though it entails moving against a pressure gradient. Any number of different mechanisms can be concocted by which water can be made to flow in a pipe. About the only thing that must be common to all these mechanisms is the fact that *something* must change along the pipe's length that results in the application of a net force. If absolutely everything is everywhere the same, there can be no reason for the fluid to move. Fluid dynamicists deal with this problem by turning the question on its head. Rather than starting with the cause and seeing what velocity it produces, they start with the velocity and define the "something" that causes it in terms of the velocity itself. This "something" is the velocity potential function, Φ, defined (for one dimension) by the relationship

$$\frac{\partial \Phi}{\partial x} = u, \qquad (5.5)$$

or, the change in the velocity potential function with distance in the x-direction is equal to the velocity in the x-direction. The concept of a velocity potential is analogous to that of the voltage potential in an electric circuit: the rate at which electric current flows is proportional to the change in voltage potential along a wire. The beauty of this concept may not be immediately apparent, but defining the velocity potential in this manner avoids the necessity of specifically saying what it is that causes flow, and one therefore has a quantity that is valid for any and all mechanisms.

The definition of the velocity potential function given above applies only for motion along a single axis. The equivalent definition for three dimensions is

$$\mathbf{v} = \mathbf{i}\frac{\partial \Phi}{\partial x} + \mathbf{j}\frac{\partial \Phi}{\partial y} + \mathbf{k}\frac{\partial \Phi}{\partial z}, \qquad (5.6)$$

where **v** is the overall velocity vector, and **i**, **j**, and **k** are unit vectors parallel to the three axes. The components of the velocity can be found by evaluating the right-hand side of this equation with respect to the x, y, and z axes.

$$\begin{array}{l} u \text{ (the magnitude of velocity} \\ \text{in the } x\text{-direction)} \end{array} = \frac{\partial \Phi}{\partial x}$$

$$\begin{array}{l} v \text{ (the magnitude of velocity} \\ \text{in the } y\text{-direction)} \end{array} = \frac{\partial \Phi}{\partial y}$$

$$\begin{array}{l} w \text{ (the magnitude of velocity} \\ \text{in the } z\text{-direction)} \end{array} = \frac{\partial \Phi}{\partial z}. \quad (5.7)$$

In discussing the velocity potential function so far, we really have not accomplished much beyond a convenient change in notation. If we knew the velocity potential function for wave motion, we could specify the water's velocity, but we are at present no closer to knowing the potential function itself. To devise an appropriate potential function, we must make sure that the function fits our assumptions regarding wave-induced flows, and we have to impose several boundary conditions.

First, we can easily show that flows must be irrotational to be described by a velocity potential. Consider rotation about the y-axis (eq. 5.3). Substituting for u and w the corresponding values in terms of the velocity potential function (eq. 5.7), we see that

$$\frac{\partial(\partial \Phi/\partial x)}{\partial z} - \frac{\partial(\partial \Phi/\partial z)}{\partial x} = 0, \qquad (5.8)$$

which must be true from the basic properties of derivatives. The same conclusion is reached for rotation about the x-axis or z-axis. The potential function thus requires irrotational flow. Conversely, only if the flow is irrotational is the potential

function valid. This is the reason behind assuming irrotational flow (Assumption 6) in constructing linear wave theory.

Next we insert the velocity potential function into the continuity equation. Doing so imposes certain limits on the velocity potential function, and these limits provide the first boundary condition necessary in specifying the velocity potential itself. When the definitions for u, v, and w (eq. 5.7) are inserted into the continuity equation (eq. 5.1), we arrive at the conclusion that

$$\frac{\partial^2 \Phi}{\partial x^2} + \frac{\partial^2 \Phi}{\partial y^2} + \frac{\partial^2 \Phi}{\partial z^2} = 0. \qquad (5.9)$$

This equation is a particular example of the Laplace equation. However, in itself the Laplace equation cannot completely define the velocity potential for wave flows, for the velocity potential function for *any* irrotational flow satisfies the Laplace equation. For instance, the velocity potential for still water does a dandy job of satisfying the Laplace equation but is of little interest to us here. To specify the velocity potential function for a water wave more fully, we must impose more boundary conditions.

The first of these additional conditions is easily established. The bottom of the ocean in most places is nearly impermeable. If the water cannot flow into or out of the bottom, the vertical velocity caused by a wave, w, must be zero at the bottom. This can be true only if

$$w = \frac{\partial \Phi}{\partial z} = 0, \quad \text{at } z = -d. \qquad (5.10)$$

Next we note that because we have decided to ignore surface tension (Assumption 5), we can assume that water at the surface of the ocean is at atmospheric pressure. If this assumption can be expressed in terms of the velocity potential, it will form the last boundary conditions required to specify Φ. The link between pressure and velocity (a function of the velocity potential) is Bernoulli's equation.

By analogy to eq. 3.13 we write

$$\tfrac{1}{2}\rho U^2 + p + \rho g \eta = \text{constant}, \quad \text{at } z = \eta,$$

where U is the overall water velocity (here, $\sqrt{u^2 + w^2}$) and η is the vertical elevation of the water surface away from its equilibrium level (see Fig. 5.1). But this equation applies only when the water motion is steady. To deal with the nonsteady flow in waves, we need an equivalent equation that takes into account how the motion changes with time. The appropriate equation is one form of the unsteady Bernoulli equation:

$$\tfrac{1}{2}\rho U^2 + p + \rho g \eta + \rho \frac{\partial \Phi}{\partial t} = K(t), \quad \text{at } z = \eta,$$

$$(5.11)$$

which is similar to eq. 3.13, but with the addition of the term $\rho \, d\Phi/dt$. This term is a measure of the change in available energy due to the change in flow through time. Here $K(t)$ signifies that the available energy is the same everywhere along a streamline at time t, but can vary from one time to the next.

We will not derive this equation here, but it is a simple matter to show that $\rho \, d\Phi/dt$ intuitively fits the bill as a measure of energy per volume in accordance with the other terms in eq. 5.11. One can think of Φ as having the dimensions of a velocity multiplied by a distance. Only if this is true can you end up with a velocity when taking the derivative of Φ with respect to x, y, or z. If Φ is the product of velocity and distance, $d\Phi/dt$ is acceleration times distance, which, when multiplied by density (mass/volume), yields a term

$$\frac{(\text{mass} \cdot \text{acceleration}) \, \text{distance}}{\text{volume}} = \frac{\text{energy}}{\text{volume}}.$$

$$(5.12)$$

For a rigorous derivation of eq. 5.11, the inquisitive reader should consult a standard text on fluid dynamics.

Using the assumption that the waves we are examining have a fixed, small wave height (Assumption 3), eq. 5.11 can be reworked to our advantage. As long as the wave height is fixed, the overall energy of the wave form is constant, and $K(t)$ does not vary with time. We have also assumed that the pressure is the same everywhere at the surface, so that p is constant and can be absorbed into $K(t)$. Similarly, ρ is constant and can be divided out, leaving the relationship

$$\tfrac{1}{2}U^2 + g\eta + \frac{\partial \Phi}{\partial t} = \text{constant}, \quad \text{at } z = \eta. \quad (5.13)$$

This equation applies only at the surface, $z = \eta$. Due to the mixture of variables, it would be difficult to use eq. 5.13 in specifying Φ. However, we can avoid this problem if we make certain assumptions allowing the equation to be "linearized" (hence the name *linear* wave theory). We showed in Chapter 4 that the velocity of fluid at the surface is proportional to the height of the wave (eq. 4.3). Thus, by assuming that wave height is very small, we are assured that the term containing U^2 (proportional to η^2) is small compared to the term containing η, and we can ignore it without unduly changing matters. This is the first aspect of linearizing the equation. Thus, to a first approximation,

$$g\eta + \frac{\partial \Phi}{\partial t} = \text{constant}, \quad \text{at } z = \eta. \quad (5.14)$$

The second step in linearizing eq. 5.13 is to assume that η is small enough so that instead of evaluating eq. 5.14 at $z = \eta$, we can evaluate it at $z = 0$ without an appreciable loss of accuracy.

We then differentiate eq. 5.14 with respect to time, yielding

$$g\frac{\partial \eta}{\partial t} + \frac{\partial^2 \Phi}{\partial t^2} = 0, \quad \text{at } z = 0. \quad (5.15)$$

We now proceed by noting that

$$w \simeq \frac{\partial \eta}{\partial t}, \quad \text{at } z = 0. \quad (5.16)$$

This approximation is the third aspect of linearizing eq. 5.13. Now $w = d\Phi/dz$ (eq.

5.7), so $d\eta/dt = d\Phi/dz$. Substituting this result into eq. 5.15, we finally arrive at an expression solely in terms of the velocity potential that incorporates the fact that the pressure is everywhere the same at the surface:

$$g\frac{\partial\Phi}{\partial z} + \frac{\partial^2\Phi}{\partial t^2} = 0, \quad \text{at } z = 0. \quad (5.17)$$

This is our third, and last, boundary condition.

We now have three equations with which the velocity potential of a wave must be consistent:

1. The flow must obey the continuity equation:

$$\frac{\partial^2\Phi}{\partial x^2} + \frac{\partial^2\Phi}{\partial y^2} + \frac{\partial^2\Phi}{\partial z^2} = 0.$$

2. The vertical velocity must be zero at the seabed:

$$\frac{\partial\Phi}{\partial z} = 0, \quad \text{at } z = -d.$$

3. At the surface the pressure is everywhere the same:

$$g\frac{\partial\Phi}{\partial z} + \frac{\partial^2\Phi}{\partial t^2} = 0, \quad \text{at } z = 0.$$

The trick now is to devise an expression for Φ which satisfies these three conditions. This is a classic problem in partial differential equations, which, through the proper machinations, leads to the result that

$$\Phi = \left\{\frac{H}{2}\right\} \sin(kx - \omega t) \left\{\frac{L}{T}\right\} \left\{\frac{\cosh(ks)}{\sinh(kd)}\right\}. \quad (5.18)$$

This expression for Φ was first published by G. B. Airy in 1845, and linear wave theory is often referred to as Airy wave theory. The mathematics leading to this equation will not be treated here, and the reader should consult Kinsman (1965) for a readable derivation.

Now that we have an expression for Φ, we are free to explore linear wave theory in earnest. As you will see, a knowledge of Φ allows us to make many useful and important predictions regarding wave motion, but to understand the results of linear wave theory one must first contend with the presence of the hyperbolic functions cosh, sinh and tanh. Because these functions may not be as familiar as their trigonometric analogs (cosine, sine, and tangent), a brief review of hyperbolic trigonometry is in order.

Hyperbolic Trigonometry

The hyperbolic sine of a value x is

$$\sinh(x) = \tfrac{1}{2}(e^x - e^{-x}).$$

When $x = 0$, the $\sinh(x) = 0$ (both e^0 and e^{-0} equal 1) (Fig. 5.2a). $\sinh(x)$ increases

Figure 5.2. The hyperbolic functions. (a) The hyperbolic cosine (cosh) and sine (sinh). (b) The hyperbolic tangent (tanh = sinh/cosh).

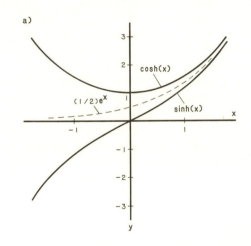

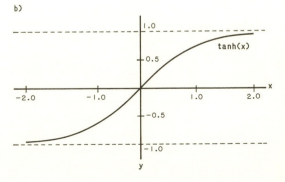

as x increases, until for large values of x, $\sinh(x)$ is approximately equal to $(\frac{1}{2})e^x$.

The hyperbolic cosine of a value x is

$$\cosh(x) = \tfrac{1}{2}(e^x + e^{-x}). \qquad (5.20)$$

When $x = 0$, $\cosh(x) = 1$. The $\cosh(x)$ also increases as x increases, again reaching values very close to $(\frac{1}{2})e^x$ for large values of x.

The hyperbolic tangent of x is defined as

$$\tanh(x) = \frac{\sinh(x)}{\cosh(x)}, \qquad (5.21)$$

in analogy to the trigonometric tangent; $\tanh(x) = 0$ at $x = 0$, and is asymptotic to 1 for large values of x (Fig. 5.2b). Table 5.1 gives the large- and small-value approximations for these functions. These approximations will be of considerable use throughout this chapter.

Table 5.1
Approximations of the Hyperbolic Functions

	x large	x small
$\sinh(x)$	$\frac{1}{2}e^x$	x
$\cosh(x)$	$\frac{1}{2}e^x$	1
$\tanh(x)$	1	x

We can now begin to examine the ramifications of linear wave theory. These findings are summarized in Tables 5.2 to 5.4.

The Surface Elevation: A First Look

If we were to be strictly logical in working our way through Airy's results, we would begin with the velocity potential and proceed to derive all the other findings from it. If we follow this route, one of the last facts to emerge is the shape of the wave form. To avoid the suspense as to what the wave looks like, we start by presenting the wave form and the terms related to it. Later, at the appropriate point in the logical

Table 5.2
Results of Linear Wave Theory

Velocity potential

$$\Phi = \left\{\frac{H}{2}\right\} \sin(kx - \omega t) \left\{\frac{L}{T}\right\} \frac{\cosh(ks)}{\sinh(kd)}$$

Surface elevation $\quad \eta = \left\{\frac{H}{2}\right\} \cos(kx - \omega t)$

Celerity $\quad C = \sqrt{\frac{g}{k} \tanh(kd)}$

Horizontal particle displacement

$$-\left\{\frac{H}{2}\right\} \sin(kx - \omega t) \frac{\cosh(ks)}{\sinh(kd)}$$

Vertical particle displacement

$$\left\{\frac{H}{2}\right\} \cos(kx - \omega t) \frac{\sinh(ks)}{\sinh(kd)}$$

Horizontal particle velocity

$$u = \left\{\frac{\pi H}{T}\right\} \cos(kx - \omega t) \frac{\cosh(ks)}{\sinh(kd)}$$

Vertical particle velocity

$$w = \left\{\frac{\pi H}{T}\right\} \sin(kx - \omega t) \frac{\sinh(ks)}{\sinh(kd)}$$

Horizontal particle acceleration

$$\frac{\partial u}{\partial t} = \left\{\frac{2\pi^2 H}{T^2}\right\} \sin(kx - \omega t) \frac{\cosh(ks)}{\sinh(kd)}$$

Vertical particle acceleration

$$\frac{\partial w}{\partial t} = -\left\{\frac{2\pi^2 H}{T^2}\right\} \cos(kx - \omega t) \frac{\sinh(ks)}{\sinh(kd)}$$

Pressure

$$p = -\rho g z + \tfrac{1}{2}\rho g H \cos(kx - \omega t) \frac{\cosh(ks)}{\cosh(kd)}$$

$$+ \text{ atmospheric } p$$

Group velocity $\quad C_g = \frac{C}{2}\left\{1 + \frac{2kd}{\sinh(2kd)}\right\}$

Average energy per area $\quad E = \tfrac{1}{8}\rho g H^2$

Rate of energy transmission $= EC_g$

Radiation stress $\quad S_{xx} = E\left\{\frac{1}{2} + \frac{2kd}{\sinh(2kd)}\right\}$

$$S_{yy} = E\left\{\frac{kd}{\sinh(2kd)}\right\}$$

$$S_{xy} = S_{yx} = 0$$

flow of the discussion, we return to this wave form and show how it is derived.

The surface elevation of a linear wave is

$$\eta = \left\{\frac{H}{2}\right\} \cos\left\{\frac{2\pi x}{L} - \frac{2\pi t}{T}\right\}. \quad (5.22)$$

This is one variation of the classic expression for a traveling cosine wave (Fig. 5.3a,b). In accordance with the way in which we have defined wave height, $H/2$ is the amplitude of the wave. We know from experience with waves that the surface elevation varies from one place to the next, and it varies from one time to the next at any given place. This variability in time and space is accounted for by the cosine function in this expression.

There are two ways to approach an understanding of this equation. First consider how the expression varies through space (i.e., along the x-axis) as time is held constant (Fig. 5.3a). In essence, we take a

Figure 5.3. Traveling waves. (a) The surface elevation as a function of distance at time = 0. (b) The surface elevation as a function of time at $x = 0$.

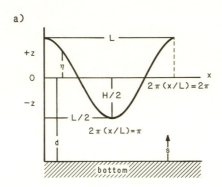

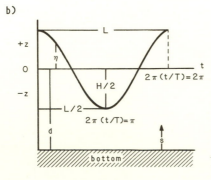

snapshot of eq. 5.22 at one instant in time. For the sake of simplicity, we define this instant to be $t = 0$. Thus the surface elevation η at position x is

$$\eta(x) = \left\{\frac{H}{2}\right\} \cos\left\{\frac{2\pi x}{L}\right\}; \quad t = 0.$$

This expression is at a maximum whenever $(2\pi x/L) = 0$, 2π, 4π, etc. In other words, wave crests occur at $x = 0$, L, $2L$, etc., all in accordance with the definition of the wave length. Conversely, at $x = L/2$, $3L/2$, etc., the cosine of $2\pi x/L$ is at a minimum, the surface elevation is similarly at a minimum, and troughs occur. The values of $\eta(x)$ given by this function are measured relative to the still water level. If the still water level is a distance d above the bottom, the crests and troughs have elevations of $d + (H/2)$ and $d - (H/2)$, respectively, relative to the bottom (Fig. 5.3).

Alternatively, one could choose to look at one point in space and examine how the surface elevation varies as a function of time (Fig. 5.3b). To keep things simple, we examine the wave at $x = 0$, in which case the expression for η becomes

$$\eta(t) = \left\{\frac{H}{2}\right\} \cos\left\{\frac{-2\pi t}{T}\right\}; \quad x = 0.$$

As expected, the wave form repeats itself with period T, reaching a maximum at $t = 0$, T, $2T$, etc., and a minimum at $t = T/2$, $3T/2$, etc.

In many cases the form of wave equations is more easily grasped if a shorthand notation is used for the values $2\pi/L$ and $2\pi/T$. Thus we define the *wave number, k*,

$$k = \left\{\frac{2\pi}{L}\right\}, \quad (5.23)$$

and the radian wave frequency, ω,

$$\omega = \left\{\frac{2\pi}{T}\right\}. \quad (5.24)$$

Using this notation, we can write the expression for surface elevation as

$$\eta = \left\{\frac{H}{2}\right\} \cos(kx - \omega t),$$

which is a more streamlined expression. One need only substitute $2\pi/L$ for k and $2\pi/T$ for ω to recover the full equation.

The Linear Velocity Potential

We now return to Airy's expression for the velocity potential function:

$$\Phi = \left\{\frac{H}{2}\right\} \sin(kx - \omega t) \left\{\frac{L}{T}\right\} \left\{\frac{\cosh(ks)}{\sinh(kd)}\right\}. \quad (5.18)$$

That this is indeed a proper solution can be verified by taking the various partial derivatives of this function and summing them up in the appropriate combinations specified by our boundary conditions, an exercise that is left to the industrious reader.

The first thing to notice about the potential function is that it can be divided into three parts. The first of these is very similar to the expression for a traveling wave as used for the surface elevation function; the only difference is that the velocity potential function varies as the sine rather than the cosine:

$$\left\{\frac{H}{2}\right\} \sin(kx - \omega t).$$

Figure 5.4 compares this term of the velocity potential to the surface elevation. Here the two expressions are plotted as functions of distance along the x-axis, both expressions being evaluated at $t = 0$. In this case the first term from the velocity potential function is

$$\left\{\frac{H}{2}\right\} \sin(kx), \quad t = 0.$$

Wherever the surface elevation is at a maximum or a minimum, the velocity potential is zero, and vice versa. Thus, the two expressions are one quarter of a wave length out of phase. If we had chosen to compare the surface elevation and the velocity potential as a function of time at one point in space, we would find that the two are out of phase by one-quarter period.

The second term in the velocity potential is simply the wave celerity, L/T.

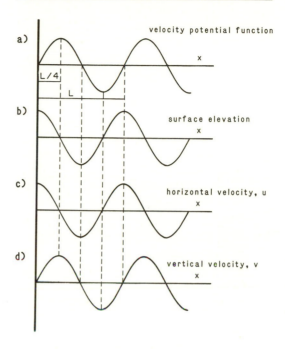

Figure 5.4. The phase relationship among the (a) velocity potential, (b) surface elevation, (c) horizontal velocity, and (d) vertical velocity.

These two expressions (the traveling sine wave and the celerity) are multiplied by the last term of the potential function, an expression containing variables for the depth of the water, d, and the length of the wave, L [remember that $k = (2\pi/L)$]. This expression also includes a variable for the distance s above the bottom at which we are to evaluate the velocity potential function (Fig. 5.1). The depth-related term is

$$\frac{\cosh(ks)}{\sinh(kd)}.$$

This portion of the velocity potential will be explored in greater detail later, but for the moment we will examine a simple example of how it works. If water depth is much greater than the wave length, kd is large and $\sinh(kd)$ is also large (see Fig. 5.2). At the surface $s = d$, in which case $\cosh(ks)$ is large and approximately equal to $\sinh(kd)$. Consequently, at the surface this depth-related expression is approximately equal to 1. At the bottom $s = 0$, in which case $\cosh(ks) = 1$ (Fig. 5.2), and the depth-related expression takes on a very small value. Thus, in deep water the velocity potential is large at the water's surface and is very small at the bottom.

Water Particle Velocities

The horizontal velocity, u, is simply the partial derivative of Φ with respect to x (eq. 5.7). Noting that $d\sin(kx - \omega t)/dx = k\cos(kx - \omega t) = (2\pi/L)\cos(kx - \omega t)$,

$$u = \frac{\partial \Phi}{\partial x} = \left\{\frac{\pi H}{T}\right\}\cos(kx - \omega t)\left\{\frac{\cosh(ks)}{\sinh(kd)}\right\}. \quad (5.25)$$

Thus the horizontal velocity is a cosine function, reaching a maximum when $x = 0$, L, $2L$, etc. Figure 5.4 compares the horizontal velocity with the surface elevation as given by eq. 5.22. The two are in phase, the horizontal velocity reaching its maximum value $(\pi H/T)$ at the surface under a crest, and its minimum value $(-\pi H/T)$ at the surface under a trough. This is exactly the result obtained in the laboratory experiments discussed in Chapter 4.

The vertical velocity, w, is the partial derivative of the potential function with respect to s, the distance from the bottom. Because $s = d + z$, taking the derivative with respect to s is equivalent to taking the derivative with respect to z as specified by 5.7. Given that $d\cosh(ks)/ds = k\sinh(ks)$,

$$w = \frac{\partial \Phi}{\partial s} = \left\{\frac{\pi H}{T}\right\}\sin(kx - \omega t)\left\{\frac{\sinh(ks)}{\sinh(kd)}\right\}. \quad (5.26)$$

The vertical velocity is a sine function and reaches a maximum when $x = L/4$, $3L/4$, etc. In other words, the maximum vertical velocity is reached at points where the surface elevation η is zero, one-quarter wave length away from the crests and troughs (Fig. 5.4). We have already seen that the horizontal velocity is zero at these points.

To compare the magnitudes of the vertical and horizontal velocities, we must eval-uate the depth-related terms. Consider the case of deep water ($d > L/2$), where kd is large. At the surface, where $s = d$, both $\cosh(ks)$ and $\sinh(kd)$ are very large and approximately equal to $(1/2)e^{kd}$ (see Table 5.1). Thus at the surface in deep water, the depth-related term for both horizontal and vertical velocities is approximately equal to 1, and

$$u \simeq \left\{\frac{\pi H}{T}\right\}\cos(kx - \omega t), \quad \text{at } s = d \quad (5.27)$$

$$w \simeq \left\{\frac{\pi H}{T}\right\}\sin(kx - \omega t), \quad \text{at } s = d. \quad (5.28)$$

The maximum magnitudes of both these two velocities are $\pi H/T$, exactly the same as calculated before (eq. 4.2), using a simpler logic based on observations in a wave tank. The two velocities are one-quarter wave length ($\pi/2$ radians) out of phase, so that when the horizontal velocity is maximum, the vertical velocity is zero, and vice versa. A moment's thought about the motion of a Ferris wheel will convince us that this is the same thing as saying that the water travels in a circular orbit.

The equality of horizontal and vertical velocities at the surface holds true for waves traveling in depths greater than $L/2$, but not if $d < L/2$. Regardless of depth, $\sinh(ks)/\sinh(kd)$ is equal to one at the surface where $s = d$. Consequently, the vertical water velocity at the surface (eq. 5.26) is the same in shallow water as it is in deep water as long as H remains constant (actually not a very good assumption, as we will see in Chapter 7). In contrast, $\cosh(ks)/\sinh(kd)$ (the depth-related term of the horizontal velocity, eq. 5.25) becomes larger than 1 as d becomes small relative to L. Thus, as linear waves move into shallow water, the horizontal velocity at the surface becomes greater than the vertical velocity, and water particles no longer travel in circular orbits. Instead, the orbits are elliptical, the long axis of the ellipse lying in the horizontal plane. When $d \ll L$, $\sinh(kd)$ approaches zero (Table 5.1), the depth-related term becomes very large, and the function is said to "blow up." Taken to the limit, we would calculate an infinite horizontal velocity when $d = 0$. This clearly cannot be true, and we conclude on logical grounds that linear wave theory is not to be trusted for waves in very shallow water.

We now explore the components of water velocity at depths below the surface (i.e., $s < d$). We begin with the case of deep water, where $d > L/2$. Consider what happens at the bottom itself. When $s = 0$, $\sinh(ks) = 0$, and consequently the vertical velocity is zero at the bottom (eq. 5.26). This had better be true, since it was one of the boundary conditions required of the velocity potential function. At $s = 0$, $\cosh(ks)$ is approximately equal to 1, while in deep water $\sinh(kd)$ is large. Thus the depth-related term of the equation for horizontal velocity (eq. 5.25) becomes very small at the bottom in deep water, but unless the water is infinitely deep it never goes all the way to zero.

To give these calculations some tangibility, consider a wave with a height of 3 m and a period of 10 s traveling in water 1000 m deep. We have not yet examined Airy's results for calculating wave length from wave period, so we will rely for the moment on the result from the wave-tank experiments: $L = gT^2/2\pi = 156$ m. At the surface, the maximum horizontal and vertical velocities are

$$u_{max} = w_{max} = \left\{\frac{\pi H}{T}\right\} = \left\{\frac{\pi 3}{10}\right\} = 0.95 \text{ m/s.}$$

At a depth of 10 m the maximum horizontal velocity is

$$u_{max} = \left\{\frac{\pi H}{T}\right\} \cosh\left\{\frac{2\pi 990}{156}\right\} \div \sinh\left\{\frac{2\pi 1000}{156}\right\}$$
$$= 0.64 \text{ m/s,}$$

or 67% of the maximum velocity at the surface. The maximum vertical velocity at a depth of 10 m is

$$w_{max} = \left\{\frac{\pi H}{T}\right\} \sinh\left\{\frac{2\pi 990}{156}\right\} \div \sinh\left\{\frac{2\pi 1000}{156}\right\}$$
$$= 0.64 \text{ m/s,}$$

likewise 67% of the surface velocity. At this depth in deep water, the maximum vertical and horizontal velocities, although less than those at the surface, are still equal: the water moves through a nearly circular orbit as each wave passes. However, because the velocity is smaller at this depth while the period is the same, the diameter of the circle must be smaller.

At a depth of one-half wave length below the surface ($s = 922$ m), the horizontal and vertical velocities are still equal but amount to only 4% of the velocities at the surface, an amount generally considered to be negligible. This conclusion is the justification for choosing $L/2$ as the cutoff point between deep and shallow water. This result leads us to believe that wave-induced velocities are likely to be unimportant for plants and animals that live at depths $> L/2$. For a typical wave with a period of 10 s, this cutoff depth occurs at approximately 75 m.

In deep water ($d > L/2$) the terms for horizontal and vertical velocity reduce to the form

$$u = \left\{\frac{\pi H}{T}\right\} \cos(kx - \omega t)e^{\{k(s-d)\}}$$
$$d > L/2$$
$$w = \left\{\frac{\pi H}{T}\right\} \sin(kx - \omega t)e^{\{k(s-d)\}} \qquad (5.29)$$

(see Table 5.1). Note that $z = s - d$ (Fig. 5.1). Inserting z into eq. 5.29, we arrive at expressions for u and w in terms of distance from still water level instead of distance from the bottom.

$$u = \left\{\frac{\pi H}{T}\right\} \cos(kx - \omega t)e^{(kz)}$$
$$d > L/2$$
$$w = \left\{\frac{\pi H}{T}\right\} \sin(kx - \omega t)e^{(kz)}. \qquad (5.30)$$

At the surface $z = 0$, and we recover the expressions for surface velocity obtained earlier (eqs. 5.27 and 5.28). Velocity decreases exponentially with depth in deep water, and the vertical and horizontal velocities remain equal to a depth where velocity is negligible. This is the mathematical basis for Figure 4.5. The various deep-water approximations for linear wave theory are summarized in Table 5.3.

In shallow water ($d < L/20$, $kd < \pi/10$), the equations for the components of velocity can again be simplified. For small values of kd, $\cosh(ks)/\sinh(kd) = 1/kd$ (Table 5.1). Thus

$$u = \left\{\frac{\pi H}{T}\right\} \cos(kx - \omega t)\left\{\frac{1}{kd}\right\}, \quad d < L/20.$$

$$(5.31)$$

As noted earlier, the horizontal velocity increases as the water gets shallower. In these shallow depths, u is constant from the surface to the sea floor, and this expression tends to blow up in very shallow water where d is small.

When $d < L/20$, $\sinh(ks)/\sinh(kd) = s/d$ and w in eq. 5.26 is approximated by

$$w = \left\{\frac{\pi H}{T}\right\} \sin(kx - \omega t)\left\{\frac{s}{d}\right\}, \quad d < L/20.$$

$$(5.32)$$

Thus the vertical water velocity in shallow water is a function of depth, decreasing to zero at the sea floor ($s = 0$) as required. The various shallow-water approximations for linear wave theory are summarized in Table 5.4.

At depths between $d < L/20$ and $d > L/2$ there are no handy simplifications to the full expressions for the velocity components (eqs. 5.25, 5.26). For this reason these depths are termed *intermediate*, and results at these depths are intermediate between those for deep and shallow water: the horizontal water velocity is somewhat increased above its value in deep water. As a result, the orbits traveled by water under waves at intermediate depths are elliptical but not nearly as flattened as those in shallow water.

Note that none of the variation in veloc-

Table 5.3
Deep-Water Approximations
for Linear Wave Theory

Range of validity $kd > \pi$

$$d > \frac{L}{2}$$

Velocity potential $\Phi = \left\{\dfrac{\pi H}{kT}\right\} e^{kz} \sin(kx - \omega t)$

Celerity $C = \sqrt{\dfrac{g}{k}}$

Wave length $L = L_0 = \dfrac{gT^2}{2\pi}$

Surface elevation $\eta = \left\{\dfrac{H}{2}\right\} \cos(kx - \omega t)$

Horizontal particle displacement

$$-\left\{\frac{H}{2}\right\} e^{kz} \sin(kx - \omega t)$$

Vertical particle displacement $\left\{\dfrac{H}{2}\right\} e^{kz} \cos(kx - \omega t)$

Horizontal particle velocity

$$u = \left\{\frac{\pi H}{T}\right\} e^{kz} \cos(kx - \omega t)$$

Vertical particle velocity $w = \left\{\dfrac{\pi H}{T}\right\} e^{kz} \sin(kx - \omega t)$

Horizontal particle acceleration

$$\frac{\partial u}{\partial t} = \left\{\frac{2\pi^2 H}{T^2}\right\} e^{kz} \sin(kx - \omega t)$$

Vertical particle acceleration

$$\frac{\partial w}{\partial t} = -\left\{\frac{2\pi^2 H}{T^2}\right\} e^{kz} \cos(kx - \omega t)$$

Pressure $p = -\rho g z + \frac{1}{2}\rho g H e^{kz} \cos(kx - \omega t)$
$$+ \text{ atmospheric pressure}$$

Group velocity $C_g = \dfrac{C}{2}$

Average energy per area $E = \frac{1}{8}\rho g H^2$

Rate of energy transmission $= \dfrac{EC}{2}$

Radiation stress $S_{xx} = \dfrac{E}{2}$

$$S_{yy} = S_{xy} = S_{yx} = 0$$

Table 5.4
Shallow-Water Approximations for Linear Wave Theory

Range of validity $kd < \dfrac{\pi}{10}$

$$d < \dfrac{L}{20}$$

Velocity potential $\Phi = \left\{\dfrac{\pi H}{k^2 Td}\right\} \sin(kx - \omega t)$

Celerity $C = \sqrt{gd}$

Wave length $L = T\sqrt{gd}$

Surface elevation $\eta = \left\{\dfrac{H}{2}\right\} \cos(kx - \omega t)$

Horizontal particle displacement

$$-\left\{\dfrac{H}{2}\right\}\left\{\dfrac{1}{kd}\right\} \sin(kx - \omega t)$$

Vertical particle displacment $\left\{\dfrac{H}{2}\right\}\left\{\dfrac{s}{d}\right\} \cos(kx - \omega t)$

Horizontal particle velocity

$$u = \left\{\dfrac{\pi H}{T}\right\}\left\{\dfrac{1}{kd}\right\} \cos(kx - \omega t)$$

Vertical particle velocity $w = \left\{\dfrac{\pi H}{T}\right\}\left\{\dfrac{s}{d}\right\} \sin(kx - \omega t)$

Horizontal particle acceleration

$$\dfrac{\partial u}{\partial t} = \left\{\dfrac{2\pi^2 H}{T^2}\right\}\left\{\dfrac{1}{kd}\right\} \sin(kx - \omega t)$$

Vertical particle acceleration

$$\dfrac{\partial w}{\partial t} = -\left\{\dfrac{2\pi^2 H}{T^2}\right\}\left\{\dfrac{s}{d}\right\} \cos(kx - \omega t)$$

Pressure $p = -\rho gz + \tfrac{1}{2}\rho gH \cos(kx - \omega t)$
          $+ \text{atmospheric pressure}$

Group velocity $C_g = C$

Average energy per area $E = \tfrac{1}{8}\rho gH^2$

Rate of energy transmission $= EC$

Radiation stress $S_{xx} = \tfrac{3}{2}E$

$$S_{yy} = \tfrac{1}{2}E$$

$$S_{xy} = S_{yx} = 0$$

ity components with depth affects the phase relationship between the water movement and the wave form. Horizontal velocity is always greatest under the crests and troughs, and vertical velocity is always greatest at still water level.

Mass Transport

Earlier in this discussion we concluded that orbits are circular in deep water because the maximum horizontal velocity equals the maximum vertical velocity. Actually, this argument has a minor error. The logic we followed says that at one instant in time a water particle at the crest moves forward with velocity u, a particle in the trough moves with velocity $-u$, and particles at the still water level have velocities w and $-w$ (equal in magnitude to u and $-u$). This is quite true and indeed is what we predict from an analogy to a Ferris wheel. However, consider what happens to one particular particle as it travels around its orbit. At the crest it moves with velocity u ($u < C$). But because it is moving in the same direction as the wave, it spends an extra bit of time under the wave crest before the wave form passes it by. The opposite is true when the particle reaches the trough. Here it moves with velocity $-u$, and because this is in a direction opposite the wave celerity, it spends less time moving at this velocity. Because the particle spends more time moving at velocity u than at $-u$, it travels a greater distance in the x-direction than in the $-x$-direction, and there is a net movement of the particle in the direction of wave motion. Furthermore, because u decreases with distance below still water level (eq. 5.29), the magnitude of a particle's velocity is greater at the top of its orbit, which amplifies the particle's net movement. This is the explanation for the phenomenon briefly noted in Chapter 4 where particles did not move in exactly circular orbits.

As long as waves have a small height

(an assumption we made in linearizing the boundary conditions for wave motion), u is much less than C, so the difference in time a particle spends under the crest and trough is small and the net transport is negligible. Thus, for real ocean waves in deep water, the orbits of water particles are very nearly, but not quite, circular.

The Surface Elevation Revisited

We are now in a position to return to the expression for the surface elevation (eq. 5.22) to see how it is calculated. From eq. 5.26 we know how fast the water is moving vertically at the surface at each point in time. By multiplying each velocity by the time spent at that velocity, we can calculate the distance traveled in the vertical direction as a function of time. Integrating this distance over time, we get

vertical displacement

$$= \int \left\{\frac{\pi H}{T}\right\} \sin(kx - \omega t) \left\{\frac{\sinh(ks)}{\sinh(kd)}\right\} dt. \quad (5.33)$$

At the surface $s = d$, so that $\sinh(ks)$ is equal to $\sinh(kd)$ and the depth-related portion of the integrand is simply equal to 1:

vertical displacement at the surface

$$= \eta = \left\{\frac{\pi H}{T}\right\} \int \sin(kx - \omega t) \, dt$$

$$= \left\{\frac{H}{2}\right\} \cos(kx - \omega t) + \text{constant}.$$

$$(5.34)$$

Taking the constant to be equal to zero (i.e., η is measured relative to still water), we arrive at the same equation for the surface elevation presented earlier (eq. 5.22).

Particle Displacements

To obtain the vertical distance traveled by a bit of fluid at points below the surface,

we integrate the entire expression for vertical velocity (eq. 5.26):

vertical displacement

$$= \left\{\frac{H}{2}\right\} \cos(kx - \omega t) \left\{\frac{\sinh(ks)}{\sinh(kd)}\right\}. \quad (5.35)$$

This expression varies with depth in exactly the same way as the velocity itself. The analogous procedure can be carried out for the horizontal displacement with the result that at the surface in deep water the displacement is

horizontal displacement $= \left\{\dfrac{H}{2}\right\} \sin(kx - \omega t);$

at $s = d$, (5.36)

as expected. Elsewhere the horizontal displacement is

horizontal displacement

$$= \left\{\frac{H}{2}\right\} \sin(kx - \omega t) \left\{\frac{\cosh(ks)}{\sinh(kd)}\right\}. \quad (5.37)$$

See Tables 5.3 and 5.4 for shallow- and deep-water approximations of these equations.

Particle Accelerations

The expressions derived for the components of velocity beneath a wave make it easy to calculate the fluid acceleration. One simply takes the partial derivative of the velocity with respect to time. Thus

$$\frac{\partial u}{\partial t} = \left\{\frac{2\pi^2 H}{T^2}\right\} \sin(kx - \omega t) \left\{\frac{\cosh(ks)}{\sinh(kd)}\right\}$$

$$\frac{\partial w}{\partial t} = -\left\{\frac{2\pi^2 H}{T^2}\right\} \cos(kx - \omega t) \left\{\frac{\sinh(ks)}{\sinh(kd)}\right\}.$$

$$(5.38)$$

These expressions vary with depth in the same manner as the velocity components.

Celerity

So far we have extracted from the velocity potential function information related solely to the movement of the water beneath a wave. What about the movement of the wave form itself? How fast does it move? Can we account for the experimental fact that waves of constant period move slower in shallow water than they do in deep water? The simplest way to answer these questions is to insert the velocity potential function (eq. 5.18) into the boundary condition requiring the pressure to be everywhere the same at the surface ($g\partial\Phi/\partial z + \partial^2\Phi/\partial t^2 = 0$, at $s = d$; eq. 5.17). Now,

$$\frac{\partial^2\Phi}{\partial t^2} = -\left\{\frac{2\pi^2 H}{T^2}\right\} \sin(kx - \omega t)\left\{\frac{L}{T}\right\}\left\{\frac{\cosh(ks)}{\sinh(kd)}\right\}$$

and

$$\frac{\partial\Phi}{\partial z} = \left\{\frac{2\pi}{L}\right\}\left\{\frac{H}{2}\right\} \sin(kx - \omega t)\left\{\frac{L}{T}\right\}\left\{\frac{\sinh(ks)}{\sinh(kd)}\right\}.$$
(5.39)

We evaluate these expressions at the surface ($s = d$) and insert them into eq. 5.17:

$$g\left\{\frac{2\pi}{L}\right\}\left\{\frac{H}{2}\right\} \sin(kx - \omega t)\left\{\frac{L}{T}\right\}\left\{\frac{\sinh(kd)}{\sinh(kd)}\right\}$$
$$= \left\{\frac{2\pi^2 H}{T^2}\right\} \sin(kx - \omega t)\left\{\frac{L}{T}\right\}\left\{\frac{\cosh(kd)}{\sinh(kd)}\right\}.$$
(5.40)

Canceling equal terms from both sides of the equation yields

$$\frac{g}{L} = 2\pi\left\{\frac{1}{T^2}\right\}\frac{\cosh(kd)}{\sinh(kd)}.$$
(5.41)

Multiplying both sides of the equation by L^2 and rearranging, we get

$$\left\{\frac{L^2}{T^2}\right\} = \left\{\frac{gL}{2\pi}\right\}\left\{\frac{\sinh(kd)}{\cosh(kd)}\right\} = \left\{\frac{gL}{2\pi}\right\}\tanh(kd).$$
(5.42)

Recall that $C = L/T$ (eq. 4.1). Thus we can rewrite eq. 5.42 as

$$C^2 = \left\{\frac{gL}{2\pi}\right\}\tanh(kd) = \left\{\frac{g}{k}\right\}\tanh(kd).$$
(5.43)

In deep water, kd is large and therefore $\tanh(kd)$ is approximately equal to 1. Thus we conclude that in deep water

$$C^2 = \frac{g}{k} = \frac{gL}{2\pi},$$

the same relationship obtained in Chapter 4 (eq. 4.9). However, the equation derived here allows us to calculate the wave celerity as the depth becomes shallow. As d decreases, $\tanh(kd)$ decreases. Thus the wave celerity decreases as waves move into shallow water, as noted in the wave tank. When $d < L/20$, $\tanh(kd)$ is approximately equal to kd, and eq. 5.43 reduces to $C^2 = gd$. Thus in shallow water the celerity is independent of the wave length and depends on the water depth, conclusions we reached in Chapter 4 for solitary waves. It is interesting to note the convergence in these properties of the two wave types.

Wave Length

The velocity potential can also be used to explain why wave length decreases in shallow water. Returning to an intermediate step in the derivation of celerity (eq. 5.42), note that:

$$\left\{\frac{L^2}{T^2}\right\} = \left\{\frac{gL}{2\pi}\right\}\tanh(kd).$$

Thus

$$L = \left\{\frac{gT^2}{2\pi}\right\}\tanh(kd) = \left\{\frac{gT^2}{2\pi}\right\}\tanh\left\{\frac{2\pi d}{L}\right\}.$$
(5.44)

This conclusion is problematical because the wave length, L, appears on both sides of the equation. It does, however, describe a unique relationship between L and d. An approximate solution for L as a function of d is given by Eckart (1952).

$$L \simeq \left\{\frac{gT^2}{2\pi}\right\}\sqrt{\tanh\left\{\frac{4\pi^2 d}{T^2 g}\right\}}.$$
(5.45)

This approximation is accurate to within about 5%.

Wave Energy

The method for evaluating wave energy from the potential function involves a straightforward integration of the terms for horizontal and vertical velocity and the surface elevation. This process has already been discussed in general terms in Chapter 4 and is left as an exercise for the industrious reader.

Group Velocity

In Chapter 4 we briefly discussed the concept of the group velocity, and noted the fact that a group of waves may move at a slower speed than individual waves. As a consequence, wave energy is transported across the ocean at the group velocity rather that at the wave celerity. A knowledge of the group velocity can be helpful, and linear wave theory can be used to calculate the group velocity:

$$C_g = \left\{\frac{C}{2}\right\}\left\{1 + \frac{2kd}{\sinh(2kd)}\right\}. \quad (5.46)$$

Unfortunately, the derivation of this expression is not straightforward, and will not be presented here. The interested reader should consult Kinsman (1965).

In deep water, $2kd$ is very small compared to $\sinh(2kd)$, and the depth-related term $[2kd/\sinh(2kd)]$ is approximately equal to zero. In this case the group velocity is half the wave celerity. In shallow water, $2kd$ is approximately equal to $\sinh(2kd)$ and the group velocity is equal to the wave celerity. The dependence of the group velocity on depth will be used in Chapter 7 to examine the change in height as waves move into shallow water.

Pressure

In dealing with real ocean waves, it is often useful to have a method of measuring the height of waves. One of the simplest of these is to place a pressure transducer on the bottom at a depth of 3–15 m. If there were no waves present on the surface, the pressure transducer would simply measure the hydrostatic pressure due to the mass of water and air above it. If the transducer lies at a depth z below the surface, this hydrostatic pressure is

hydrostatic pressure

$$= -\rho gz + \text{atmospheric pressure.} \quad (5.47)$$

This is easily seen by calculating the mass of a column of water z meters high. If the column has a cross-sectional area S, the volume of the column is zS. The mass of this volume of water is ρzS, and under the acceleration gravity this mass exerts a force ρgzS. This force acts over the area S, so the pressure (force per area) is ρgz. Below the water's surface, z is negative, so the pressure is $-\rho gz$. The weight of a column of atmosphere is piled on top of the water, so the total pressure at the transducer is that due to the water plus atmospheric pressure.

As a wave crest moves over the transducer, the column of water above the transducer increases in height by $H/2$. Thus one might expect that the total pressure due to the water above the transducer would be

$$\begin{array}{c}\text{total}\\\text{pressure}\end{array} = -\rho gz + \left\{\frac{\rho gH}{2}\right\} + \begin{array}{c}\text{atmospheric}\\\text{pressure}\end{array}$$

$$(5.48)$$

However, if we were actually to use this equation, we would find that the wave heights calculated from the measured pressure would be smaller than those actually present. To understand why, we will derive an expression for the pressure from the velocity potential.

We start by choosing an equation that relates pressure, surface elevation, and the velocity potential—Bernoulli's equation for unsteady flow (eq. 5.11). We linearize eq. 5.11, noting that the term $(\frac{1}{2})\rho U^2$ is small, and conclude that

$$p + \rho gz + \rho\frac{\partial\Phi}{\partial t} = \begin{array}{c}\text{atmospheric}\\\text{pressure}\end{array} \quad (5.49)$$

where $K(t)$ has been set equal to atmospheric pressure, the pressure known to act at the surface. Solving for pressure and substituting the full expression for $d\Phi/dt$, we see that

$$p = -\rho g z + \left\{\frac{\rho \pi H L}{T^2}\right\} \cos(kx - \omega t) \left\{\frac{\cosh(ks)}{\sinh(kd)}\right\}$$

$$+ \text{ atmospheric pressure.} \qquad (5.50)$$

We know from eq. 5.42 that

$$C^2 = \left\{\frac{L^2}{T^2}\right\} = \left\{\frac{gL}{2\pi}\right\}\left\{\frac{\sinh(kd)}{\cosh(kd)}\right\} \quad \text{or}$$

$$\left\{\frac{1}{T^2}\right\} = \left\{\frac{g}{2\pi L}\right\}\left\{\frac{\sinh(kd)}{\cosh(kd)}\right\}.$$

Substituting this value of $1/T^2$ into eq. 5.50 and canceling terms, we can conclude that

$$p = -\rho g z + \left\{\frac{\rho g H}{2}\right\} \cos(kx - \omega t) \left\{\frac{\cosh(ks)}{\cosh(kd)}\right\}$$

$$+ \text{ atmospheric pressure.} \qquad (5.51)$$

At the wave crest, $\cos(kx - \omega t) = 1$, so

$$p(\text{under the crest})$$

$$= -\rho g z + \left\{\frac{\rho g H}{2}\right\}\left\{\frac{\cosh(ks)}{\cosh(kd)}\right\}$$

$$+ \text{ atmospheric pressure.}$$

This equation is very similar to that obtained earlier (eq. 5.48), the only difference being the depth-related term $\cosh(ks)/\cosh(kd)$. This difference, however, explains why a subsurface transducer measures a lower pressure than might be expected. The depth-related term is equal to one only when $s = d$—in other words, directly at the surface. As the depth of the transducer is increased (as s becomes smaller), the observed pressure is less than the hydrostatic pressure that would be measured if the water above it were not moving. For example, consider a wave with height of 2 m and a period of 10 s. Such a wave has a length of about 150 m in water with a depth of 10 m (eq. 5.45). If a column of *stationary* fluid equal to half the wave height were added to the water above it, a pressure transducer located on the bottom would measure an increase in pressure of

$$\Delta p = \left\{\frac{\rho g H}{2}\right\}$$

$$= (1025 \text{ kg/m}^3) \cdot (9.81 \text{ m/s}^2) \cdot (1 \text{ m})$$

$$= 10^4 \text{ N/m}^2.$$

However, the actual increase in pressure measured is

$$\Delta p = \left\{\frac{\rho g H}{2}\right\}\left\{\frac{\cosh(ks)}{\cosh(kd)}\right\}$$

$$= (1025 \text{ kg/m}^3) \cdot (9.81 \text{ m/s}^2)$$

$$\cdot (1 \text{ m})\left\{\frac{\cosh(0)}{\cosh(2\pi 10/150)}\right\}$$

$$= 4.6 \cdot 10^3 \text{ N/m}^2,$$

only 46% of the hydrostatic value.

By now we should be convinced that linear wave theory can be a useful tool. If we know what Φ is, we can calculate almost any important flow parameter for any position beneath a wave. We will make extensive use of this ability throughout the rest of this book.

Other Wave Theories

Linear wave theory is but one of many theories devised to describe wave motion. Several of these alternative theories have proven useful in the study of the wave-swept environmment.

Stokes's Theory for Waves of Finite Height

Recall that the derivation of linear wave theory required that the height of the wave be small relative to the wave length. Only by making this assumption could our boundary conditions be linearized. However, in circumventing this initial problem, the assumption of small wave height leads to problems when we try to apply linear theory to actual ocean situations. For instance, one often wants to know the pertinent details about the highest waves that are likely to appear. These are the waves that contain the most energy and can impose the largest forces on the shoreline, but these are also the waves that most blatantly violate the assumption of infinitesimal wave height required by linear wave theory. Fortunately, the problems of the nonlinear terms in the equations governing wave motion have been dealt with to a certain extent. The first approximate solution allowing waves of finite height was devised by G. G. Stokes in 1847, and a series of similar theories has been developed since.

The results of these theories are expressed as a series approximation to an exact solution. For example, the series expression for the surface profile is given in the general form

$$\eta = A \cos(\theta) + A^2 B_2(L, d) \cos(2\theta)$$
$$+ A^3 B_3(L, d) \cos(3\theta)$$
$$+ \ldots + A^n B_n(L, d) \cos(n\theta), \quad (5.52)$$

where θ is $(kx - \omega t)$. A is $H/2$ for the first and second orders and may be less than $H/2$ for higher orders. The functions B_2 through B_n are functions of wave length and water depth. The trick in formulating wave theories of this general form is to choose the appropriate values for the constants A, and the functions $B(L, d)$. The higher the order of the series approximation (i.e., the larger the multiple of the angle term θ contained in the series), the closer the series comes to an exact solution. These series approximations have been carried out to a very high order, but the simple second-order expansion (Stokes II) contains all the results pertinent to this exploration, and it is this approximation that is shown in Table 5.5.

Table 5.5
Results of Stoke's Finite
Amplitude Wave Theory (2nd Order)

Velocity potential

$$\Phi = \left\{\frac{H}{2}\right\} \sin(kx - \omega t) \left\{\frac{L}{T}\right\} \frac{\cosh(ks)}{\sinh(kd)}$$
$$+ \left\{\frac{3H}{16}\right\} \sin 2(kx - \omega t) \left\{\frac{L}{T}\right\} \frac{\cosh(2ks)}{\sinh(2kd)}$$

Surface elevation

$$\eta = \left\{\frac{H}{2}\right\} \cos(kx - \omega t)$$
$$+ \left\{\frac{\pi H^2}{8L}\right\} \cos 2(kx - \omega t)[2 + \cosh(2kd)] \frac{\cosh(ks)}{\sinh^3(kd)}$$

Horizontal particle displacement

$$-\left\{\frac{H}{2}\right\} \sin(kx - \omega t) \frac{\cosh(ks)}{\sinh(kd)} + \left\{\frac{\pi H^2}{8L}\right\} \left\{\frac{1}{\sinh^2(kd)}\right\}$$
$$\cdot \left\{1 - \frac{3 \cosh(2ks)}{2 \sinh^2(kd)}\right\} \sin 2(kx - \omega t)$$
$$+ \left\{\frac{\pi H^2}{L}\right\} \left\{\frac{Ct}{2}\right\} \frac{\cosh(2ks)}{\sinh^2(kd)}$$

Vertical particle displacement

$$\left\{\frac{H}{2}\right\} \cos(kx - \omega t) \frac{\sinh(ks)}{\sinh(kd)}$$
$$+ \left\{\frac{3\pi H^2}{16L}\right\} \cos 2(kx - \omega t) \frac{\sinh(2ks)}{\sinh^4(kd)}$$

Horizontal particle velocity

$$u = \left\{\frac{\pi H}{T}\right\} \cos(kx - \omega t) \frac{\cosh(ks)}{\sinh(kd)}$$
$$+ \left\{\frac{3\pi^2 H^2}{4}\right\} \cos 2(kx - \omega t) \frac{\cosh(2ks)}{\sinh^4(kd)}$$

Vertical particle velocity

$$w = \left\{\frac{\pi H}{T}\right\} \sin(kx - \omega t) \frac{\sinh(ks)}{\sinh(kd)}$$
$$+ \left\{\frac{3\pi^2 H^2}{4TL}\right\} \sin 2(kx - \omega t) \frac{\sinh(2ks)}{\sinh^4(kd)}$$

Horizontal particle acceleration

$$\frac{\partial u}{\partial t} = \left\{\frac{2\pi^2 H}{T^2}\right\} \sin(kx - \omega t) \frac{\cosh(ks)}{\sinh(kd)}$$
$$+ \left\{\frac{3\pi^3 H^2}{T^2 L}\right\} \sin 2(kx - \omega t) \frac{\cosh(2ks)}{\sinh^4(kd)}$$

Vertical particle acceleration

$$\frac{\partial w}{\partial t} = -\left\{\frac{2\pi^2 H}{T^2}\right\} \cos(kx - \omega t) \frac{\sinh(ks)}{\sinh(kd)}$$
$$- \left\{\frac{3\pi^3 H^2}{T^2 L}\right\} \cos 2(kx - \omega t) \frac{\sinh(2ks)}{\sinh^4(kd)}$$

Pressure

$$p = -\rho g z + \tfrac{1}{2}\rho g H \cos(kx - \omega t) \frac{\cosh(ks)}{\cosh(kd)}$$
$$+ \frac{3}{4} \rho g H \left\{\frac{\pi H}{L}\right\} \left\{\frac{1}{\sinh(2kd)}\right\} \left\{\frac{\cosh(2ks)}{\sinh^2(kd)} - \frac{1}{3}\right\}$$
$$\cdot \cos 2(kx - \omega t) - \frac{1}{4} \rho g H \left\{\frac{\pi H}{L}\right\} \left\{\frac{1}{\sinh(2kd)}\right\}$$
$$\cdot [\cosh(2ks) - 1] + \text{atmospheric pressure}$$

The wave form predicted by Stokes II is not exactly sinusoidal due to a second-order term that varies with $2(kx - \omega t)$ rather than with $(kx - \omega t)$. The deviation from a sinusoidal profile becomes increasingly apparent as the depth decreases—the finite amplitude waves are more peaked (Fig. 5.5). In very shallow water the second-order term becomes large and the predicted wave form deviates from reality. Thus, like linear theory, second-order Stokes theory blows up in shallow water.

A close examination of the expression for horizontal particle displacement (Table 5.5) reveals a term that is not periodic:

$$\text{net displacement} = \left\{\frac{\pi H^2}{4L}\right\}\left\{\frac{\cosh(2ks)}{\sinh^2(kd)}\right\}\omega t.$$
(5.53)

This term increases steadily through time. In other words, this second-order theory makes explicit the fact noted earlier—that water particles do not travel through exactly closed orbits. There is some transport of water in the direction of wave propagation, and the magnitude of the transport decreases with increased depth. Dividing eq.

Figure 5.5. A comparison of the surface elevations predicted by linear and Stokes II wave theories. The second-order approximation has steeper peaks and more flattened troughs.

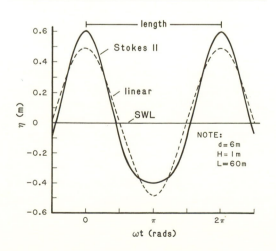

5.53 by t, we arrive at an expression for the average velocity with which water is transported in the direction of wave propagation:

$$u = \left\{\frac{\pi H^2}{4L}\right\}\left\{\frac{\cosh(2ks)}{\sinh^2(kd)}\right\}\omega.$$
(5.54)

This transport velocity gets larger as the water depth decreases.

Near a coast, inshore mass transport by waves must be offset by a seaward-flowing countercurrent. If it were not, the ocean would empty itself onto the beach. In a simple scenario we can assume that the seaward return flow is the same at all depths, and by setting the velocity of the return flow to the average inshore velocity we assure ourselves that net transport is zero. Integrating eq. 5.54 between $s = 0$ and $s = d$ and dividing by d, we see that the average inshore mass transport velocity is

$$\bar{u} = \frac{\pi H^2 \omega \sinh(2kd)}{16d \sinh^2(kd)}.$$
(5.55)

Thus near a coast we predict that the net mass transport velocity is (Fig. 5.6a)

$$\text{net } \bar{u} = \frac{\pi H^2 \omega \cosh(2ks)}{4L \sinh^2(kd)} - \frac{\pi H^2 \omega \sinh(2kd)}{16d \sinh^2(kd)}.$$
(5.56)

The transport velocity calculated from Stokes II is only a first approximation and must be modified before being applied to real flows in shallow water. Longuet-Higgins (1953) devised a second approximation for the mass transport that takes into account the viscosity of water. In deep water $(d \gg L/2, kd \gg \pi)$,

$$\bar{u} \simeq \left\{\frac{\pi H^2}{4T}\right\}k^2 d\left[3\left\{\frac{z}{d}\right\}^2 + 4\left\{\frac{z}{d}\right\} + 1\right].$$
(5.57)

Like Stokes's solution, the transport velocity is greatest at the surface and is zero at the sea bed $(-z = d)$ (Fig. 5.6a), but there is a small negative velocity below $-z = d/3$. When $-z = 2d/3$, flow is maximal in the direction opposite wave propagation and equal to one-third the velocity at the surface.

Figure 5.6. Mass transport in waves. (a) The transport in deep water predicted by Stokes II (eq. 5.56) and Longuet-Higgins (eq. 5.58). (b) Transport in shallow water (eq. 5.59).

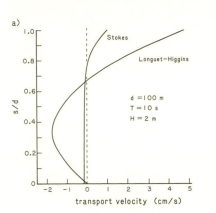

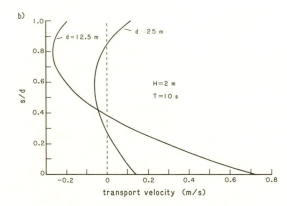

At intermediate and shallow depths, the Longuet-Higgins solution predicts that

$$\bar{u} = \left\{\frac{\pi H^2}{8T}\right\}\left\{\frac{k}{\sinh^2(kd)}\right\}\left[2\cosh\left\{2kd\left[\frac{-z}{d}-1\right]\right\}\right.$$
$$+ 3 + kd\sinh\left\{2kd\left[3\left\{\frac{z}{d}\right\}^2 + 4\left\{\frac{z}{d}\right\} + 1\right]\right\}$$
$$\left. + 3\left\{\frac{\sinh(2kd)}{2kd} + \frac{3}{2}\right\}\left\{\frac{z^2}{d^2}-1\right\}\right]. \quad (5.58)$$

In shallow water ($d < L/20$), the transport velocity (in the direction of water propagation) is maximal near the sea bed, reaching a maximum of

$$\bar{u}_{max} = 0.688\left\{\frac{\pi H^2}{T}\right\}\left\{\frac{k}{\sinh^2(kd)}\right\} \quad (5.59)$$

a few millimeters above the bottom (Fig. 5.6b).

Mass transport by waves approaching the shore may be biologically important. For instance, the switch from maximum shoreward transport near the surface in deep water to maximum transport near the bottom in shallow water may have practical consequences for strategies of larval dispersal. Larvae returning to the shore may find it advantageous to sink to the bottom once the water is shallow and ride the benthic mass transport. However, this strategy will only work as far as the surf zone. Inside the line of the breakers there is often a seaward-directed current near the bottom (the *undertow*; Longuet-Higgins 1983). The undertow is one mechanism by which water carried inshore by the surf is returned to the sea.

The changes in wave shape and water transport predicted by Stokes II for shallow water give some hint as to how sinusoidal, deep-water waves are transformed into waves resembling solitary waves in shallow water. However, the Stokes approximations are subject to the same malady suffered by linear wave theory—they blow up in shallow water. A general rule of thumb is that if $d/L < 1$, Stokes wave theories are useful only when $H/d \ll (kd)^2$ (Sarpkaya and Isaacson 1981). In other words, the theory is useful only for waves of appreciable height [(H/d of the same order as $(d/L)^2$] when the water is deep.

There are certain aspects of Stokes second order theory that are precisely the same as for linear wave theory. The wave celerity and the wave energy are predicted to be exactly the same for waves of finite amplitude as they are for waves of very small height. These two properties will figure heavily in our further discussions of waves. Higher-order Stokes theories predict energies and celerities that are slightly different. Fifth-order Stokes theories (commonly referred to as Stokes V) have found considerable use in ocean engineering practice and the study of nearshore wave processes. Fenton (1985) provides a review of these theories and the latest revisions. Unfortunately, Stokes fifth-order theory is so unwieldy that for practical use a computer is required.

Solitary Wave Theory

We have already encountered many of the results of solitary wave theory in the discussion of waves in a laboratory wave tank (Chapter 4). A thorough theoretical treatment of the fluid motion in a solitary wave results in a more precise description (Table 5.6, coupled with Fig. 5.7) but in no surprises.

The shape of the wave form is that of the square of a hyperbolic secant function

$$\eta = H \operatorname{sech}^2(q), \quad (5.60)$$

where $q = ([3H/4d]^{1/2}[x - Ct]/d)$. An examination of Figure 5.8a shows that $\operatorname{sech}^2(q)$ is a maximum (= 1) at $q = 0$ and asymptotes to zero as q goes to infinity. The larger the H/d, the larger is q, and the faster $\operatorname{sech}^2(q)$ approaches zero. In other words, the shape of the wave form depends on the ratio of wave height to water depth—solitary waves are steeper in shallower water.

The wave celerity predicted by the second approximation to solitary wave theory is given in Table 5.6. This is very close to

$$C = \sqrt{gd(1 + H/d)} = \sqrt{g(d + H)}, \quad (5.61)$$

the value derived earlier (eq. 4.14)

Figure 5.7. The variation in constants M and N used to calculate particle velocities beneath solitary waves (see Table 5.6) (redrawn from Munk 1949).

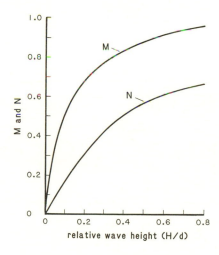

Table 5.6
Solitary Wave Theory,
2nd Approximation

Surface elevation

$$\eta = H \operatorname{sech}^2 \left\{ \sqrt{\frac{3}{4} \frac{H}{d}} \frac{x - Ct}{d} \right\}$$

Celerity[a]

$$C = \sqrt{gd} \left\{ 1 + \frac{H}{2d} - \frac{3H^2}{20d^2} + \cdots \right\} \simeq \sqrt{g(d + H)}$$

Horizontal particle velocity[b]

$$u = NC \left\{ \frac{1 + \cos\left\{\frac{Mz}{d}\right\} \cosh\left\{\frac{M[x - Ct]}{d}\right\}}{\left\{\cos\left\{\frac{Mz}{d}\right\} + \cosh\left\{\frac{M[x - Ct]}{d}\right\}\right\}^2} \right\}$$

Vertical particle velocity

$$w = NC \left\{ \frac{\sin\left\{\frac{Mz}{d}\right\} \sinh\left\{\frac{M[x - Ct]}{d}\right\}}{\left\{\cos\left\{\frac{Mz}{d}\right\} + \cosh\left\{\frac{M[x - Ct]}{d}\right\}\right\}^2} \right\}$$

Pressure

$$p \approx \rho g(\eta - z) + \text{atmospheric pressure}$$

[a] Laitone 1959.
[b] N and M are obtained from Figure 5.7.

The particle trajectories and velocity distributions for a steep solitary wave in shallow water are given in Figure 5.8b,c. Note that in shallow water ($H/d > 0.4$) the maximum water velocity at the sea bed is about 30% of the wave celerity. In the absence of any return flow, water particles are essentially stationary at a distance $x = 10d$ from the crest. Thus waves need to be spaced with their crests only $20d$ apart to behave as if they were truly

Figure 5.8. Aspects of the solitary wave. (a) The wave profile is described by sech^2 (q), where q is a function of H and d (see text). (b) Particle trajectories in a solitary wave (redrawn from Paul D. Komar, *Beach Processes and Sedimentation*, 1976, p. 58, by permission of Prentice-Hall, Englewood Cliffs, N.J.). (c) Velocities in a solitary wave (redrawn from Munk 1949 by permission of the New York Academy of Sciences).

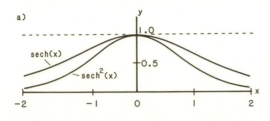

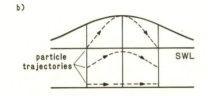

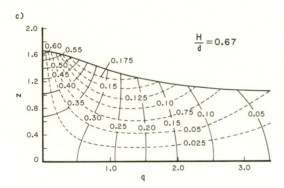

"solitary." In sufficiently shallow water, this criterion is easily met by ocean waves.

The energy of a solitary wave is, to a first approximation, evenly divided between kinetic and potential energy. The total energy per unit crest width is

$$\text{energy} = \left\{ \frac{8}{3\sqrt{3}} \right\} \rho g H^{3/2} d^{3/2}. \quad (5.62)$$

Because a solitary wave has no wave length, this energy is not directly comparable to that calculated earlier for periodic waves. In a solitary wave, energy is transmitted with the same velocity as the wave form. This is because all the water in a truly solitary wave is above the still water level and is moving forward. There is none of the backwards motion associated with the trough of an oscillatory wave, and consequently no kinetic energy is associated with portions of the ocean not contained in the wave crest. We examine the consequences of this change in the pattern of energy transmission when we deal with wave shoaling (Chapter 7).

Solitary wave theory is most applicable to waves in very shallow water. A reasonable rule of thumb is that when the Ursell number (HL^2/d^3) is greater than 58, an oscillatory wave approximates a solitary wave.

Stream Function Wave Theory

Waves need not be described only in terms of a velocity potential function; they can also be described by a *stream function*. A stream function is a close relative to a velocity potential function in that it is a general method for describing the pattern of fluid flow. However, instead of being defined relative to the difference in conditions along the direction of flow, the stream function, Ψ, is defined in terms of differences across the direction of flow. Thus, u, the velocity in the x-direction is

$$u = \frac{\partial \Psi}{\partial z}, \quad (5.63)$$

and

$$w = -\frac{\partial \Psi}{\partial x}. \quad (5.64)$$

This of course assumes that waves are two-dimensional, i.e., there is no flow in the y-direction. By plugging the stream function into the continuity equation (eq. 5.1), we can show that those areas of a fluid that have the same value for the stream function lie on a streamline. Hence the name.

Stream function theory (as approached by Dean 1974) begins by specifying the shape of a streamline in a wave. This is relatively simple, since the water's surface is a streamline. Thus the shape of the wave profile sets the stream function, and the calculation proceeds from there. The equation for the stream function is expressed as a series expansion, and the coefficients of the series are chosen so as to provide the best fit to the specified boundary conditions, these generally being the same as used earlier. The advantage of this method over that used for Stokes theories is that rather than specifying each coefficient starting with the first order and moving on to the second order, etc., all coefficients are manipulated at once in the same way one would manipulate coefficients in a polynomial to provide the best least-squares fit to a curve. The resulting series expansion may thus provide a more accurate estimate of the real flows. The recent stream function theory of Rienecker and Fenton (1981) has been shown to provide accurate estimates of wave flows for waves of all heights and for all depths, from deep water inshore to the point where waves break.

However, there are two drawbacks to stream function theory. The first is due to the method by which the various coefficients are determined—each coefficient in and of itself has no particular meaning. The coefficient of the first-order term is different in a fifth-order expansion than the first-order coefficient in a tenth-order expansion. Thus the expansions themselves have no heuristic value. Second, the calculations involved in this theory are not something you bash out on your pocket calculator. To simplify the practical use of his theory, Dean (1974) has provided an extensive set of tables enumerating the water flow conditions in various wave situations. If the particular case you are examining happens to match one of the situations tabulated by Dean, these tables are the simplest method of quickly ac-

quiring an accurate answer regarding subsurface velocities and accelerations for waves in shallow water. However, if a particular case falls between Dean's tabulated values, a time-consuming interpolation must be carried out. These tables (a sizable tome) have been published by the U.S. Government and are available at a reasonable price. Alternatively, the theory of Rienecker and Fenton is easily developed as a computer program. Once the time has been invested in writing the program, this theory can be very conveniently used.

Edge Waves

The waves discussed so far have been assumed to propagate more or less perpendicular to the shoreline as incident or reflected waves. A different type of wave, an *edge wave*, propagates along the shore. As the name implies, edge waves are found only near the water's edge, being trapped there by refraction. In contrast, the energy of incident waves can freely move away from shore, and these wave forms are often referred to as "leaky" waves.

Edge waves can exist in many modes, differing in the manner in which their amplitude varies offshore. The simplest of these modes, mode 0, is used here as an example (Table 5.7). The inquisitive reader should consult Holman (1983) or Eckart (1951) regarding other modes.

Aside from their direction of travel, edge waves differ from incident gravity waves in several respects. First, their amplitude is largest at the shoreline, just the opposite of incident swell, which break and decrease in height near the shore (see Chapter 7). Thus if edge waves have appreciable heights, the water velocities and accelerations they cause may be important in the flow regime of the swash zone, and these flows have been implicated in the formation of cusps and bars on sandy beaches (for a review see Holman 1983).

Table 5.7
Edge Wave Theory (0 mode),
1st Approximation

Velocity potential

$$\Phi = \left\{\frac{HTg}{4\pi}\right\} \exp(-kx\cos\theta) \exp(kz\sin\theta) \sin(ky - \omega t)$$

Surface elevation

$$\eta = \left\{\frac{H}{2}\right\} \exp(-kx\cos\theta) \cos(ky - \omega t)$$

Celerity

$$C = \left\{\frac{gT}{2\pi}\right\} \sin\theta$$

Horizontal particle velocities

$$u = \left\{\frac{-\pi H}{T}\right\}\left\{\frac{1}{\tan\theta}\right\}$$
$$\cdot \exp(-kx\cos\theta) \exp(kz\sin\theta) \sin(ky - \omega t)$$

$$v = \left\{\frac{\pi H}{T}\right\}\left\{\frac{1}{\sin\theta}\right\}$$
$$\cdot \exp(-kx\cos\theta) \exp(kz\sin\theta) \cos(ky - \omega t)$$

Vertical particle velocity

$$w = \left\{\frac{\pi H}{T}\right\} \exp(-kx\cos\theta) \exp(kz\sin\theta) \sin(ky - \omega t)$$

Note: θ is the slope of the sea bed; the y-axis is parallel to the shoreline and x is positive to seaward.

The amplitude of edge waves decreases rapidly with distance from the shore; for instance in the mode 0 wave with x-positive away from the shoreline,

$$\eta(x, y, t) = \left\{\frac{H}{2}\right\} \exp(-kx\cos\theta) \cos(ky - \omega t),$$
$$(5.65)$$

reaching a negligible amplitude only $L/2$ from the shoreline. Here θ is the bottom slope.

The properties of edge waves are highly dependent on the slope of the bottom (Table 5.7). The wave length and celerity both increase as the slope becomes steeper:

$$L = \left\{\frac{gT^2}{2\pi}\right\} \sin\theta, \qquad C = \left\{\frac{gT}{2\pi}\right\} \sin\theta. \quad (5.66)$$

The orbital motion of water in an edge wave is similar to that under a linear wave,

except that particles travel in orbitals that are nearly horizontal. If the slope of the bottom is gentle, the orbitals are approximately circular. As the bottom slope becomes steeper, the orbitals stretch out to form ellipses with their long axes parallel to the shore. The steeper the slope, the slower the particle velocities in the x and y-directions:

$$u = \left\{\frac{-\pi H}{T}\right\}\left\{\frac{1}{\tan\theta}\right\} \exp(-kx\cos\theta)$$
$$\cdot \exp(kz\sin\theta) \sin(ky - \omega t)$$

$$v = \left\{\frac{\pi H}{T}\right\}\left\{\frac{1}{\sin\theta}\right\} \exp(-kx\cos\theta)$$
$$\cdot \exp(kz\cos\theta) \cos(ky - \omega t). \quad (5.67)$$

The vertical velocity is small relative to the horizontal velocities except on very steeply sloping shores:

$$w = \left\{\frac{\pi H}{T}\right\} \exp(-kx\cos\theta) \exp(kz\sin\theta)$$
$$\cdot \sin(ky - \omega t). \quad (5.68)$$

Edge waves may occur with a variety of periods. One typical form is driven by the incident swell and has a period precisely twice that of the swell (Guza and Davis 1974). Other edge waves may have very long periods not tightly coupled to the incident surf.

Although edge waves have received considerable interest in recent years, it is only very recently that their existence has been unequivocally demonstrated outside of the laboratory (see Holman 1983 for a review). No studies have been made of edge waves on rocky shores, and until the mode, amplitude, and period of these edge waves are known, it is difficult to assess their biological importance.

Other Theories

The wave theories discussed here are only a few of the theories that exist. For example, an elegant, exact theory exists for capillary waves (Crapper 1957). Cnoidal wave theory bridges the gap between shallow-water linear wave theory and solitary wave theory (Korteweg and deVries 1895; Wiegel 1964; Fenton 1979). There are theories for trochoidal, vocoidal, and hyperbolic waves. However, these theories have proven to be of little practical utility in the study of the wave-swept environment. The interested reader should consult standard texts dealing with wave theory for a review of these alternative approaches to waves (e.g., Lamb 1945; Kinsman 1965; Sarpkaya and Isaacson 1981; Wiegel 1964).

Which Theory to Use?

One's choice of a wave theory depends very much on the purpose to which the theory is to be applied. For example, in many cases one wants an intuitive and qualitatively accurate understanding of some aspect of wave motion and can accept some quantitative inaccuracy in the answer. In such a case, linear wave theory is by far the best choice. The theory is easily understood and manipulated, and provides answers which (to a first approximation) accurately describe the real world. In fact, when various theories are compared to empirical measurements made on real ocean waves, linear theory gives values that are as close and sometimes closer to those observed in nature than are the predicted values of Stokes V or other high-order theories (Dean 1974). This is true even in shallow water, where linear theory, theoretically, should not be accurate. On these bases, linear wave theory should be considered the starting point for any understanding of wave motion.

In very shallow water the first approximation of solitary wave theory can serve the same function as linear wave theory. It is easily understood and manipulated, and therefore a good heuristic tool.

In cases where one requires as accurate an estimate as possible of a certain flow

parameter, Stokes V can be used for deep water, and a computer program based on the stream function theory of Rienecker and Fenton (1981) can be used at any depth.

All of these theories have a common failing—the waves must be two-dimensional and periodic, and one must know the wave height and period before the theory can be applied. In the real ocean, the height and period vary from wave to wave, and it is not a simple matter to choose an appropriate set of parameters. The problem of specifying wave height and period in a chaotic sea is dealt with in Chapter 6.

Sources of Information

The texts referred to in this chapter provide the best comprehensive introduction to wave theories. However, new theories and interpretations are continually being published, and one is advised to keep abreast of the current literature. Periodicals such as the *Journal of Fluid Mechanics*, the *Journal of Geophysical Research* (the "Oceans" section), the *Journal of Physical Oceanography*, and the *Journal of Waterway, Port, Coastal and Ocean Engineering* are excellent sources of information.

Appendix 5.1

The Continuity Equation

Consider a small cubic volume (Fig. 5.9) having dimensions dx, dy, and dz, known in fluid-dynamic parlance as a control volume. We let water flow into the cube across three of its sides, and then we calculate what must happen at the other three sides to keep the mass of water in the cube constant. Because water is virtually incompressible, the maintenance of a constant mass is equivalent to the maintenance of a constant volume.

Consider the side of the cube with area $dy\,dz$. At this side, the water has a velocity

Figure 5.9. The principle of continuity revisited. The mass of fluid entering the control volume must be equal to the mass leaving. The variation in velocity across the control volume is shown for one direction; analogous variations may occur for v and w.

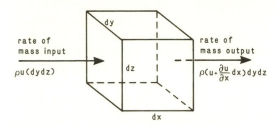

$u(= dx/dt)$ in the x-direction so that volume enters the cube at rate $u\,dy\,dz$. This particular velocity may not be maintained as the fluid moves across the control volume. This can be taken into account by allowing the velocity to change at a rate $\partial u/\partial x$. Thus the velocity at the side of the cube opposite from where fluid enters has changed by an amount $(\partial u/\partial x)\,dx$. In this one direction, then, water enters the control volume at rate $u\,dy\,dz$ and leaves at the rate of $u\,dy\,dz + [(\partial u/\partial x)\,dx]\,dy\,dz$. The same reasoning can be applied to water flowing with velocity v across side $dx\,dz$ and velocity w across side $dx\,dy$.

If the volume (and hence, the mass) of water in the cube stays the same, the rate at which water enters the control volume must equal the rate at which it leaves. This can be stated mathematically by subtracting the rate at which water leaves from the rate at which it enters, and setting the difference to zero:

$(u\,dy\,dz + v\,dx\,dz + z\,dx\,dy)$

$\quad -\left\{(u\,dy\,dz + \dfrac{\partial u}{\partial x}\,dx\,dy\,dz) + (v\,dx\,dz\right.$

$\quad + \dfrac{\partial v}{\partial y}\,dy\,dx\,dz) + (w\,dx\,dy + \dfrac{\partial w}{\partial z}\,dz\,dx\,dy)\Bigg\} = 0.$

$$\text{(5.69)}$$

Upon simplification this becomes

$$\left\{\frac{\partial u}{\partial x} + \frac{\partial v}{\partial y} + \frac{\partial w}{\partial z}\right\} dx\,dy\,dz = 0. \quad \text{(5.70)}$$

Since we know that $dx\,dy\,dz$ (the volume of the cube) is not zero, this can be true only if

$$\frac{\partial u}{\partial x} + \frac{\partial v}{\partial y} + \frac{\partial w}{\partial z} = 0. \quad \text{(5.71)}$$

This simple differential equation, then, is a statement of the conditions that must be met for the mass of an incompressible fluid to be conserved, and it is known as the *continuity equation.*

Appendix 5.2

Irrotational Motion

Our task here is to verify intuitively that eq. 5.3 indeed describes irrotational movement. Consider an arrow moving in a circular path around the origin. As it moves, its head always points away from the origin (Fig. 5.10a). The direction in which the arrow points changes: at the top of its path it points up, at the bottom it points

Figure 5.10. Rotational vs. irrotational movement. (a), (b) Rotational movement: the arrow changes direction as it moves and the components of velocity vary between the arrow's head and tail. (c), (d) Irrotational movement: the arrow always points in the same direction.

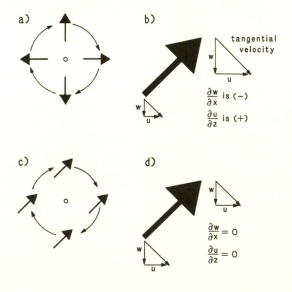

down, and in between it points to the sides. If we just look at the arrow's orientation about its center and ignore how it moves relative to the origin, it is clear that the arrow rotates about its center. Let us examine this motion at one arbitrary position to see how u and w vary between the head of the arrow and its tail (Fig. 5.10b). Because the head of the arrow is farther from the origin, it has a greater tangential velocity than the tail, shown by the longer velocity vector at its head. At both the head and the tail the u and w components of the tangential velocity vectors are shown. u is greater at the head of the arrow, and the head is higher than the tail, so the change in u with an increase in z (that is, $\partial u/\partial z$) is positive. w has a larger magnitude in the negative direction at the head than at the tail, so $\partial w/\partial x$ is negative. Thus when we subtract $\partial w/\partial x$ from $\partial u/\partial z$, the result is a positive number. By the criterion of eq. 5.3 the motion of the arrow would not qualify as irrotational, as indeed it should not. Now con-

sider the motion of the arrow shown in Figure 5.10c. Again the arrow moves in a circular path around the origin, but the arrow always points in one direction—it does not rotate. If we now examine u and w at an arbitrary position, a different picture emerges. If the arrow does not rotate, both the head and the tail must have the same u. Thus $\partial u/\partial z$ is zero. Similarly, w must be the same at the head and the tail; and $\partial w/\partial x$ is also zero. Inserting these values into eq. 5.3, we see that $\partial u/\partial z - \partial w/\partial x$ indeed equals zero, as it should.

This is only an informal varification of the validity of eq. 5.3, and should not be taken too seriously. The fact that both $\partial u/\partial z$ and $\partial w/\partial x$ are zero for the irrotational arrow does not mean that this is a general requirement for irrotationality. Here it is a result of the arrow being a solid object. For a water particle, which can change shape, irrotationality requires only that $\partial u/\partial z$ and $\partial w/\partial x$ be of the same magnitude and the same sign.

CHAPTER 6

~ ~ ~ ~ ~ ~ ~ ~ ~ ~ ~

The Random Sea

In the last two chapters we have considered the fluid motions accompanying water waves. This analysis has assumed that waves in deep water have a sinusoidal shape and that one wave is much like another, with wave height and period essentially the same. Such a depiction of the water surface is shown schematically in Figure 6.1a, but it is not an accurate portrait. A brief visit to the seashore will convince anyone that no two waves are exactly alike, and that successive waves can be quite different in both their height and period. A more typical picture of the ocean surface is given in Figure 6.1b, which is drawn from a record of the water-surface level off the California coast. Both the wave height and period are variable, and at times it is difficult to tell where one wave ends and the next begins. This, then, is the real world, and it leaves us with two large questions to answer: Why does the real ocean surface have these characteristics? And how can we reconcile the ideas we have derived for sinusoidal waves with this kind of ocean surface?

Wind-Generated Waves

In general, the waves that we see on the surface of the ocean are produced by the wind. As wind blows across a water surface, some of its energy is transferred to the fluid, producing waves. There are several mechanisms by which this energy transfer is accomplished. For example, localized variations in air pressure associated with turbulence in the wind may push and pull on the water's surface, forming small waves. As the wind blows over the crests of preexisting waves, its velocity increases, resulting in a decrease in air pressure above and downwind of the wave crest (see the discussion of lift in Chapter 11). The resulting force acts over a distance as the wave propagates, adding to the wave's energy. Various other mechanisms of energy transfer have been proposed, and their relative importance is a matter of some controversy. Regardless of the precise mechanism by which wind raises waves, the net result is the production of waves with a large range of heights and lengths, traveling in all directions. This wind-generated, chaotic ocean surface (often produced well out at sea in the midst of a storm) is the source of the waves that break on shore.

The size of waves produced by the wind depends on three factors: the wind speed, the length of time the wind blows, and the *fetch*, the length of the stretch of water affected by the wind. For a given wind speed, the wave height can be limited either by the length of time the wind blows, the

Figure 6.1. Sinusoidal vs. actual surface waves. (a) A monochromatic, sinusoidal wave train; wave height, length, and period are well defined. (b) A record of sea surface elevations; wave height, length, and period vary (redrawn from U.S. Army Corps of Engineers 1984).

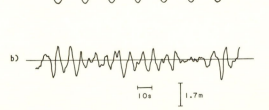

a)

b)

⊢——⊣ 10s ⊢ 1.7m

Table 6.1
Conditions in Fully Developed Seas
(from Bascom 1980)

Wind Velocity (m/s)	Fetch (km)	Time (hours)	Average Wave Height (m)	Sig. Wave Height (m)
5.1	18	2.4	0.3	0.4
7.7	63	6	0.8	1.1
10.3	139	10	1.5	2.4
12.9	297	16	2.7	4.3
15.4	519	23	4.3	6.7
20.6	1315	42	8.5	13.4
25.7	2630	69	14.6	23.8

fetch, or both. If neither the time nor the fetch limits the wave height, the waves reach an equilibrium height, and the sea is said to be "fully developed." Table 6.1 gives empirical measurements for the wave heights observed in fully developed seas as a function of the wind speed. These are the largest waves that the wind can produce; for the winds found in hurricanes and typhoons (> 60 knots, or about 30 m/s), the waves can reach monstrous proportions. For instance, instruments mounted on offshore oil rigs have recorded waves 24 m high. The highest wave reliably recorded was sighted by the watch officer of the HMS *Ramapo* during a storm in the South Pacific on February 6, 1933. By sighting from the bridge to the crow's-nest and on to the wave crest, Lt. F. C. Margraff estimated the wave to be 34 m high (Bascom 1980). Fortunately, these huge waves are rare. This is largely a result of the long time and fetch required for the sea to become fully developed at these high wind velocities. Hurricane winds need nearly three days to raise waves to their equilibrium height and require a fetch of more than 2250 km. Storms of this duration and size are seldom seen.

Several semiempirical methods are available for predicting wave heights from meteorological data. These are somewhat complex and will not be dealt with here. Readers interested in wave generation and forecasting should consult Pierson et al. (1955), Kinsman (1965), or the U.S. Army Corps of Engineers' *Shore Protection Manual* (1984).

Wave Propagation

Waves produced by storms often outlive the disturbance that produced them and propagate to a different section of the ocean. It is these propagated waves that we often see, and their size and shape are different from those in storms where waves are actively being produced.

Why should waves be any different after they have propagated? The most obvious reason is the dispersive nature of water waves; waves of different wave lengths travel with different celerities (eq. 4.9). Although a localized storm produces waves with a range of lengths, these waves are spatially sorted out as they travel. The longer waves (which travel fastest) arrive at a distant location first, spending the least time in transit. Conversely, the shorter waves arrive last and spend the greatest time in transit. In calculating the time it takes a wave to travel a certain distance, we must remember that wave energy is transmitted at the group velocity rather than at the celerity of individual waves. In deep water the group velocity is half the celerity (Table 5.2).

As waves travel they lose some of their energy to viscous processes. The loss depends not so much on the distance traveled as on the length of travel time. The height of a deep-water wave at time t is

$$H(t) = H_i \exp\left\{\frac{-32\pi^4 vt}{g^2 T^4}\right\}$$

$$H(t) = H_i \exp\left\{\frac{-8\pi^2 vt}{L^2}\right\}, \qquad (6.1)$$

where H_i is the initial wave height at time $t = 0$ (Komar 1976). As before, T is the wave period, L the wave length, and v the kinematic viscosity of water. The factors T^4 and L^2 in the denominators of these equations mean that short period, short wave-length waves lose energy to viscosity much faster than do waves of longer periods and wave lengths, and shorter wavelength waves are more attenuated when they reach the shore. Waves with periods above 10 s lose relatively little height in propagation. For example a wave with $T = 1$ s would lose half its height in traveling 16 km, while a wave with $T = 10$ s would have to travel about 16 million km to lose half its initial height. These easily propagated waves of long wave length and long period are the *swell* described earlier. Swell more closely approximate our idealized sinusoidal wave train of Figure 6.1a than do the waves shown in Figure 6.1b.

The swell produced by a distant storm can interact with the waves created locally by wind. These local waves (the *seas* referred to in Chapter 4) are simply added onto the swell. This combination of locally produced seas and propagated swell is perhaps the most typical condition for the ocean surface in coastal waters.

Specifying the Random Sea

Having arrived at a reasonable, if cursory, explanation for the origin of waves and the complex wave pattern we typically observe, we are now faced with devising some means of reconciling this observed pattern with our theoretical understanding of wave motions. The crux of the problem is that each wave is different from the rest. One cannot speak of *the* wave height or *the* wave length. Instead, we are forced to deal with what appears to be a continuous distribution of wave heights and lengths. The reconciliation between wave theories and the random sea requires the use of a specialized set of statistics.

The Mean Square Amplitude

Recall from Chapter 3 that the amplitude of a wave is its maximum vertical deviation from mean sea level. Each wave has a positive amplitude (the crest) and a negative amplitude (the trough). A linear wave has half its wave form above sea level and half below, and the amplitude is simply one-half of the wave height. We make the assumption that this is true for real ocean waves as well.

To describe the "typical" sea surface shown in Figure 6.1b, we must have some means of measuring the average wave amplitude of a complex wave form. The first step is to measure the surface elevation at a large number of evenly spaced points. The average of these elevations is, by definition, the mean sea level. Each elevation is then expressed as a deviation from mean sea level, A_i, (a positive number for elevations above still water level, negative for points below). Each deviation is then squared (thereby automatically ending up with nothing but positive values), and squared deviations are averaged,

$$\text{mean square deviation} = \frac{1}{N} \sum_{i=1}^{N} A_i^2, \quad (6.2)$$

where N is the number of surface elevations measured. This statistic is formally known as the *mean square amplitude*, $\bar{A}^2$; it is the same as the *variance* of the surface elevation. Because it is the sum of squared lengths, it has the units m². To return to units of amplitude, we simply take the square root of $\bar{A}^2$ to arrive at the *root mean square (rms) wave amplitude*, A_{rms}. The rms amplitude is the same as the standard deviation of the sea surface about mean sea level. By doubling A_{rms} we get a measure of the rms wave height, H_{rms}.

The Rayleigh Distribution

In itself, the root mean square wave amplitude is not very useful. It gives us a general measure of how "wavy" a section of ocean is, but it is certainly not true that every wave in that section of ocean has the rms amplitude. In general, we are more concerned with the probability of encountering a wave of at least a certain amplitude than with knowing the amplitude of some hypothetical "average wave." Fortunately, a knowledge of the rms wave height allows for the calculation of these sorts of probabilities.

The amplitudes of ocean waves have been shown to follow a Rayleigh distribution (Longuet-Higgins 1952; Fig. 6.2a). There are few waves with small or very large amplitudes, the most probable wave amplitude being somewhat less than the rms amplitude. This distribution is described by the following equation:

$$P(A) = \left\{ \frac{2A}{A_{rms}^2} \right\} \exp - \left\{ \frac{A}{A_{rms}} \right\}^2, \quad (6.3)$$

where $P(A)$ is the probability density for amplitude A. In other words, the probability that a wave has an amplitude within range $A \pm dA/2$ is $P(A)\,dA$. The total area under the probability density curve, $\int_0^\infty P(A)\,dA$, is the probability that a wave has a height between zero and infinity, and is therefore equal to 1.

The probability that a wave chosen at random has an amplitude greater than a certain value A' is

$$P_{(A > A')} = 1 - \int_0^{A'} P\,dA = \exp - \left\{ \frac{A'}{A_{rms}} \right\}^2. \quad (6.4)$$

This cumulative probability curve is shown in Figure 6.2b. Even on the calmest of days there are always some waves present on the ocean, thus the probability that there are waves with amplitude greater than zero is 1. The probability that a particular wave chosen at random is greater than the rms amplitude is $\exp - (1) = 0.37$. Thus 37% of waves have amplitudes greater than A_{rms}. The higher the wave we specify, the smaller

Figure 6.2. The Rayleigh distribution. (a) The Rayleigh probability density distribution. (b) The Rayleigh exceedance distribution.

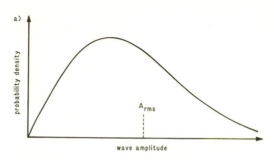

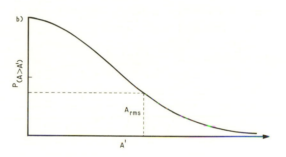

the probability that a wave chosen at random exceeds that height.

The Rayleigh distribution is satisfyingly simple to use. Unlike the Gaussian (normal) distribution where we must know both an arithmetic mean and a standard deviation, the Rayleigh distribution is scaled by a single factor, A_{rms}.

Knowing that wave amplitudes follow a Rayleigh distribution, we can specify all sorts of interesting probabilities by simply measuring the root mean square wave height. Take a tangible example: You have established an experimental site on the shore and know from past experience that when a wave exceeds a certain height, H, the surge reaches your site. As you sit on the rock, counting snails in your quadrat, you might reasonably ask the question, "Will the next wave get me wet?" By specifying the wave conditions using A_{rms}, this question can be answered:

$$P_{(A > H/2)} = \exp - \left\{ \frac{H/2}{A_{\mathrm{rms}}} \right\}^2.$$

Significant Wave Height

You are left, however, with the practical problem of measuring the rms wave height. If you perchance have a recording wave meter (see Chapter 19), this poses no problem. The instrument provides a chart recording similar to Figure 6.1b, and measurements are made as outlined above. Unfortunately, most casual wave observers do not have recording wave meters and are therefore faced with trying to estimate A_{rms} from visual sightings. This is a notoriously error-prone method of measuring wave amplitudes. In a choppy sea it is often difficult to tell where one wave stops and another begins, and it requires a great deal of practice to estimate wave heights accurately. A partial solution to this problem was devised by W. H. Munk during World War II. In those days oceanographers were assigned the problem of forecasting surf conditions for the beaches where amphibious assaults were to be made. Providing these forecasts involved, in part, knowing the sea conditions near the beach, and the only reliable way to get this information was to send someone out to look. The observer came back with an estimate along the lines of, "Well, the waves looked to be about 6 feet high." What precisely did this mean? By comparing visual observations with accurate wave records, Munk (1944) determined that the wave height as estimated by a skilled observer was approximately equal to the average height of the one-third-highest waves. In other words, when you look at waves, in an attempt to make some order out of the visual chaos your brain ignores two-thirds of the waves present (those with the smallest amplitudes) and averages what is left. This visually estimated wave height is called the *significant wave height*, H_s or $H_{1/3}$. For our purposes we define a significant wave amplitude, A_s, equal to half the significant wave height. It has been shown (Longuet-Higgins 1952) that

$$A_s = 1.412\, A_{\mathrm{rms}},$$
$$A_{\mathrm{rms}} = 0.71 A_s. \qquad (6.5)$$

Thus, from a visual estimation of wave height, it is possible to calculate a rough estimate of A_{rms}, and thereby to gain access to the Rayleigh distribution. This conclusion should be treated with caution. Upon first observing waves, we have a strong tendency to overestimate their height. When we are standing at sea level, or, worse yet, swimming, a 1 m wave appears quite large and a 2 m wave looks as big as a house. Only when we have carefully watched a number of waves pass a fixed object marked with a meter scale can we train our eyes to be reasonably accurate.

We must also use caution in applying the Rayleigh distribution to extreme situations. The U.S. Army Corps of Engineers (1984) has shown that the Rayleigh distribution overestimates by about 5% the height of waves occurring with probabilities of 0.01, and overestimates by about 15% those with probabilities of 0.0001 or less. There can also be problems with assuming a Rayleigh distribution in shallow water. Thornton and Guza (1983) measured wave heights at various places in the surf zone on a sandy beach and found that as waves steepen and break, the nonlinearity of their behavior can cause their height distribution to deviate from the Rayleigh distribution. However, once waves have broken, they again become Rayleigh-distributed. This is a surprising result, one not easily accounted for by theory, and we will use it cautiously here. In summary, we can use the Rayleigh distribution with some confidence for waves outside the surf zone and, with some reservations, for broken waves in the surf zone, but we must be careful in using this distribution to describe breaking wave heights.

The Rayleigh distribution will be used

extensively in Chapter 16 in the discussion of structural wave exposure.

The Wave Spectrum

Useful as the above approach may be as a means for coping with distributed wave amplitudes, it is only a partial fulfillment of our needs. An examination of Figure 6.1b (our typical wave record) clearly shows that not only do wave amplitudes vary, but wave periods vary as well. The use of an rms amplitude and the probabilities calculated from the Rayleigh distribution tell us nothing about which wave heights are associated with which wave lengths. However, most calculations regarding wave-induced water motions require knowing both the wave amplitude and the wave period. Solitary waves (which are aperiodic) are an important exception to this rule, but this exception is not sufficient to alleviate the general need to know something about both the amplitude and the period of real ocean waves.

Information about the relationship between amplitude and period is best conveyed by the *wave power spectrum* or, more simply, the *wave spectrum*.

We begin by examining the additive properties of linear waves in a laboratory wave tank. Recall that waves are produced in this tank by sinusoidally moving a paddle. We can vary the frequency with which the paddle oscillates, but only one frequency can be produced at a time. We now replace the paddle with a more sophisticated device, one capable of oscillating at several different frequencies simultaneously. Devices appropriate for this task are described in Chapter 19.

We first use our wave maker to produce waves of a single frequency, causing the surface of the tank to look like Figure 6.3a. This is the familiar sinusoidal (also known as monochromatic) wave train discussed in Chapter 4. Alternatively, we can produce waves of a different, higher frequency, as shown in Figure 6.3b. If we

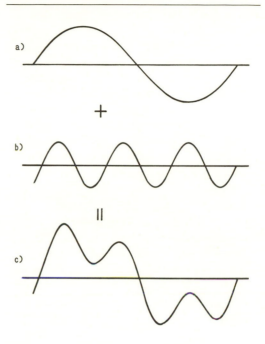

Figure 6.3. The surface elevations of water waves are additive. The sum of the wave forms (a) and (b) is the wave form shown in (c).

produce both wave forms simultaneously, we find that the surface of the tank at one instant in time looks like Figure 6.3c. At points where the peaks of the two wave forms coincide, the amplitude of the combined wave is equal to the sum of the amplitudes of the component waves. Similarly, where troughs coincide, the negative amplitude is greater than before. At points where the trough of one wave coincides with the crest of another, the resulting amplitude is again the sum of the two, and the surface deviation is less than that present in either wave form alone. The surface resulting from the superposition of two wave trains of different frequencies is far more complex than that of a single frequency wave train, and looks more like Figure 6.1b.

We can state these findings in a more formal manner. A sinusoidal wave form can be expressed in both time and space by the equation

$$\eta(x,\ t) = \alpha \cos\left\{\frac{2\pi x}{L} - \frac{2\pi t}{T}\right\}, \quad (6.6)$$

where $\eta(x, t)$ is the surface elevation at position x at time t, and α is the wave amplitude. For the present discussion, we need not worry about both time and space; instead, we will take the perspective of an observer watching waves pass a piling, and examine how the water's surface varies as a function of time alone. It is from this perspective that figures such as 6.1a and b are produced—a recording wave meter sits at one spot on the bottom and measures the amplitude of waves as they pass overhead. If we define our fixed spot in the ocean to be $x = 0$, eq. 6.6 can be rewritten as

$$\eta(t) = \alpha \cos \omega t, \qquad (6.7)$$

where ω (the radian frequency) $= 2\pi/T$. Since $\cos(\omega t) = \cos(-\omega t)$, we are justified in ignoring the negative sign from eq. 6.6. We can shift the phase of this wave by subtracting a fixed value ϕ from ωt; ϕ can take on values between 0 and 2π. Thus

$$\eta(t) = \alpha \cos(\omega t - \phi). \qquad (6.8)$$

The surface elevation produced by two waves acting simultaneously is simply the sum of the surface elevations for the individual waves:

$$\eta(t) = \alpha_1 \cos(\omega_1 t - \phi_1) + \alpha_2 \cos(\omega_2 t - \phi_2). \qquad (6.9)$$

Any arbitrary surface-elevation pattern can be represented by the appropriate sum of several component wave forms,

$$\eta(t) = \sum_{i=1}^{N} \alpha_i \cos(\omega_i t - \phi_i), \qquad (6.10)$$

where N is the number of wave forms of different frequency that we add. For example, the wave form shown in Figure 6.4a, which is very reminiscent of our typical wave record, is formed from the component waves shown beneath it.

It would be possible to continue the analysis of wave patterns using eq. 6.10 as it stands, but it will be useful to restate this equation in a more conventional form. Consider Figure 6.5. Here a sine wave and a cosine wave of equal amplitudes and the

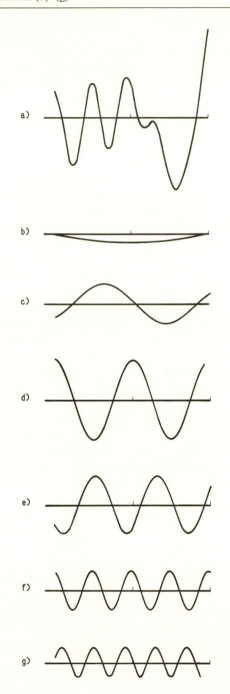

Figure 6.4. Complex surface elevations. The wave form shown in (a) is the sum of the component waves shown in (b)–(g).

Figure 6.5. The addition of a sine wave (a) and a cosine wave (b) of the same frequency results in a wave with the same frequency but a different phase (c).

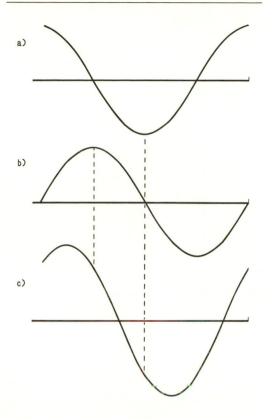

same frequency have been added together. The result is a wave with an amplitude slightly larger than either component and, more importantly, with a phase that has been shifted from either component. Thus, one effect of adding sine and cosine waves of the same frequency is a shift in the phase of the resulting wave. By varying the relative amplitudes of the component sine and cosine waves, we can produce phase shifts from 0 to 2π. Given this fact, we can rewrite eq. 6.10 as

$$\eta(t) = \sum_{i=1}^{N} \{\alpha_i \cos \omega_i t + \beta_i \sin \omega_i t\}, \qquad (6.11)$$

where the term containing $\sin(\omega t)$ substitutes for ϕ in eq. 6.10. Here η is measured relative to mean sea level. If η is measured relative to some other reference point, we need to add a term for the average eleva-

tion, $\bar{\eta}$, to eq. 6.11 to account for the shift in reference. Thus, in general,

$$\eta(t) = \bar{\eta} + \sum_{i=1}^{N} \{\alpha_i \cos \omega_i t + \beta_i \sin \omega_i t\}. \qquad (6.12)$$

This is the standard expression for the Fourier series. By appropriate manipulation of amplitude coefficients, we can use this series to describe practically any arbitrary wave form.[1]

The fact that a complicated surface pattern can be the result of the addition of a series of component waves of different frequencies has many practical consequences. In the present context it provides a means for calculating the contributions of waves of various frequencies to the overall mean square wave amplitude. Because the square of wave amplitude is a measure of wave energy (Table 5.2), we have a method whereby we can determine what portions of the overall wave energy are associated with what frequencies. The frequency of a wave is the inverse of its period. Thus our knowledge of the energy associated with a given frequency gives us the energy (and power; see Chapter 7) associated with a certain wave period.

How do we calculate the energy contributions of different wave-frequency components? We begin by calculating the mean square amplitude, $\bar{A}_f^2$, for a single frequency, f. By definition, the mean square amplitude is

$$\bar{A}_f^2 = \frac{1}{2\pi} \int_0^{2\pi} \alpha_f^2 \cos^2\theta \, d\theta$$

$$= \frac{\alpha_f^2}{2\pi} \left\{ \frac{2\pi}{2} + \frac{\sin 4\pi}{4} \right\}$$

$$= \frac{\alpha_f^2}{2}, \qquad (6.13)$$

---

[1] To be described by a Fourier series, a function, $f(t)$, must fulfill Dirichlet's conditions: the integral $\int f(t) \, dt$ must be finite over the period of interest and $f(t)$ must be piecewise continuous and piecewise monotonic. We will not concern ourselves with these technicalities. By their physical nature, water waves fulfill Dirichlet's conditions.

where α_f is the amplitude of the wave form. The same result can be obtained for the sine component of a wave form—the mean square amplitude is one-half the square of the overall amplitude. By applying this fact, we see that the mean square amplitude of one frequency component of a complex wave form is

$$\overline{A_f^2} = \frac{\alpha_f^2}{2} + \frac{\beta_f^2}{2} = (\tfrac{1}{2})(\alpha_f^2 + \beta_f^2). \quad (6.14)$$

The overall mean square amplitude for a complex wave formed of N equally spaced frequency components is

$$\overline{A_{\text{total}}^2} = (\tfrac{1}{2}) \sum_{i=1}^{N-1} (\alpha_{f_i}^2 + \beta_{f_i}^2) + \alpha_{f_N}^2, \quad (6.15)$$

where the subscript, f_i, is used to denote the different frequencies.[2] Note that the highest frequency (f_N) is a special case in what is otherwise a simple summation of the contributions from individual frequencies.

The Periodogram. The additive nature of mean squares allows one to construct a periodogram. In (Fig. 6.6a) the individual mean squares are plotted for the various component frequencies of the complex wave form shown in Figure 6.5a. Because the overall mean square wave height is the sum of these individual mean square amplitudes, the height of the bars is an unbiased representation of the relative contribution of various frequencies to the overall mean square amplitude. This is even more clearly seen if the ordinate is normalized by dividing by the total mean square amplitude (Fig. 6.6b). The normalized contributions made by various frequencies sum to 1, and it is easy to see that the component with a frequency of 0.1 Hz contributes 44% of the overall mean square amplitude, while the component with frequency 0.2 Hz contributes only 11%.

[2] This result (a consequence of Parseval's theorem) is less intuitive and less easily proved. The interested reader should consult Jenkins and Watts (1968) or Chatfield (1984).

Figure 6.6. The periodogram quantifies the contribution of component waves of different frequencies to the overall mean square wave amplitude. (a) The periodogram for the wave form shown in Figure 7.5a. (b) The periodogram normalized to the total mean square wave amplitude (the variance of surface elevation).

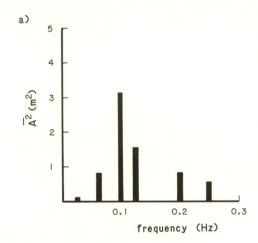

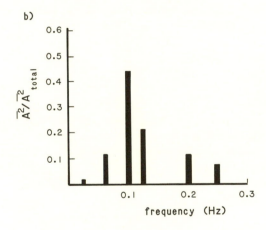

This sort of analysis is useful in a different way from that of the Rayleigh distribution. For example, from the periodogram shown in Fig. 6.6a, we know that waves with a frequency of 0.1 Hz ($T = 10$ s) account for nearly half the wave energy in this case and have a mean square amplitudes of about 3 m. Thus, for this component frequency, the rms amplitude (A_{rms}) is about $\sqrt{3} = 1.7$ m, and the rms wave height is $2 A_{\text{rms}} = 3.4$ m. In deep water these waves have a length of 156 m (eq.

4.8). With this information regarding wave height and length, we can use linear wave theory to provide a rough estimate of what the water motion is at any depth. This estimate will not be exact because waves of this frequency are not the only ones present, but in this case where one frequency predominates, the estimate will not be far wrong. We could then use the mean square amplitude calculated here for one frequency component to calculate from the Rayleigh distribution the probability of seeing a wave of at least a certain height with this particular frequency.

The Intensity. Although the wave form shown in Figure 6.4a superficially resembles that of our typical sea, this resemblance is somewhat misleading. Whereas this figure is formed from six distinct frequency components, the wave form of the sea is formed from a continuous spectrum of frequencies. The presence of a continuous spectrum leads to problems in calculating a periodogram. The overall mean square amplitude $\overline{A_{total}^2}$, is a sum of the mean square amplitudes of the component frequencies. Because $\overline{A^2}$ is proportional to wave energy, we know a priori that it must have a finite value. No wave has an infinite energy. This poses no problem when only a few frequencies are present; each contributes a measurable portion of the overall mean square amplitude. However, as the number of frequency components increases, $\overline{A_{total}^2}$ does not (lest the total wave energy tend toward infinity), and the contribution of each component must decrease. For a continuous spectrum of frequencies that, by definition, is made up of an infinite number of components, each component frequency contributes an infinitesimal amount to the overall mean square amplitude. A periodogram for a sea surface with a continuous spectrum of frequencies would look like a bare set of axes, a totally uninformative picture.

How can we avoid this problem? The simplest procedure is to define a new term

for the contribution of each frequency to the overall "waviness" of the surface. At each frequency the *intensity* is

$$I(f_i) = \frac{N}{2}(\alpha_{f_i}^2 + \beta_{f_i}^2) \qquad i = 1, 2, \ldots, N-1$$

$$(6.16)$$

$$I(f_N) = N(\alpha_{f_N}^2 + \beta_{f_N}^2),$$

where N is the total number of frequency components contributing to the surface elevation. As N increases, $(\alpha_f^2 + \beta_f^2)$ at each frequency decreases in a manner so that the intensity remains finite. By using this simple stratagem we can plot an equivalent to the periodogram for as many frequencies as we care to examine. Given this definition of the intensity,

$$\overline{A_{total}^2} = \sum_{i=1}^{N} I(f_i) \frac{1}{N}. \qquad (6.17)$$

$(1/N)$ is a measure of how small the difference in frequency, Δf, is between the component waves contributing to the surface elevation. In the limit as N goes to infinity, $1/N$ becomes df, and we can rewrite eq. 6.17 in integral form:

$$\overline{A_{total}^2} = \int_0^\infty I(f)\, df. \qquad (6.18)$$

In this form $I(f)$ is given a new symbol, $\Gamma(f)$, and is called the *power spectral density function*, or more simply, the *power spectrum*. Thus

$$\overline{A_{total}^2} = \int_0^\infty \Gamma(f)\, df. \qquad (6.19)$$

In other words, $\Gamma(f)$ is defined to be a function of frequency such that the *area* under the function [equal to the integral of $\Gamma(f)$ over its entire range] is equal to the overall mean square amplitude (Fig. 6.7a). In essence we have recalculated the periodogram so that the contribution of each frequency to the $\overline{A_{total}^2}$ is proportional to the area of the graph beneath $\Gamma(f)$ at that frequency, rather than to the value of $\overline{A^2}$ corresponding to that frequency. Defined in this manner, $\Gamma(f)$ has the units m² s.

In one sense this new calculation does

Figure 6.7. The power spectrum. (a) The area under the spectrum is the total mean square amplitude (the variance of surface elevation). The area under one portion of the curve is the contribution of that frequency range to the overall mean square wave amplitude. (b) A shift in the height of the wave spectrum reflects a change in wave amplitude. Shifts along the frequency axis reflect changes in the frequency of component waves contributing to a complex surface elevation.

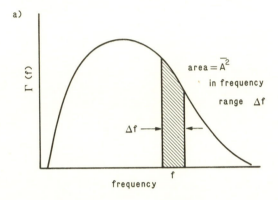

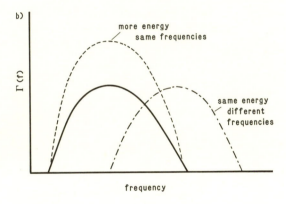

quencies does correspond to an area on the graph. Thus if we know the shape of the power spectrum, we can quantify the contribution of any range of frequencies to the overall mean square amplitude. For example, the range of frequencies (Δf) shown in Figure 6.7a accounts for about 15% of the area under the spectrum, and therefore for 15% of $\overline{A^2_{total}}$. From eq. 6.13 we can then write

$$0.15\overline{A^2_{total}} = (\tfrac{1}{2})A^2_f$$
$$A_f = \sqrt{0.3\overline{A^2_{total}}},$$

where A_f is the average amplitude of the waves in the frequency range, Δf. More generally, $\Gamma(f)$, the value of the spectrum at a point, when multiplied by Δf, a small range of frequencies, yields an area such that

$$\Gamma(f)\,\Delta f = (\tfrac{1}{2})A^2_f$$
$$A_f = \sqrt{2\Gamma(f)\,\Delta f}. \qquad (6.20)$$

Be warned that this is an approximation; it gets less and less exact as Δf is allowed to get wider.

Figure 6.7b shows a spectrum with roughly the same shape as Figure 6.7a, but with double the area under the curve. This doubling of area means that $\overline{A^2_{total}}$ is doubled, corresponding to a 1.4-fold increase in the rms wave height. Thus a shift of the spectrum up or down on the ordinate corresponds to a difference in the "waviness" of the sea surface. Shifting of spectral peaks to the left or right (Fig. 6.7b) corresponds to a change in the frequency pattern of the sea surface.

It is possible to normalize this kind of graph in exactly the same fashion as for the periodogram. If each value of $\Gamma(f)$ is divided by $\overline{A^2_{total}}$, the area under the normalized curve must equal 1. In this case the curve is known as the *normalized power spectral density function*.[3] Figure 6.8 is

not change anything. Although a particular frequency may have a finite value for $\Gamma(f)$, because it is a discrete point on the abscissa no measurable area is associated with this single frequency. Thus we arrive at the same conclusion as before: each individual frequency in a continuous spectrum contributes an infinitesimal amount to the overall mean square amplitude. But with this new function we are not confined to looking at individual frequencies. From Figure 6.7a it is clear that a range of fre-

---

[3] There is some variation among authors in the names used for these functions. For example, Jenkins and Watts (1968) refer to what we have here termed the "power spectral density function" as simply the

Figure 6.8. The wave spectrum for broken waves (turbulent bores), recorded along the rocky coastline of Hopkins Marine Station, Pacific Grove, California. The major spectral peak indicates that the bores have a period of 12.5 s. The bores with this period have an average amplitude of 0.55 m (as calculated using eq. 6.20), corresponding to an average height of 1.1 m. Spectral estimates were averaged in groups of ten, i.e., there are twenty degrees of freedom.

Figure 6.9. Daily wave spectra recorded in Monterey Bay, April 1985. Reproduced from the Coastal Data Information Program, U.S. Army Corps of Engineers and the California Department of Boating and Waterways.

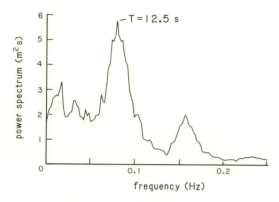

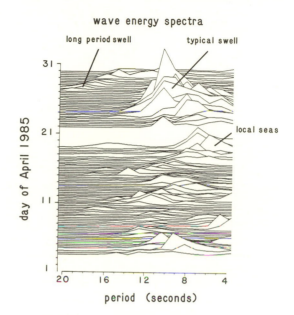

the normalized spectral density curve for waves breaking on a rocky shore during a storm in December 1985. The spectrum tells us that the peak energy is associated with waves with frequencies around 0.1 Hz ($T = 10$ s). Thus the waves produced by this storm behave (for the most part) as if they have wave lengths of approximately 150 m.

Figure 6.9 is taken from the records of the U.S. Army Corps of Engineers' Coastal Data Information Program for the month of April 1985. Four spectra are calculated for each day, and the compilation clearly shows how the wave spectrum changes from day to day. On April 18 and 19 local seas with periods of 4–8 s accounted for most of the wave energy. By April 24 these seas had been replaced by swell with periods centered around 10 s. On April 27–28 the spectrum was dominated by very long period swell (17–19 s), presumably from a distant storm. Note that the

abscissa in this graph is slightly different from what we have examined so far—it is linear in period rather than in frequency, tending to expand the low-frequency end of the spectral curve while compressing the high-frequency end.

The mathematical process we have used here to examine the frequency components of the random sea is useful in other contexts as well. In general, this process is known as "spectral analysis," and it is a standard tool for exploring the predictability of a series of events.

Measuring the Wave Spectrum

Knowing that a complex sea surface is composed of a spectrum of waves of different frequencies does not make it readily apparent how, given a wave record, we should go about deciding which ranges of frequencies contributed which amplitudes. In practice, we arrive at the wave spec-

---

"power spectrum," and the "normalized power spectral density function" as the "power spectral density." Here we follow the convention used by Chatfield (1984).

trum through a simple, but somewhat laborious, process. We start with a record of the sea-surface elevations, such as that in Figure 6.1b, and measure the derivations from mean sea level at a series of N equally spaced points. This is equivalent to measuring the surface elevation at one point on the sea surface at a series of equally spaced times, producing a record in the *time domain*. The appropriate interval between measurements depends on the frequency of the waves in which we are interested. The shortest wave period we can discern is equal to twice the interval at which we sample. Conversely, if we want information about waves with very long periods, we need to sample for a long time. The longest wave period we can say anything about is equal to the length of our time series.

Once we have obtained a time series of N surface elevations, we transform this information about η as a function of time to information about η as a function of frequency. In other words, we transform the data from the time domain to the *frequency domain* via the Fourier series. We know (eq. 6.12) that by specifying the amplitudes of the sine and cosine waves at each of many frequencies we can tailor the Fourier series to reproduce any arbitrary wave form. The task, then, is to correctly specify α_i and β_i in eq. 6.12 to best fit a given time series.

The time period for which we have information is equal to $N\Delta$, where N is the total number of data points and Δ is the time between samples. Noting the fact that frequency = 1/period, we define a fundamental frequency, $f = 1/(N\Delta)$. We will examine the surface elevation at even multiples (harmonics) of this fundamental frequency. In other words, we define α_i and β_i at each frequency f_i, where f_i is $i/(N\Delta)$ and i is any integer between 1 and q, the highest harmonic that can be discerned. As noted before, the shortest period we can discern is 2Δ, so the highest frequency we can deal with is $1/(2\Delta)$, known as the

Nyquist frequency. Consequently, $q = N/2$.

By a derivation that we will not delve into here (see Jenkins and Watts 1968; Chatfield 1984), we can show that the appropriate values for α_{f_i} and β_{f_i} are

$$\alpha_{f_i} = \frac{2}{N} \sum_{j=0}^{N-1} \eta_j \cos(2\pi f_i[\Delta j]).$$
$$\dots i = 1, 2, \dots q-1$$

$$f_q = \frac{1}{N} \sum_{j=0}^{N-1} \eta_j \cos(2\pi f_q[\Delta j])$$

$$\beta_{f_i} = \frac{2}{N} \sum_{j=0}^{N-1} \eta_j \sin(2\pi f_i[\Delta j])$$
$$\dots i = 1, 2, \dots q-1$$

$$\beta_{f_q} = 0,$$

$$(6.21)$$

where q sets the upper limit to the frequency for which we calculate α_i and β_i.

Let us briefly examine these equations to see how we would go about actually carrying out the calculation. Consider the equation defining α_{f_2}, the second harmonic of the fundamental frequency, for a time series consisting of $N = 10$ points taken at intervals of $\Delta = 1$ s:

$$\alpha_{f_2} = \frac{2}{N} \sum_{j=0}^{N-1} \eta_j \cos(2\pi f_2[\Delta j]).$$

For the moment, consider only the cosine term in this series. In this example, $f_2 = 2/(N\Delta) = 0.2$ Hz. Now, 2π times frequency (in cycles per second) is the radian frequency, ω (eq. 6.7). The variable j in the expression for α_{f_i} is just a tool for counting through the sample; Δj is equal to the time t (from the beginning of our time series) at which a certain value of surface elevation (η_j) was taken. For instance, when $j = 5$, we are dealing with the sample taken 5 s after the beginning of the series. Thus the term $\cos(2\pi f_2 \Delta j)$ is equivalent to $\cos(\omega t)$, the same sort of expression encountered in eq. 6.12. For each time in the series, we multiply the measured surface elevation (η_j) by the appropriate term for $\cos(\omega t)$. Adding all N of these values together and multiplying by $(2/N)$ yields α_{f_2} for frequency f_2 (in this example 0.2 Hz). To calculate α_{f_i} for a different frequency, we repeat the whole procedure using a different i.

Once the various α_{f_i} and β_{f_i} have been calculated, it is a simple procedure to calculate the intensity at each frequency:

$$I(f_i) = \frac{N}{2}(\alpha_{f_i}^2 + \beta_{f_i}^2) \qquad i = 1, 2, \ldots, q - 1$$

$$I(f_q) = N\alpha_{f_q}^2, \qquad\qquad (6.22)$$

where $q = N/2 = 5$. These intensities are estimates of the true power spectrum.

In actual practice, the computation of the wave spectrum is somewhat more complicated than presented here. As noted, we must choose the sampling procedure with care to avoid spurious results (known as *aliasing*), where wave components above the Nyquist frequency bias measurements at lower frequencies. Furthermore, spectral estimates as calculated here provide an imperfect picture of the true spectrum. In particular, when we examine frequencies near the fundamental frequency, only a few cycles are present in a time series, and there is a substantial possibility for error in estimating the true spectrum. As a result, it is common to find that spectral estimates fluctuate wildly at low frequencies. This problem is circumvented by "smoothing" the spectral estimates in a process known as *apodizing* or *applying a spectral window*. There are many methods for apodizing; the simplest is to average each spectral estimate with estimates at several adjacent frequencies. Thus the smoothed estimate at frequency f_i is

$$I(f_i) = \frac{1}{(2j + 1)} \sum_{m=i-j}^{i+j} I(f_m), \quad (6.23)$$

where $J = 2j + 1$ is the size of the group averaged at each frequency. The larger the group of estimates used in this running average, the smoother the spectrum. Herein lies the art of spectral estimation. If J is chosen too small, the detail shown by the spectral estimate may be spurious.[4] If J is chosen too large, the spectrum is smoothed to the point where important detail is lost. Readers interested in developing expertise in spectral analysis should consult the excellent introductions by Denman (1975) and Chatfield (1984) or standard texts such as Jenkins and Watts (1968), Box and Jenkins (1970), or Bendat and Piersol (1971).

As one might imagine, the number of computations involved in calculating a spectrum rises rapidly with the number of data points in the time series. For oceanographical time series that may consist of tens of thousands of points, the computational load is immense. To ease this load, an efficient algorithm (the Fast Fourier Transform, or FFT) has been devised. Most spectra are now calculated using FFT methods rather than using the computation outlined here. Again the reader is referred to standard texts on spectral analysis for a discussion of this technique.

Summary

Although the ocean's surface is exceedingly complex, the statistical techniques outlined in this chapter allow us to derive some order from this complexity. The Rayleigh distribution provides a tool for predicting the height distribution of waves. In Chapter 16 we use these predictions to study the structural exposure of wave-swept organisms. The power spectrum allows us to calculate equivalent wave heights and lengths for a complex ocean surface, and thereby provides a method by which we can apply wave theory to real ocean waves.

[4] The parameter $2J$, twice the group size, is the degrees of freedom of the spectral estimate. This parameter can be used in a chi-squared calculation of confidence interval size. See Chatfield (1984) for details.

CHAPTER 7

~ ~ ~ ~ ~ ~ ~ ~ ~ ~ ~

Breaking and Broken Waves

$\mathbf{I}$n Chapter 5 we examined waves as they exist in deep and shallow water, but we did not follow one particular wave as it moves inshore and breaks. The process of moving from deep to shallow water is known a *shoaling*, and a simple theory for wave shoaling is the next topic of discussion.

Wave Shoaling

We begin by examining the rate at which energy is conducted by waves. Consider a vertical plane 1 m wide lying parallel to the wave crests and extending from the bottom of the ocean through the surface. As waves move ashore, how much energy is conducted through this plane in a given period of time? From Table 5.2 we know that the total energy per area of surface in an oscillatory wave is

$$E = (\tfrac{1}{8})\rho g H^2.$$

This energy is transported at the group velocity, which in deep water is half the wave celerity (eq. 5.51). Thus in one second $C/2$ square meters of wave area move through this 1 m-wide plane. The energy per time, Q, passing through the plane is

$$Q = (\tfrac{1}{2})EC. \qquad (7.1)$$

This is the power (also known as the *wave energy flux*) of one meter of wave crest in deep water.

In shallow water the group velocity is equal to the wave celerity (eq. 5.51), and the energy flux per meter of wave crest is

$$Q = EC. \qquad (7.2)$$

In general, then, the wave power is pro-

portional to the energy per square meter times the wave celerity, and the constant of proportionality, n, is equal to C/C_g, which varies between $\tfrac{1}{2}$ (in deep water) and 1 (in shallow water):

$$Q = nEC. \qquad (7.3)$$

From eq. 5.51 we see readily that

$$n = (\tfrac{1}{2})\left\{1 + \frac{2kd}{\sinh(2kd)}\right\}. \qquad (7.4)$$

When $d > L/2$, $\sinh(2kd)$ is much greater than $2kd$, and n approaches $\tfrac{1}{2}$ as required. When $d < L/20$, $\sinh(2kd) \simeq 2kd$, and $n \simeq 1$.

Unless there is some process operating that drains energy from a wave as it moves inshore, the energy flux remains constant. On moderate sloping shores appreciable energy is lost only if the wave breaks, in which case the action of viscosity draws energy from the wave and converts it to heat. Thus before a wave breaks, its power inshore is approximately equal to its power offshore. If we denote the offshore (deep-water) values for energy and celerity using the subscript O, we can equate eqs. 7.1 and 7.3:

$$(\tfrac{1}{2})E_O C_O = nEC,$$

$$\frac{E}{E_O} = \frac{1}{2}\frac{1}{n}\frac{C_O}{C}. \qquad (7.5)$$

As described by linear wave theory (Table 5.2), the total energy of the wave is

$$E = (\tfrac{1}{8})\rho g H^2, \qquad E_O = (\tfrac{1}{8})\rho g H_O^2$$

where H_O is the wave height in deep water. Inserting these values for E and E_O into eq. 7.5, we see that

$$\frac{H}{H_O} = \sqrt{\frac{E}{E_O}} = \sqrt{\frac{1}{2}\frac{1}{n}\frac{C_O}{C}}. \qquad (7.6)$$

Now, recall that $C_O = gT/2\pi$ and $C = (gT/2\pi)\tanh(kd)$ (Table 5.2). Substituting these values for the celerity and the expression calculated earlier for n (eq. 7.4) into eq. 7.6, we see that

$$\frac{H}{H_O} = \sqrt{\frac{\sinh(2kd)}{\sinh(2kd) + 2kd}\frac{1}{\tanh(kd)}}. \qquad (7.7)$$

The height of a wave near the shore thus depends on H_O, the height of the wave when it was in deep water, and kd, the water depth relative to wave length (recall that $k = 2\pi/L$). As the water gets shallower (in other words, as kd decreases) the term $(1/n)$ gets smaller and the term C_O/C gets larger. Except for a narrow range of intermediate depths, C_O/C increases faster than $1/n$ decreases, and H/H_O is larger than 1 (Fig. 7.1). Thus waves approaching the shore increase in height. In very shallow water the wave height can be several times the height in deep water, provided of course that the wave does not break. This approach to an estimation of wave shoaling relies on linear theory and therefore cannot be expected to be accurate in

Figure 7.1. Relative wave length, height, and steepness as waves move from deep water ($d > L_O/2$, right-hand side of graph) to shallow water ($d < L_O/2$, left-hand side of graph).

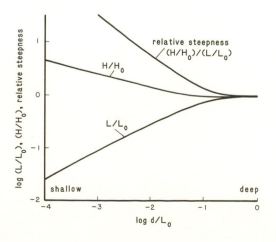

very shallow water. Similar calculations using Stokes theories, cnoidal theory, or solitary wave theories provide estimates for wave shoaling that are more accurate and slightly different (they predict that the waves are higher).

Recall from Chapter 5 that wave length decreases as waves move inshore. Consequently, at the same time that waves are growing in height, their length is decreasing. In other words, shoaling waves become steeper. This effect is clearly shown in Figure 7.1, where relative wave height (H/H_O), relative wave length (L/L_O) (eq. 5.45), and wave steepness (relative wave height/relative wave length) are plotted.

Refraction

The calculation presented here tacitly assumes that a section of wave crest one meter long in deep water stays one meter long as the wave moves inshore. However, the length of a wave crest may change due to refraction as a wave moves ashore, and the power of one meter of deep-water crest may be spread over a different length of shallow-water crest. As a result, the shallow-water wave need not have the height predicted by eq. 7.7. To see this we restate eq. 7.2 to account for the length of crest. The energy flux for a length, y_O, of a deep-water wave crest is

$$Q = (\tfrac{1}{2})E_O C_O y_O.$$

The flux in this same section of the wave when it has traveled into shallow water is

$$Q = nECy,$$

where y is the new length of the wave crest. By analogy to eq. 7.5, we can then write

$$\frac{E}{E_O} = \frac{1}{2}\frac{1}{n}\frac{C_O}{C}\frac{y_O}{y} \qquad (7.8)$$

and

$$\frac{H}{H_0} = \sqrt{\frac{\sinh(2kd)}{\sinh(2kd) + 2kd} \frac{1}{\tanh(kd)} \frac{y_0}{y}} \quad (7.9)$$

The expression $\sqrt{y_0/y}$ is known as the *refraction coefficient*, K_R, and can be less than, equal to, or greater than 1. For instance, consider a series of waves with parallel crests approaching the shore (Fig. 7.2). We mark a series of equally spaced points on these deep-water crests, knowing from eq. 7.2 that the intervals between marks contain equal wave powers. Due to refraction, the wave crests are bent as they move ashore. In a bay, this results in a stretching of the wave crests. $y > y_0$, and K_R is less than 1. At a headland or point, the opposite effect occurs—the waves wrapping around the sides of the headland crowd the waves approaching the point. In this case, $y > y_0$, $K_R > 1$, and the waves approaching the point increase more in height than they would in a perpendicular approach to a linear shoreline.

The results of wave refraction are directly analogous to the results of light refraction. A bay can be thought of as a concave lens that spreads the waves that travel into it. A headland can be thought of as a convex lens that tends to focus wave energy.

An extreme example of this principle is reported by Bascom (1980). On April 20–24, 1930, a portion of the breakwater protecting the harbor at Long Beach, California, was destroyed by unusually large waves. Blocks weighing from 4 to 20 tons were broken off the tip of the breakwater and washed away. This destruction was peculiar in that large waves occurred only at the breakwater; the sea offshore was calm, the winds were light, and beaches to the north and south reported no large surf. The answer to this puzzle was discovered some seventeen years later. M. P. O'Brien noticed that the bottom topography to the southwest of Long Beach forms an underwater headland extending out a dozen miles or more. Most of this headland is at a depth of greater than 100 m, and is therefore in deep water for typical ocean waves having a period of 10 s or so. However, the waves that destroyed the breakwater were timed to have periods of 20–30 s, corresponding (eq. 4.8) to wave lengths of 625–1405 m. For these long-wave-length swells, the underwater headland was effectively in shallow water, and the waves were therefore refracted. A refraction diagram showed that the underwater headland focused the swell precisely at the tip of the breakwater, resulting in waves of an unusual height and in the perplexing pattern of destruction on an otherwise normal day. The long-period swell was thought to be the result of an unusually large storm somewhere in the Southern Hemisphere.

The evaluation of the refraction coefficient for actual bottom topographies is a tedious and time-consuming business. As with many such tasks these days, refraction calculations are best done by computers. An introduction to calculation of refraction effects can be found in the

Figure 7.2. Refraction of waves. Wave energy is diffused in a bay and focused on a headland.

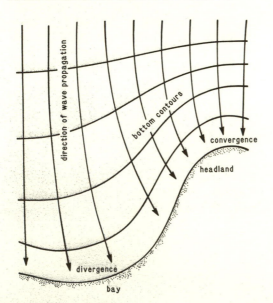

U.S. Army Corps of Engineers' *Shore Protection Manual* (1984).

Wave Breaking

We have seen (Chapter 4) that a solitary wave traveling into shallower water peaks up and becomes unstable when the velocity of water at the crest becomes equal to the wave celerity. McGowan (1894) calculated that this point of instability is reached when

$$\frac{\text{breaking height}}{\text{breaking depth}} = \frac{H_b}{d_b} = 0.78. \quad (7.10)$$

The definition of H_b and d_b are given in Figure 7.3a. This value has been widely cited and is a useful rule of thumb. However, this relationship is accurate only when the slope of the bottom is very gradual. Galvin (1972) cites data showing that for steep slopes H_b/d_b can be as high as 1.4. This increase in H_b/d_b is in part due to the fact that the change in wave shape leading to instability and breaking takes a finite amount of time. A wave traveling over a steeply sloping bottom may begin to break when $H_b/d_b = 0.78$, but by the time breaking is fully underway the wave has traveled to shallower water. Values of H_b/d_b for several values of the tangent of bottom slope θ are shown in Figure 7.3b. The steeper the slope, the greater the ratio of breaking height to breaking depth, up to $\tan(\theta) = 0.2$ (a slope of 1:5). The ratio H_b/d_b also depends on the steepness of the wave (H_O/L_O) before it begins to shoal. The steeper the wave offshore, the deeper the water in which it breaks (Fig. 7.3b).

We noted earlier that a shoaling wave can in theory reach a height several times its deep-water height, limited only by breaking. The maximum relative height (H_b/H_O) a wave can reach is a function of both the bottom slope and deep-water wave steepness, H_O/L_O (Fig. 7.3c). The

Figure 7.3. (a) Wave height (H_b) and water depth (d_b) at breaking (redrawn from Galvin 1972 by permission of Academic Press). (b) Breaker height relative to water depth (H_b/d_b) as a function of offshore wave steepness (H_O/L_O) and bottom slope (tan θ). The steeper the waves, the greater the depth at which they break. The steeper the bottom slope, the higher the breaker (redrawn from U.S. Army Corps of Engineers 1984). (c) Breaker height (H_b) relative to off-shore height (H_O) as a function of offshore wave steepness (H_O/L_O) and bottom slope (tan θ). The steeper the wave, the less height it attains as it shoals; the steeper the bottom slope, the more height a wave attains as it shoals (redrawn from U.S. Army Corps of Engineers 1984.)

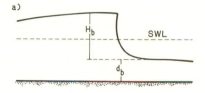

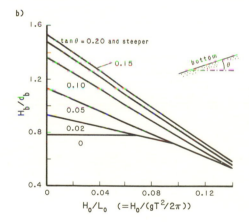

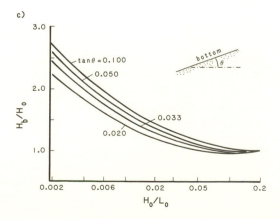

steeper the slope, the greater H_b/H_O. The steeper the wave, the less H_b/H_O.

Figure 7.3 is based primarily on laboratory studies, and caution should be used in applying its quantitative predictions to waves on rocky shores. A reasonable approach to dealing with the problems of breaking waves is to assume that waves break when their height (measured from trough to peak) is equal to the water depth below the trough. This may underestimate the breaking height on steep shores and overestimate it for more gently sloping beaches, but in either case it will not be too far wrong.

Having a method for estimating the wave height and water depth at breaking is only one part of dealing with breaking waves. A visit to the shore quickly reveals that all waves do not look the same when they break. What determines the shape (as opposed to the height) of a breaking wave? The manner in which a wave breaks depends on three factors: (1) The height of the wave at breaking, H_b; (2) the period of the wave, T; and (3) the slope of the bottom, expressed as the $\tan(\theta)$ (Fig. 7.3). These three factors have been combined by Galvin (1972) to form a breaking parameter, B_b:

$$B_b = \frac{H_b}{gT^2 \tan(\theta)}. \qquad (7.11)$$

When B_b is greater than about 0.068, waves are of a type referred to as *spilling*. The wave is nearly symmetrical at breaking; the leading slope of the wave looks nearly the same as trailing slope except for a small section near the crest. Breaking is initiated when a small amount of water at the crest arches over and begins to spill down the front of the wave, hence the name. An example of a typical spilling wave is shown in Figure 7.4a. These sorts of waves have relatively short periods and travel over beaches with shallow slopes. Their height at breaking is relatively large, implying that their deep-water steepness was low (see Figure 7.3).

Waves with a B_b between 0.068 and 0.003 are of the sort termed *plunging*. As noted previously, these are the sort of waves one finds pictured in surfing magazines. Breaking is initiated when water at the crest arches over the front of the wave. In this respect, plunging waves are similar to spilling waves, but the size of the plunging jet is much increased. A classic plunging breaker forms a tube parallel to the crest as water from the crest arches far over the front of the wave to land well down the forward slope (Fig. 7.4b). Eq. 7.11 predicts that plunging waves have periods that are longer than those of spilling waves and occur on beaches with steeper slopes. Their height at breaking is less than for spilling waves, implying that their deep-water steepness is larger.

Figure 7.4. The categories of breaking waves (redrawn from Galvin 1968 by permission of the American Geophysical Union).

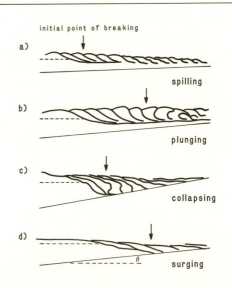

At a B_b of less than 0.003, waves do not break in the normally accepted sense. The crest of the wave does not become turbulent and foamy, and the wave simply flattens out and surges up the beach (Fig. 7.4d). These waves are consequently known as *surging* breakers, and are typical of very long-period waves and steep shores. In the limit, where the beach is vertical (a sea wall extending down into deep water, for instance) a surging breaker is entirely reflected from the shore. The effects of wave reflection will be discussed in greater detail later in this chapter when we consider inshore wave processes.

For breakers with a B_b near to but greater than 0.003, Galvin has coined the term *collapsing*. These breakers are similar to plunging breakers, but instead of only the crest arching over, most of the vertical front surface of the peaked wave moves ahead of the crest, and then collapses. A typical collapsing breaker is shown in Figure 7.4c.

The B_b values given here for the transitions between the various categories of waves are somewhat arbitrarily drawn. There is a smooth continuum in breaker type from spilling to plunging to collapsing, depending on how much of the front surface of the wave arches forward at breaking.

The information presented here may prove useful in forming a general impression of what happens when a wave breaks, but there are often exceptions to the rules of thumb. A series of waves of different heights (and therefore different celerities), each with a height too small to break, may suddenly come into phase in shallow water, producing a wave of exceptional height and unusual breaking characteristics. One such exception is the "sand buster" referred to by Peregrine (1983). This is a wave that for some reason manages to survive with appreciable height into very shallow water, at which point the entire front face becomes vertical and is dumped onto the beach as a solid mass of water. The present understanding of wave-breaking processes is such that there is simply no adequate substitute for empirical observations at each particular site.

Figure 7.5. The broken wave—a turbulent bore. (a) Profile of the bore; velocity within a turbulent bore varies chaotically. (b), (c) Creation of a bore in a laboratory wave tank (see text).

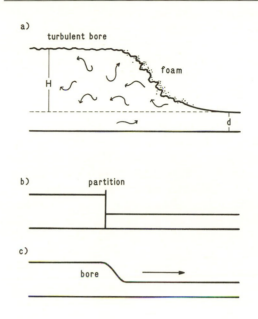

Postbreaking Flows

Although the process of breaking substantially changes the shape of a wave, it does not completely absorb the wave's energy, and the wave in its new form continues to move shoreward. In many cases, it is this postbreaking flow that has the greatest contact with organisms on the shore and therefore is of considerable interest here.

Four aspects of postbreaking flow require our attention: (1) The propagation of the broken wave; (2) The introduction of turbulence into the flow; (3) Local variations in sealevel within the surfzone; and (4) Longshore and rip currents. These aspects are considered in turn.

Propagation

Regardless of the manner in which a wave breaks, the shape of the broken wave resembles that shown in Figure 7.5a. The front face of the wave is steep and foaming, and there is often a distinct "toe" where this foam encounters the water in front of the wave. The back surface of the wave has the gradual slope characteristic of a solitary wave, although here the slope carries a train of foam. This wave form closely resembles a type of wave not yet considered— a *bore*.

A bore is the pattern of fluid motion that results when two bodies of stationary fluid with different surface levels are suddenly joined. This process is easily observed in the laboratory. Divide a long, narrow tank into two compartments by a partition (Fig. 7.5b) and fill the two compartments with water, allowing the water in the left-hand compartment to rise to a higher level. Then remove the partition suddenly. (Fig. 7.5c). As might be expected, a wave form, a bore, travels to the right. To be strictly correct, the left-hand compartment would have to be large enough for its water level not to be substantially lowered by the bore's movement.

A bore is analogous to the front half of a solitary wave. The entire wave form is above the water into which it propagates, and water is transported forward. In a true bore, the back half of the wave is missing; all the water behind the crest remains at crest height. In this respect the wave form produced by a breaking wave is not strictly a bore; a distinct crest is still present. However, because the broken wave is propagating into shallower water, the water depth in front of the broken wave is always less than that behind, and the approximation to a bore is sufficient for our purposes.

How fast does a bore move? The idealized bore shown in Figure 7.5 is identical to the wave form used to calculate the celerity of a solitary wave (Fig. 4.7), and we can immediately conclude that the celerity of a bore is the same as that of a solitary wave. To a first approximation,

$$C \simeq \sqrt{g(d + H)}. \qquad (7.12)$$

This approximate relationship holds true even for a bore moving into water with zero depth. In other words, a wall of water moving over a dry surface moves with approximately the same celerity as a bore moving in very shallow water. In practice, turbulence acts to slow a propagating bore in shallow water, and eq. 7.12 may overestimate the celerity by 10%–20%.

The propagation of broken waves as bores is most obvious on beaches with gradually sloping bottoms. Here waves break a substantial distance from the shoreline, and the travel of the broken waves can be clearly seen before the bores are dissipated on the beach. On sufficiently steep shores, waves break directly onto the beach itself, in which case a bore may never form or may be limited to the swash as the wave runs up the beach.

The velocity distribution in a bore is similar to that in a solitary wave. For a bore moving into still water, water begins to move forward as the crest approaches. In a true bore the water continues to move forward as long as the bore is present, transporting water in the direction of propagation. The velocity is highest at the surface and decreases slightly with depth, but it is still appreciable near the bottom. In the surf zone, bores often move shoreward against the seaward-moving backwash of the previous wave. In this situation all velocities in the bore are shifted by the imposed countercurrent. In the case of a bore moving over a dry surface ($d = 0$), all the water at the leading edge of the bore travels with a speed equal to the bore's celerity.

Turbulence

Consider what happens when the crest of a wave arches over during breaking. The water in the jet travels at approximately the wave celerity and strikes the wave form somewhere on its front surface at a point where the water is moving back and up toward the crest. Thus, at the moment the jet touches down, two streams of water moving in opposite directions are brought directly into contact, forming what is termed a *shear layer* or *mixing layer*. The velocity gradient at the point of contact is very great, and as a result substantial viscous forces are brought to bear.

The result of this velocity gradient is a rotation of fluid in the area where the jet impacts the front surface of the wave. Consider the situation shown in Figure 7.6a A solid cube (shown here in cross section) is placed between two streams of water moving in opposite directions. As a result, velocity gradients are established between the top and bottom surfaces of the cube and the water flowing past them. Because water has viscosity, the gradient at the top results in a force tending to push the top of the cube to the right (eq. 3.8). The opposite situation is present on the bottom surface of the cube, and this surface is pushed to the left. Under the influence of these oppositely directed forces, the cube rotates clockwise. The water near the interface between the breaking jet and the front surface of a wave acts in the same manner as the cube: it rotates, forming a series of vortices or eddies (Fig. 7.6b).

The vortices formed in the vicinity of the jet tend to propagate. The rotation of each vortex forms a velocity gradient between fluid in the vortex and fluid outside. This velocity gradient produces secondary vortices of a smaller dimension, which in turn produce tertiary vortices, and so forth. Soon after the jet impacts the front surface of the wave, much of the water in front of the wave crest has been stirred into chaotic motion by the cascade of vortices and is said to be *turbulent*. As the bore propagates, the turbulence streams out behind the crest in a plume.

This turbulent motion amounts to a chaotic fluctuation of water velocity superimposed on the overall motion of the fluid as the bore moves shoreward. This fluctuation makes it extremely difficult to predict the motion of a particle of fluid at a

Figure 7.6. The shear layer created by a breaking wave causes rotation of the water. (a) Rotation of a solid cube due to the shear forces exerted (see text). (b) The analogous rotation of a volume of water—a vortex or eddy.

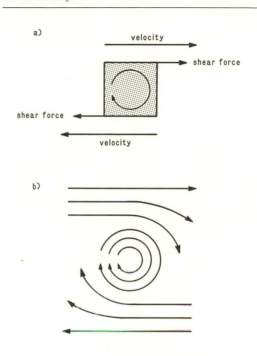

certain point in the wave form. To see how this works, we return to the wave tank and propagate a bore moving in the positive *x*-direction. Using the same logic as before (Chapter 4), we bring this wave form to a halt relative to our stationary frame of reference by imposing a current with velocity equal to the bore's celerity but moving in the opposite direction. If the bore is not turbulent, it appears as shown in Figure 7.5c. Water moves smoothly along streamlines, and if we pick a point anywhere in the fluid we find that the velocity of water flowing past that point remains the same through time. However, if the bore is turbulent, the picture is different. The velocity at a point varies unpredictably as turbulent vortices move past. Thus, one cannot pick one point and time and say with any certainty what the velocity will be. However, if we measure the velocity at a certain point at a large number of times and average these velocities, we find that the average velocity is nearly the same as the steady velocity in the nonturbulent bore. Thus, even though the turbulent bore is chaotic on a fine temporal scale, its motion averaged over time is still recognizable as that of a bore. This concept will be treated in greater detail in Chapter 9.

The turbulent nature of broken waves has two consequences of importance. First, turbulent flow quickly consumes wave energy. Wherever a shear gradient exists, a force is exerted in moving one particle of viscous fluid past another, and work is done by the fluid; this work appears in the form of heat. A breaking wave therefore expends much of its mechanical energy in heating the water. As mechanical energy is expended, the energy remaining (per meter of wave crest) decreases, and the wave height decreases proportionally (Table 5.2). Thus broken waves get smaller and smaller as they move ashore, in contrast to the process of shoaling, where unbroken waves increase in height. Thornton and Guza (1982) have shown that the root mean square height of bores in the surf zone of a sandy beach is linearly proportional to the water depth:

$$H_{rms} = \gamma d. \qquad (7.13)$$

For the beaches they examined, $\gamma = 0.42$, but γ is likely to vary from one beach to the next.

The random motion of turbulent water also has an important effect on the acceleration of water relative to a fixed point (such as an animal) on the shore. Take, for instance, the situation shown schematically in Figure 7.7. A turbulent eddy with diameter $D = 0.1$ m rotates ten times per second ($T = 0.1$). This vortex is moved by a bore at a velocity of $u = 5$ m/s. When the eddy first contacts the stationary point P, P experiences a flow with a component $\pi D/T = 3.1$ m/s in the positive y-direction. At 5 m/s it takes only 0.02 s for the diameter of the eddy to be carried past point

P. When the trailing edge of the eddy is at P, P experiences a velocity of 3.1 m/s in the negative y-direction. Thus the velocity in the y-direction has changed by 6.2 m/s in 0.02 s, an acceleration of 315 m/s². The velocity in the x-direction has remained 5 m/s through this entire period, and therefore there is no acceleration in this direction.

An example of flow in a turbulent bore running up a rocky beach is given in Figure 7.7c. The very rapid accelerations are probably due to the advection of turbulent eddies past the fixed point at which the velocity was measured. As we will see in Chapter 11, the acceleration of the water in broken waves can place large forces on wave-swept organisms.

The turbulence associated with a bore depends to a large extent on the type of breaker that created the bore. A spilling breaker causes turbulence that is often confined primarily to a layer of water at the surface near the crest of the bore, and the water below the surface moves much as it would if the wave had not broken. The arching jets of plunging and collapsing waves tend to inject turbulence much farther down into the water in front of the crest, often reaching the bottom. When a wave breaks directly onto a steep beach, turbulence is guaranteed to reach the bottom. Thus organisms living on steep rocky shores are very likely to experience turbulent flow (and the consequent large accelerations) with every breaking wave. The biological consequences of this will be discussed in Chapter 17.

A third important aspect of turbulence—its role in the transport of larvae and gametes—will be discussed in Chapter 10.

Despite the action of viscosity in decreasing wave height, many broken waves arrive at the shoreline with an appreciable height and therefore still have appreciable energy and momentum. Several important processes are associated with these broken waves.

Figure 7.7. Convective acceleration. (a), (b) As an eddy is moved past point *P*, the lateral velocity changes rapidly (see text). (c) Velocity and acceleration in a turbulent bore. Values for velocity and acceleration were calculated from the force exerted on a limpet (redrawn from Denny 1985 by permission of the American Society of Limnology and Oceanography).

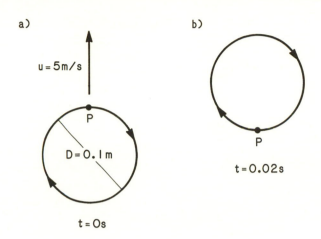

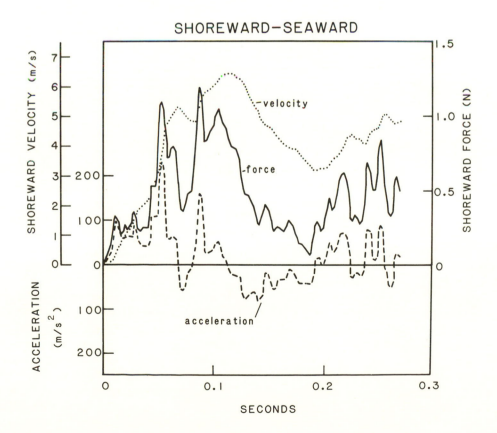

Figure 7.8. (a) The movement of a block on a frictionless plane—an analogy to run-up on a beach (see text). (b) Measured run-up (z_{max}) compared to predicted run-up (redrawn from van Dorn 1976 by permission of American Society of Civil Engineers).

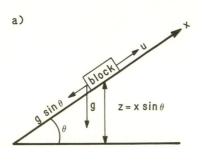

a)

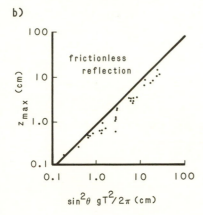

b)

Run-up

When bores reach the shore before losing all of their mechanical energy to turbulence, a volume of water is delivered to the shoreline with an appreciable velocity. This volume of water moves up the shore against the acceleration of gravity, gradually slowing down. In this respect, a bore running up the beach is roughly analogous to a block sliding up a frictionless ramp against the pull of gravity, and we can use this analogy to calculate the height to which the bore will run up on the beach.

Recall from introductory physics that x, the distance along a ramp traveled by a projectile block, is a function of the block's initial velocity, u_0, and the magnitude of

the acceleration tending to decrease this velocity

$$x = u_0 t - \left\{ \tfrac{1}{2} g \sin \theta t^2 \right\}, \qquad (7.14)$$

where θ is the angle at which the ramp slopes up and $g \sin \theta$ is the component of the gravitational acceleration that acts along the ramp (Fig. 7.8a). In this analogy we substitute a volume of water for the block and assume that the water swashes up and down the beach in time with the incoming waves. Given this assumption, the water is at its highest point on the beach one-half a wave period after reaching the shoreline. Thus, at $t = T/2$, x is maximum, and therefore $dx/dt = 0$. Taking the derivative of eq. 7.14 and substituting $T/2$ for t, we see that

$$u_0 - \left\{ g \sin \theta \, \frac{T}{2} \right\} = 0$$

$$u_0 = g \sin \theta \, \frac{T}{2}. \qquad (7.15)$$

Upon substituting this value for u_0 into Eq. 7.14 and again setting $t = T/2$, we calculate the maximum distance the water flows up the ramp (the beach):

$$x_{max} = \left\{ g \sin \theta \, \frac{T}{2} \frac{T}{2} \right\} - \left\{ \frac{1}{2} g \sin \theta \left[\frac{T}{2} \right]^2 \right\}$$
$$= (\tfrac{1}{8}) g \sin \theta T^2. \qquad (7.16)$$

The height the water rises above sea level (the run-up) is $z_{max} = x_{max} \sin \theta$. Thus

$$z_{max} = (\tfrac{1}{8}) g \sin^2 \theta T^2. \qquad (7.17)$$

Note that

$$(\tfrac{1}{8}) g \sin^2 \theta T^2 \simeq \sin^2 \theta \left\{ \frac{gT^2}{2\pi} \right\} = \sin^2 \theta L_0, \quad (7.18)$$

where L_0 is the deep-water wave length (eq. 4.8). From this calculation we predict that the vertical run-up is greater the more steeply the beach slopes (the greater $\sin \theta$) and the longer the wave period (or wave length). Van Dorn (1976) has shown that this relationship is qualitatively correct: a plot of z_{max} versus $\sin^2(\theta)(g T^2/2\pi)$ yields a straight line (Fig. 7.8b).

Although the derivation presented here is useful in providing a feeling for how run-up works, there are problems with its practical application. For instance, if the beach is a vertical wall, eq. 7.18 predicts a run-up equal to the deep-water wave length, 156 m for a wave with a period of 10 s. In reality, the maximum height water reaches on a vertical wall is roughly equal to the wave height. In practice, eq. 7.18 is useful for beach slopes less than about 1:5.

Because eq. 7.18 is derived assuming no friction between the water and the beach, the values it predicts should always be larger than the actual run-up. How much less than z_{max} the actual run-up is depends on beach roughness, beach permeability, and wave steepness. Information regarding the effect of these parameters is provided by empirical studies, which are discussed in depth in the U.S. Army Corps of Engineers' *Shore Protection Manual* (1984). Figure 7.9 gives actual values for run-up as a function of the slope of the beach. The value for run-up is normalized to H_0, the height of the wave in deep water, and curves are shown for waves of different deep-water steepnesses. The steeper the wave in deep water, the less the run-up, probably due to the fact that steep waves break in deeper water and therefore lose more energy to turbulence before they reach the shore.

These curves are derived from laboratory experiments in which waves were allowed to run up a smooth, impermeable beach. In general, the rougher the surface over which the water flows, the less the run-up. As we will see in Chapter 9, roughness on the beach creates turbulence in the flow near the substratum and thereby augments the rate at which the moving water does work against viscosity, work that ends up as heat. Any mechanical energy lost to heat is energy unavailable to move water upward against the force of gravity. To convert the run-up values of Figure 7.9 to those present on actual shores, we need to account for the rough-

ness of the substratum. This is done by multiplying the predicted run-up by the empirically determined coefficient of roughness tabulated in Table 7.1. A reasonable value for a densely settled rocky intertidal beach would seem to be 0.6–0.8. As with all extrapolations from laboratory studies, the figures cited here for run-up should be taken with a large grain of salt; the precise local topography of each site is likely to have profound effects on the actual run-up.

The run-up height can have effects that are important for wave-swept animals. The most obvious of these is the determination of the maximum height on the shore at which marine organisms can live. For example, animals such as barnacles must be immersed with reasonable frequency in order to feed and to avoid desiccation. For a given wave height and deep-water wave steepness, these organisms can survive at a higher point on the shore (measured relative to mean sea level) on a steeply

Figure 7.9. Measured run-up height (z_{max}) relative to offshore wave height (H_0) as a function of beach slope (redrawn from U.S. Army Corps of Engineers 1984).

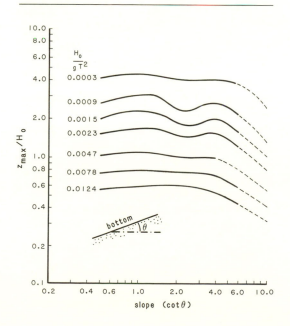

Table 7.1

Roughness Coefficients for Run-up on Rough Surfaces (from U.S. Army Corps of Engineers 1984)

Surface Characteristics	Placement	Roughness Coefficient
Smooth, impermeable		1.00
Concrete blocks	Fitted	0.90
Basalt blocks	Fitted	0.85–0.90
Grass		0.85–0.90
Rounded quarry stone	Random	0.60–0.65

Figure 7.10. Water velocity as a function of height on the shore for various beach slopes.

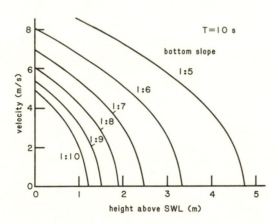

$$z = \left\{ g \sin^2 \theta \frac{T}{2} \right\} t - \frac{1}{2} g \sin^2 \theta t^2. \quad (7.19)$$

The derivative of x with respect to time yields velocity as a function of time:

$$dx/dt = u = \left\{ g \sin \theta \frac{T}{2} \right\} - g \sin \theta t. \quad (7.20)$$

Knowing both u and z as functions of time, we can plot u as a function of z. An example is shown in Figure 7.10 for a wave with a 10 s period running up beaches with different slopes. The steeper the slope, the faster the water is moving when it reaches a certain height above still water level. The differences can be dramatic. For instance, at a height of 2 m above SWL on a slope of 1:7, an organism would see a maximum velocity of about 3 m/s. For a slope of 1:8, the maximum run-up is less than 2 m; at this beach slope, an organism 2 m above SWL would not even get wet. These calculations of water velocity versus height should not be taken too seriously. They suffer from the same problems cited earlier when we calculated x_{max}.

The *Shore Protection Manual* provides methods for approximating run-up for beaches with composition that varies from one level to the next, and for beaches where the inclination varies.

sloping rock beach than on a gently sloping beach. Conversely, at a given height on the shore, organisms living on a gently sloping beach are subjected to less energetic water flows than organisms on a steep beach. For example, in an intermediate step in our calculation of run-up (eq. 7.15), we calculated that $u_0 = g \sin(\theta) T/2$. This allows us to calculate distance traveled along the beach as a function of time:

$$x = \left\{ g \sin \theta \frac{T}{2} \right\} t - \frac{1}{2} g \sin \theta t^2.$$

In terms of height above still-water level.

Wave Reflection

After water has run up a beach, it slides back seaward under the influence of gravity. In effect, the wave is reflected from the shore and a new wave is propagated seaward. There are two important effects caused by the reflection of waves from a beach. First, because the swash is in time with the incoming wave, there is the possibility of constructive interaction between the reflected wave and the shoreward-moving incident waves. This effect is seen most vividly when the beach is vertical and extends into deep water; the sea walls that protect harbors are a good example.

If the surface of the sea wall is smooth, very little mechanical energy is lost to turbulence as the wave runs up the "beach," and the reflected wave has an amplitude approximately equal to the incident-wave amplitude. The reflected wave travels seaward with the same velocity that the incident wave travels shoreward. The result of thus superimposing two waves traveling in opposite directions is a *standing wave* or *clapotis* (Fig. 7.11). Wherever the peaks or troughs of reflected and incident waves coincide (an *antinode*), the amplitude of the surface elevation is the sum of the two and is therefore doubled. As long as the celerities of incident and reflected waves are the same, the positions of the antinodes remain fixed. The shoreline forms one of the antinodes. Halfway between antinodes are *nodes*, points where the water level does not change. Water flow is vertical beneath an antinode and horizontal beneath a node. If the height of the standing wave is equal to or greater than the breaking height, incident waves can break as they enter the region where they interact with the reflected wave. At times this can lead to a spectacular "water jump" some distance offshore.

If the beach is not vertical or if it is rough, energy is lost to viscosity during run-up. As a result the reflected wave has a smaller amplitude than the incident wave. Battjes (1974, as cited in the U.S. Army Corps of Engineers' *Shore Protection Manual* [1984]) proposed that the reflection coefficient (height of the reflected wave/height of the incident wave) is a function of a "wave similarity index,"

$$\xi = \frac{\tan \theta}{\sqrt{\dfrac{H_i}{L_0}}}, \qquad (7.21)$$

where H_i is the height of the incident wave in shallow water. This function is shown for various substrata in Figure 7.12. On gradually sloping beaches [$\tan(\theta)$ small] the amplitude of the reflected wave may be so small that the effects of its interaction with incident waves are negligible.

Radiation Stress

Propagating waves transport momentum as well as energy, with interesting consequences for inshore water movement. To see how this works, consider a small cube of fluid having dimensions dx, dy, and dz (Fig. 7.13). We calculate how much momentum flows into and out of this cube as water flows in the x-direction. First, we examine the upstream face of the cube. The rate at which volume moves across this face is $(dx/dt)\, dy\, dz = u\, dy\, dz$. Thus the rate at which mass enters the cube is $\rho u\, dy\, dz$, and the rate at which momentum (mass × velocity) enters the cubes is $\rho u^2\, dy\, dz$. If this were the only flux of momentum into or out of the cube, the momentum of fluid in the cube would increase at this rate. Recall that the rate of change of momentum is a force (eq. 3.3). Thus the rate at which momentum

Figure 7.11. Motion in a standing wave (clapotis) (redrawn from U.S. Army Corps of Engineers 1984).

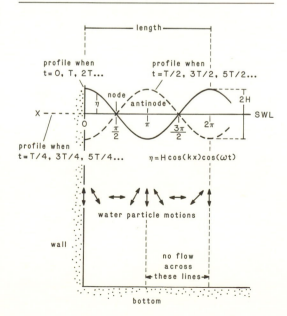

Figure 7.12. The reflection coefficient (H_r/H_i) as a function of the wave similarity index. The steeper the bottom slope, the better the wave is reflected. The steeper the wave, the less energy is reflected. (Redrawn from U.S. Army Corps of Engineers 1984 after Battjes 1968.)

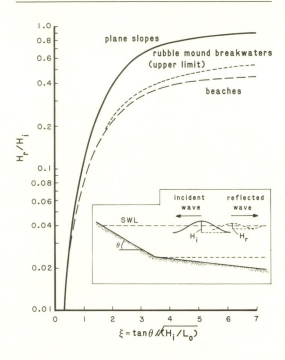

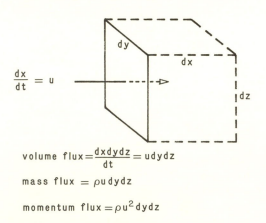

Figure 7.13. Calculation of the momentum flux due to water velocity (see text).

enters the cube (the momentum flux) is the equivalent of a force acting in the positive x-direction on the upstream face of the cube. The force per area (termed a *stress*) acting on the upstream face ($dy\,dz$) is thus ρu^2. Now, if water flows out of the cube through the downstream face at the same rate it enters the cube, the momentum flux per area across the downstream face is $-\rho u^2$, the negative sign indicating that momentum flows out. The momentum flux due to water leaving exerts a force tending to push the cube in the negative x-direction. The two forces offset, and the net force on the cube is zero. However, this is true only if the momentum flux does not change across the cube. If less momentum flows out of the cube than in, a net force is exerted equal to the change in momentum flux per change in x.

This same idea applies to the momentum transported by waves. If the momentum flux changes from one place to another, a force is exerted. Longuet-Higgins and Stewart (1963, 1964) have calculated the momentum flux due to surface waves. Some of the total momentum flux is due to the hydrostatic pressure exerted by still water, but the momentum flux in excess of that in still water is termed the *radiation stress*. In other words, radiation stress is the rate at which waves transport momentum across an area. Like wave energy, radiation stress is proportional to the square of wave height and therefore can be conveniently quantified in terms of the wave energy, E. The radiation stress in the direction of wave propagation is

$$S_{xx} = E\left\{\frac{1}{2} + \frac{2kd}{\sinh(2kd)}\right\}. \quad (7.22)$$

In deep water, the ratio $2kd/\sinh(2kd)$ is zero so that $S_{xx} = (\frac{1}{2})E$. In shallow water, $2kd/\sinh(2kd) = 1$, so that $S_{xx} = (\frac{3}{2})E$. There is also a radiation stress perpendicular to the direction of wave propagation (in the y-direction):

$$S_{yy} = E\left\{\frac{kd}{\sinh(2kd)}\right\}. \qquad (7.23)$$

Thus in deep water, where kd is negligible relative to $\sinh(2kd)$, $S_{yy} = 0$, and in shallow water, $S_{yy} = E/2$. The flow of x-directed momentum in the y-direction is zero, so that

$$S_{xy} = 0. \qquad (7.24)$$

This last conclusion is true only if the x-axis is chosen to be parallel to the direction of wave propagation, and the y-axis is parallel to the wave crests. In some cases it is more convenient to place the x-axis perpendicular to the shoreline and the y-axis parallel to the shoreline. Such a transformation of axes requires a change in the expressions for S_{xx}, S_{yy}, and S_{xy} (Basco 1982):

$$S_{xx} = E\left[\left\{\frac{1}{2} + \frac{kd}{\sinh(2kd)}\right\}\cos^2\alpha + \frac{kd}{\sinh(2kd)}\right]$$

$$S_{yy} = E\left[\left\{\frac{1}{2} + \frac{kd}{\sinh(2kd)}\right\}\sin^2\alpha + \frac{kd}{\sinh(2kd)}\right]$$

$$S_{xy} = E\left\{\frac{1}{2} + \frac{kd}{\sinh(2kd)}\right\}\sin\alpha\cos\alpha,$$

$$(7.25)$$

where α is the angle between the x-axis and the direction of wave propagation (Fig. 7.15). Note that when $\alpha \neq 0$, S_{xy} is nonzero. In other words, when waves do not approach perpendicular to the shoreline, some momentum flux is directed along the shore.

These various radiation stresses have important consequences for nearshore flows. First, we examine the effect of wave breaking. Consider the situation shown in Figure 7.14. Small-amplitude waves are moving shoreward across a gently sloping bottom. The x-axis lies in the direction of wave propagation and is perpendicular to the shore. We examine the momentum flux into and out of a small section of the flow bounded by two vertical planes and separated by a distance

Figure 7.14. The variation in momentum flux as a wave moves inshore. To maintain momentum balance as waves lose height after breaking, the average surface elevation (n) must rise. (Redrawn from Longuet-Higgins and Stewart 1964 by permission of Pergamon Press.)

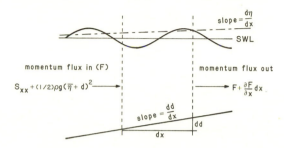

dx. Because there is no external force acting on the water, we know that the momentum flux into the section must equal the flux out.

The total momentum flux into the section is the radiation stress plus the flux due to hydrostatic pressure:

$$S_{xx} + \int_{-d}^{\eta} \rho g(\bar{\eta} - z)\, dz = S_{xx} + (\tfrac{1}{2})\rho g(\bar{\eta} + d)^2, \qquad (7.26)$$

where $\bar{\eta}$ is the time-averaged surface elevation. If the momentum flux changes at a rate

$$\frac{d}{dx}\{S_{xx} + (\tfrac{1}{2})\rho g(\bar{\eta} + d)^2\}, \qquad (7.27)$$

the flow of momentum out of the segment is

$$\{S_{xx} + (\tfrac{1}{2})\rho g(\bar{\eta} + d)^2\}$$
$$+ \frac{d}{dx}\{S_{xx} + (\tfrac{1}{2})\rho g(\bar{\eta} + d)^2\}\, dx. \quad (7.28)$$

Unless the flux of momentum out of the segment equals the flux in, the net momentum of the fluid in the segment will change. Here we examine the case in which the flux out is less than the flux in because of a decrease in wave height across the segment. As a result, the momentum of the fluid in the segment increases either by increasing its mass or its velocity, or both. In practice, water would accumulate in the segment until the change, $d/dx[S_{xx} + (\frac{1}{2})\rho g(\bar{\eta} + d)^2]\,dx$, is balanced by the change in hydrostatic pressure across the section:

$$\frac{d}{dx}\left\{S_{xx} + (\tfrac{1}{2})\rho g(\bar{\eta} + d)^2\right\}dx = \rho g(\bar{\eta} + d)\frac{dd}{dx}\,dx. \tag{7.29}$$

Working through the math, we arrive at the conclusion that

$$\frac{dS_{xx}}{dx} + \rho g(\bar{\eta} + d)\frac{d\bar{\eta}}{dx} = 0. \tag{7.30}$$

Now, this calculation assumes that the wave amplitude is small. In this case $\bar{\eta} \ll d$, so that $(\bar{\eta} + d) \simeq d$. Using this approximation, we see that

$$\frac{d\bar{\eta}}{dx} = -\left\{\frac{1}{\rho g d}\right\}\frac{dS_{xx}}{dx}. \tag{7.31}$$

Thus, the average surface elevation can change as a function of the change in S_{xx}. Earlier in this chapter we saw that wave height in the surf zone is a linear function of water depth, $H_{rms} = \gamma d$ (eq. 7.13). Furthermore, we know from Chapter 5 (Table 5.1) that the energy of a wave is a function of its height

$$E = \tfrac{1}{8}\rho g H^2. \quad \text{(from Table 5.1)}$$

In other words, in shallow water, wave energy is a function of water depth. We know from eq. 7.22 that in shallow water the radiation stress S_{xx} is equal to (3/2) the wave energy. Thus S_{xx} decreases as water depth decreases. Combining all these facts, we arrive at the conclusion that in the surf zone,

$$S_{xx} = (\tfrac{3}{2})E = (\tfrac{3}{16})\rho g \gamma^2 d^2. \tag{7.32}$$

Taking the derivative of eq. 7.32 with respect to x, and inserting the result in eq. 7.31, we arrive at our goal:

$$\frac{d\bar{\eta}}{dx} = -(\tfrac{3}{8})\gamma^2\frac{dd}{dx}. \tag{7.33}$$

In the surf zone, the average surface ele-

vation is increased by the radiation stress; this is the set-up. The greater the bottom slope $(-dd/dx)$, the greater the rate at which the average surface elevation rises toward shore, and the greater the total set-up. On sandy beaches the increase in surface elevation is typically 10%–20% of the breaking-wave height, which can amount to a rise in average sea level of 0.4 to 0.6 m. Note that this argument assumes that waves break. On very steeply sloping shores, waves reflect rather than break and eq. 7.33 is not valid.

For present purposes, set-up is important in two respects. First, it can affect surf conditions near the shore. As we have seen, the height, and therefore the celerity, of bores in the surf zone is a function of the local water depth. An increase in depth due to set-up can result in larger waves reaching inshore areas. Second, as we have seen, set-up is accompanied by a rise in the local hydrostatic pressure near the shoreline. Because the water level is higher in the surf zone than it is offshore, water tends to flow away from the beach. Some of this seaward transport may occur at the bottom, forming an "undertow" (Longuet-Higgins 1983).

Nearshore Currents

When waves approach a beach so that their direction of propagation is not perpendicular to the shoreline, the influx of momentum in the guise of the radiation stress S_{xy} causes currents to flow. As wave momentum flows into the surf zone, the momentum of water there is increased. This increase is manifested as an increase in velocity. Unless there is some mechanism that limits the momentum of water in the surf zone, the velocity of the longshore current increases as long as wave momentum is delivered to the shore. In reality, the longshore-current velocity increases until momentum is lost to turbulence and to bottom friction at the same rate at which it is delivered by waves. Once this equilibrium condition is reached, the longshore current flows at a constant velocity. Longshore currents of 0.5–1.0 m/s are common, and storms can be accompanied by currents of 2 m/s or more.

When swimming on a wave-swept shore, we often find ourselves being carried down the beach by the longshore current. At some point the longshore current turns seaward as a *rip current* (Fig. 7.15). The seaward flow in a rip can be unnerving to uninitiated swimmers, but it is harmless once we realize the pattern of flow. The rip current is quite localized, and by swimming across the current we can avoid being swept out to sea.

Figure 7.15. Longshore currents. (a) Waves approaching the beach at an oblique angle. (b) Waves approaching perpendicular to the beach (redrawn from Shepard and Inman 1950 by permission of the American Geophysical Union).

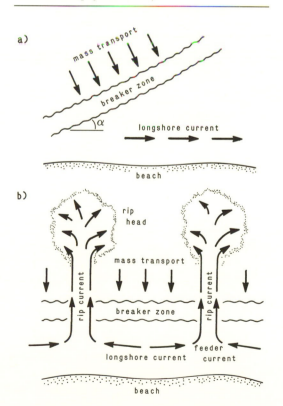

Longshore currents are of practical and economic importance because they transport sediment. For instance, any man-made structure that changes the pattern of the longshore current can cause the local depletion or build-up of sand. The Army Corps of Engineers has spent many years studying sand transport in an attempt to control sedimentation. The triumphs and failures of this effort make interesting reading (see Bascom 1980; U.S. Army Corps of Engineers' *Shore Protection Manual* 1984). In a biological context, longshore currents may be important in dispersing planktonic larvae over a scale of meters to a few kilometers, and the seaward flow in a rip current may be responsible for delivering planktonic larvae to offshore currents that can carry them tens or hundreds of kilometers.

When waves approach the beach so that their crests are nearly parallel to the shore, radiation stress may drive longshore currents that flow in opposite directions along alternating sections of the shore. Where two oppositely directed currents flow into each other, the flow turns seaward as a rip current (Fig. 7.15b). This pattern of flow in rip currents is determined by longshore variations in wave height. These variations may be due to differences in shoaling as a result of bottom topography, or to interaction with edge waves (Chapter 5).

Unfortunately, studies of set-up, longshore currents, and rips have been confined to sandy beaches. Virtually nothing is known about set-up and the resulting water flows on rocky shores or coral reefs. For a more thorough discussion of longshore and rip currents on beaches, consult Komar (1976, 1983).

Surf Beat

Offshore effects of the radiation stress can also have consequences for inshore flows. Consider the situation shown in Figure 7.16. Two groups of deep-water waves

Figure 7.16. The variation in wave height between wave "sets" can result in a variation in mean sea level—the surf beat.

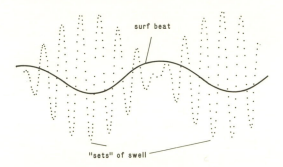

with slightly different wave lengths (and therefore celerities) have interacted so that their amplitudes are periodically modulated, forming the sort of wave "sets" familiar to surfers. Because the wave height varies from place to place, the radiation stress also changes, leading to a spatial variation in average sea level analogous to the set-up. This periodic change in sea level is itself a surface wave, one with a very long wave length. The period of the wave is equal to the interval between wave sets, typically 20 s to 5 min.[1] These long wave length, long-period waves are known as *surf beat*, one form of *infra-gravity wave*. Because their period is so long, they are difficult to distinguish visu-

ally among the swell and are usually noticed as a gradual rise and fall in the depth of water in the surf zone. The wave lengths of surf beat are sufficiently long so that they do not break on the shore (see eq. 7.11) and are reflected, forming standing waves. The antinode of the standing wave is at the shoreline, and the first node is one-half wave length offshore (usually well beyond the surf zone). Thus, in the nearshore the amplitude of the surf beat is highest at the shore and decreases seaward, just the opposite of broken waves (eq. 7.13). In this respect, surf beat are similar to edge waves (Chapter 5). Because the amplitude of surf beat (and therefore their energy) is greatest where that of the swell is least, these waves may be important in near-shore sediment transport. The biological importance of surf beat has not been explored, but a brief speculation on the subject is offered in Chapter 17.

---

[1] Because the wave period is imposed by the interval between sets of swell, surf beat traveling shoreward are forced to have this period and to move at the celerity of the sets. In this respect they are different from seas and swell, which are free to move at the celerity determined by their wave lengths. Surf-beat waves are "set free" when they reflect from the shore. The distinction between *free* and *forced* waves is discussed in Chapter 8.

Tides

Tidal fluctuations in sea level have profound effects on the lives of wave-swept organisms. Certainly for those organisms in the intertidal zone, the tides are "the heartbeat of the ocean" (Defant 1958). However, in addition to their biological relevance, tides are also of economic importance, determining when waters will be navigable and when tidal currents flow. In response to this economic importance, the governments of various nations go to a great deal of trouble each year to predict the tides for the world's shores. Unlike information about wave patterns or fluid-dynamic forces, accurate information about the tides is as close as the local bait and tackle shop, and most newspapers in coastal cities report tidal predictions along with the weather.

Because practical information on the tides is so easily available, we need not discuss the fine points of how this information is derived. What follows is a brief introduction to the physics of the tides, an hors d'oeuvre to the subject of tidal analysis. A more thorough analysis can be found in any of a number of excellent texts on the subject (e.g., Clancy 1968; Defant 1958; Godin 1972).

The tides can be studied from three points of view. The simplest explanation is provided by the equilibrium model first developed by Sir Isaac Newton. Aspects of tidal fluctuations that cannot be explained by the equilibrium model are approached by various dynamic models. Although these two viewpoints serve a valuable heuristic purpose, neither is an accurate method for predicting the tides at a certain point on the shore, and for a practical approach to predicting the tides we must turn to harmonic analysis.

The Equilibrium Model for the Tides

We begin with yet another application of Newton's three laws of motion. From the first law we know that a mass moves in a straight line unless acted upon by some force. Consequently, if a mass is to travel in a circular path, it must continually be coaxed into changing its direction. To accomplish this, the mass must be acceled sideways, and, from the second law, we know that this acceleration requires a force. The direction of this force changes: if the mass travels in a circular path, the applied force must always be directed from the mass to the center of the circle. Such a center-directed force is termed *centripetal*. A car on a Ferris wheel is an example of this sort of motion. As the car moves in a circle, the spokes of the wheel pull on it with the necessary centripetal force. From Newton's third law we know that this centripetal force is accompanied by an equal and opposite force (the *centrifugal* force). It is this force that alternately adds and subtracts from the force of gravity to provide the thrill of the ride.

Although a Ferris-wheel car moves in a circle around the hub of the wheel, each car does not rotate around its center of mass; the cars stay upright. As a result, every part of the car travels the same distance around the hub (see Fig. 8.1) and is subject to the same centripetal acceleration. This equality of acceleration applies to direction as well as magnitude; the acceleration for each part of the car is di-

Figure 8.1. A Ferris-wheel car acts as an analogy to a nonrotating earth. As the wheel revolves, the top of the car is acted upon by a centripetal force directed at the hub of the wheel (the solid dot). The bottom of the car moves in a circle that is displaced downward (dotted circle) and is acted upon by a centripetal force directed at the center of this displaced circle (open dot). The two centripetal forces have the same magnitude and are parallel.

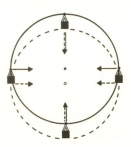

Figure 8.2. The centripetal acceleration due to gravitational attraction causes the earth and the moon to revolve about their common center of mass. Because the earth is much more massive than the moon, the center of mass lies within the earth. (Redrawn from Paul D. Komar, Beach Processes and Sedimentation, 1976, pg. 124, by permission of Prentice-Hall, Englewood Cliffs, N.J.)

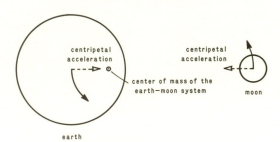

rected parallel to the line connecting the car's center of mass and the wheel's hub.

Now consider the moon and the earth. As the moon and earth whirl through space, each travels in a circular path around their common center of mass (Fig. 8.2). For the sake of simplicity, the moon and earth will be treated as if they were cars on a Ferris wheel and are assumed not to rotate. The earth is much more massive than the moon (Table 8.1), so the

center of mass for the earth-moon system actually lies within the earth. As a result, the earth's movement around the system's center of mass is something of a wobble, but a circular path nonetheless. For the moon and earth to travel in their circular paths, there must exist a centripetal force, and assuming the earth does not rotate, every part of the planet feels the same centripetal acceleration. This centripetal force is supplied by gravity. As Newton so insightfully determined, the force of gravity between two objects is proportional to the product of their masses and inversely proportional to the square of the distance between them. This means that with respect to the earth and moon, a force is tending to pull them together:

$$\text{centripetal force} = G \frac{m_e m_m}{R^2}, \quad (8.1)$$

where m_e is the mass of the earth, m_m the mass of the moon, and R the distance from the center of the earth to the center of the moon. G is the universal gravitational constant ($6.67 \cdot 10^{-11}$ Nm^2/kg^2). The distance between the earth and moon is such that this gravitational attraction is just sufficient to keep them traveling in their circular paths. However, this equality between gravitational attraction and centripetal acceleration applies only to the objects taken as a whole. Unlike the cen-

Table 8.1
Facts about the Earth, Moon, and Sun
(from Komar 1976)

Diameter of the earth	12,753 km
Diameter of the moon	3,479 km
Mass of the earth	$5.98 \cdot 10^{24}$ kg
Mass of the moon	$7.34 \cdot 10^{22}$ kg
Mass of the sun	$1.96 \cdot 10^{30}$ kg
Average distance between the centers of the earth and moon	384,329 km
Average distance between the centers of the earth and sun	149,360,000 km

tripetal force, the gravitational force acting on a bit of the earth varies from one spot to another. On the side of the earth facing the moon, a bit of mass (be it rock or water) is closer to the moon, making the gravitational attraction higher than it is for a bit of mass on the side of the earth opposite the moon.

The stage is now set for a first look at the tides. We pretend for the moment that the earth is a perfectly smooth sphere, and that it is completely covered by water. This water is acted upon by the same forces that act on the solid earth: the water is everywhere attracted toward the center of the earth by the gravitational attraction of the earth's mass. Similarly, it is attracted toward the moon by the gravitational attraction of the moon's mass. And finally, the water is acted upon by the centripetal force. For simplicity, we will examine the acceleration (force/mass) acting on masses of water rather than calculate the forces directly. Thus at the earth's surface the acceleration due to the earth's mass can be expressed as

earth's gravitational acceleration, $g = G\dfrac{m_e}{R_e^2}$,

$$(8.2)$$

where R_e is the radius of the earth.

Figure 8.3. The equilibrium model of the tides—the accelerations acting on a mass of water. (Redrawn from Paul D. Komar, Beach Processes and Sedimentation, 1976, pg. 126, by permission of Prentice-Hall, Englewood Cliffs N.J.)

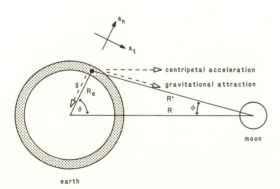

Consider a point P on the earth's surface. This point is a distance R_e from the center of the earth and a distance R' from the center of the moon (Fig. 8.3). The gravitational acceleration of the earth acting at P is directed toward the center of the earth, and the gravitational acceleration of the moon is directed toward the center of the moon. As we have noted, the centripetal acceleration is directed parallel to the line connecting the centers of the moon and the earth. Now, one could manipulate these three vectors to arrive at the net acceleration acting on a mass at P, but in the long run it is more instructive to calculate separately the components of this net acceleration: one component acting normal to the earth's surface, a_n, and one acting tangential to the surface, a_t.

Through an examination of the geometry shown in Figure 8.3, the component acting normal to the earth's surface is seen to be

$$a_n = -g - G\frac{m_m}{R^2}\cos\theta + G\frac{m_m}{R'^2}\cos(\theta + \phi).$$

$$(8.3)$$

This expression is graphed in Figure 8.4a. The normal component of the net acceleration is greatest at points directly facing or away from the moon and is least where θ is 90°. However, this net acceleration is very small compared to the acceleration of gravity. At maximum, each kilogram of water experiences a change in force of only $1.1 \cdot 10^{-7}$ N due to the presence of the moon, compared to the constant force of 9.81 N it feels due to the gravitational acceleration from the earth's mass.

The tangential component of the net acceleration is

$$a_t = G\frac{m_m}{R'^2}\sin(\theta + \phi) - G\frac{m_m}{R^2}\sin\theta. \quad (8.4)$$

Because the acceleration due to the earth's gravity acts normal to the earth's surface, it does not enter this computation. The variables R' and ϕ can be eliminated by noting that

Figure 8.4. The equilibrium model of the tides. (a) The normal (a_n) and tangential (a_t) accelerations acting on a mass of seawater as a function of θ (Figure 8.3). (b) The tangential acceleration creates tidal bulges and valleys.

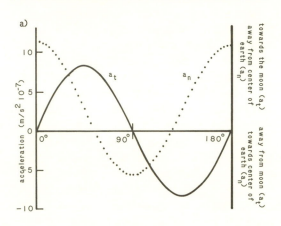

a)

b)

$$a_t = \frac{3}{2}\, g\, \frac{m_e}{m_m}\, \frac{R_e^3}{R^3} \sin 2\theta. \qquad (8.6)$$

This result is graphed in Figure 8.4a. The maximum tangential acceleration occurs when θ is 45° or 135° and is zero when θ is 0°, 90°, or 180°. The tangential acceleration is slightly smaller than the normal acceleration, but in this case the small force is not opposed by any other force, and therefore the force due to tangential acceleration is capable of moving the water on the earth's surface. This is the force responsible for the tides. The direction of the tangential "tractive" force is away from points at $\theta = 90°$. Thus water has a tendency to flow toward and away from the moon, forming tidal bulges and leaving tidal valleys behind (Fig. 8.4b). The bulges continue to grow in height until the increased hydrostatic pressure under the bulge is sufficient to offset the tractive force. At this point the water ceases to flow, and equilibrium is reached.

We now relax our original assumptions and allow the earth to rotate. What are the consequences? First, there is a centripetal acceleration due to the rotation of the earth about its center. This acceleration has no component tangential to the earth's surface and is very small compared to the acceleration of gravity, so for the moment we ignore it. We will return to this centripetal acceleration when we discuss the dynamic model of the tides. Second, one might reasonably assume that the spinning earth interacts with its coating of water. A frictional force is indeed present and it acts to gradually slow the earth's rotation. However, the frictional force cannot be accounted for in this model, and for the time being we ignore it as well.

The third and most important effect of the earth's rotation is to move areas on the surface relative to the tidal bulges of water (which remain fixed in relation to the moon). For example, if the earth rotated about an axis perpendicular to the plane of the moon's orbit, a point on the earth

$R'^2 = R_e^2 + R^2 - 2R_eR \cos\theta$ (the law of cosines)

and

$$\phi = \arctan\left\{\frac{R_e \sin\theta}{R - R_e \cos\theta}\right\}.$$

Thus

$$a_t = G\left\{\frac{m_m}{R_e^2 + R^2 - 2R_eR \cos\theta}\right\}$$

$$\cdot \sin\left\{\theta + \arctan\frac{R_e \sin\theta}{R - R_e \cos\theta}\right\}$$

$$- G\frac{m_m}{R^2} \sin\theta. \qquad (8.5)$$

A series expansion of this equation can be written. When terms containing a high order of the expression R_e/R (which are very small) are dropped, eq. 8.5 can be seen to approximate

Figure 8.5. The equilibrium model of the tides. (a) As the earth rotates, point *P* is carried past a tidal bulge (high tide) and a tidal valley (low tide). (b) Because the earth's axis is not perpendicular to the plane of the moon's orbit, a point on the earth experiences tides daily with different amplitudes.

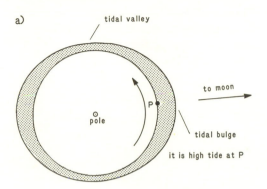

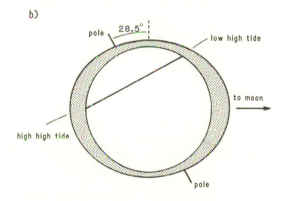

would move cyclically from a tidal bulge to a tidal valley and back again (Fig. 8.5a). Thus the water level at a point fixed to the solid earth would be high twice a day and low twice a day—a tidal fluctuation.

In this hypothetical example, the two tides would have the same amplitude. Because the tidal bulges are highest at points directly facing or away from the moon (in this example, at the equator), points on the equator would experience the largest tidal fluctuations, and fluctuations would decrease in amplitude toward the poles. However, the axis about which the earth spins is not perpendicular to the plane of the moon's orbit; it is tilted by about 28.5° (Fig. 8.5b). Consequently, as the earth

rotates, a point on its surface changes its latitude relative to the moon's orbital plane by 2 × 28.5° = 57°. For points not on the equator, this results in encountering different parts of the two tidal bulges during the course of a day's rotation. For example, a point at 57°N latitude (relative to the earth's equator) passes under one tidal bulge well north of the moon's orbital plane (where the bulge is small) and under the other bulge at the orbital plane where it is maximal. The tidal fluctuations therefore vary semi-diurnally—there is a high high tide and a low high tide, and similarly a high low tide and a low low tide. This is indeed the pattern of the tides for many areas on the earth.

The orientation of the earth's axis is fixed—it does not change as the earth is orbited by the moon nor as the earth orbits the sun. It is for this reason that day length varies through the year. Similarly, the relation of a point on the earth to the tidal bulges varies through the course of a lunar orbit. As a result, the inequality of the tides is maximal twice per month and minimal twice per month.

The equilibrium model for the tides also explains the tidal period. Consider a point on the earth directly facing the moon. According to the equilibrium model, this point experiences a high tide. Twenty-four hours later, this point has made one complete rotation about the earth's axis. However, during this time the moon has moved around its orbit in the same direction as the earth rotates (Fig. 8.6). Because it takes the moon twenty-eight days to complete one orbit, it has moved 1/28th of an orbit (12.9°) in the time it takes the earth to complete one rotation. Thus, our point must rotate an additional 12.9° before it is again directly facing the moon and experiences a high tide. This extra rotation takes about 50 minutes. Consequently, the tidal day is 24 hours and 50 minutes long, and the time of high (or low) tides moves forward 50 minutes per day.

Figure 8.6. The tidal period. During one 24-hour rotation of the earth about its axis, the moon has moved about 12.9° along its orbit. It takes the earth an additional 50 minutes to make up the difference.

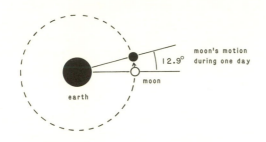

So far we have considered only the dynamics of the earth and the moon. The earth also moves in a nearly circular path around the sun, and the same mechanics apply. Although the sun is 27 million times as massive as the moon, it is 390 times as far away. From eq. 8.6 one can calculate that the tractive force due to the sun is only about 46% as strong as that due to the moon. Of course, the sun's effect on the tides has a 24-hour period rather than the Lunar period of 24 hours and 50 minutes.

The forces due to the sun and moon are additive. When the moon and the sun are in line (new moon or full moon), the tractive forces from both sources act in the same direction, and the tides have an exceptionally large amplitude. These are the *spring tides*, which happen twice a month. When the moon is one-quarter or three-quarters full, its tractive forces are opposed by those of the sun, resulting in small tidal amplitudes, the *neap tides*.

One final tidal phenomenon can be explained by the equilibrium model. Neither the orbit of the earth around the sun nor of the moon around the earth is exactly circular; rather, they are elliptical. Thus the distance from earth to sun and earth to moon varies. When the moon is nearest to the earth (perigee), the tidal amplitude is increased. This happens about once per month. When the moon is nearest the earth and the earth is nearest the sun, the tides are exceptionally high. This happens once a year.

The Dynamic Model

The equilibrium model of the tides is simple and apparently tidy. Unfortunately, there are many aspects of the tides that it cannot explain. For example:

1. On the west coast of Central America it is often too hot for even a diligent biologist to be out studying intertidal life during daytime low tides. However, the tides at night are just as low as those during the day, and the air temperature is much more conducive to scientific exploration. Light is no problem: all one has to do is wait for the moon to be high in the sky and go out and work the low tide. But wait. If the full moon is overhead, the equilibrium model predicts that the tide should be high, not low.

2. Monterey, California, is almost due south of the Golden Gate at San Francisco. The equilibrium model would predict that high tide should occur at the two spots simultaneously. In fact, the high tide at Monterey occurs about 45 minutes before the high tide at the Golden Gate.

3. The time of high tide can vary substantially over even smaller distances. For instance, within the limited confines of San Francisco Bay the time of the tides can vary by almost by hour.

4. According to the equilibrium model, Hawaii, at latitude 20°N should have approximately the same tidal fluctuations as Salina Cruz, Mexico (16°N). However, a tidal fluctuation is almost nonexistent in Hawaii (< 0.6 m), while at Salina Cruz it reaches 2 m.

These and other common examples indicate that the equilibrium model of the tides cannot answer all questions. A different, complementary approach was introduced by the great French mathematician Simon Laplace in the late 1700s and has been expanded and revised many times since. This alternative approach views the tides through the dynamics of water movement rather than through the equilibrium of forces.

Laplace's basic insight was to treat the tides as waves in the same way that we have dealt with water waves in this book. The tidal bulges are the crests of the waves, and the tidal valleys are the troughs. Thus the tidal wave has a wave length ($L = 20,000$ km) equal to half the circumference of the earth.[1] By the definition of shallow water (Chapter 4), tidal waves are shallow-water waves. In fact, they are very well-behaved shallow-water waves. Because their wave height (a few meters at most) is much less than their wave length, tidal waves in the open ocean can be treated accurately using linear wave theory. Thus we can immediately predict that the celerity of a tidal wave ought to be

$$C = \sqrt{gd}, \quad \text{(Table 5.4)}$$

where d is depth of the ocean. For example, consider a hypothetical canal circling the globe at the equator. In response to the tractive forces, the water in the canal piles up on opposite sides of the earth and tidal waves are propagated. If the canal completely circled the globe, these waves would behave in much the same fashion as the tidal bulges predicted by Newton. But this model fails to take into account one major aspect of the earth—dry land. Only in the Antarctic Ocean does the sea extend completely around the earth. Elsewhere the oceans are interrupted by the continents. Thus, if our hypothetical canal is going to be at all analogous to the real earth, it cannot extend straight around the globe. For instance, a canal at the equator would have ends at Ecuador and Southeast Asia or Gabon and Brazil.

[1] The term *tidal wave* is used here strictly in connection with the tides. The sort of large, potentially destructive waves that are caused by seismic disturbances are known as *tsunamis*.

The presence of ends on the canals drastically changes the behavior of the tidal waves. As a tidal wave runs into the end of a canal it is reflected, forming the sort of standing wave discussed in Chapter 7. This kind of fluid motion is easily demonstrated in a wave tank, or simpler yet, in a bathtub. To carry out this experiment, fill a tub about one-quarter full, float a board at one end, and move the board up and down; a wave is generated that reflects from the opposite end of the tub. If you oscillate the board at too high a frequency, several waves can be present at once. However, we confine ourselves to the simple situation where the wave length is greater than the length of the tub, so that at any one time only a single crest and a single trough are present. This simple mode of oscillation is analogous to a tidal wave in the ocean.

Once the standing wave is fully developed, water motion such as that shown in Figure 8.7 can be clearly seen. The water at the ends of the tub rises and falls. When the water at one end is high, water at the other end is low. In the center of the tub is a *nodal line* where the water level is constant. At the nodal line the water motion is strictly horizontal, while at the ends of the tub the water moves strictly in a vertical direction. The velocity of the water is out of phase with the surface elevation. When the surface is maximally displaced from the still water level, the fluid motions in one direction have just come to a halt and are preparing to move back the other way. At this instant, the water's velocity is zero throughout the tub (Fig. 8.7a, c). Conversely, when the standing wave is halfway between maximal excursions, the surface throughout the tub is at the still-water level. At this instant, the water velocities are at their maximum as water is rapidly transferred from one end of the tub to the other (Fig. 8.7b).

These water motions are analogous to the motions of the standing wave that would be established in a canal in the

Figure 8.7. A standing wave. When the wave is at its maximum excursion (a, c) the water is stationary. At midexcursion (i.e., when the water is at still-water level [SWL], the water velocities are greatest. At the nodal line the velocity is horizontal.

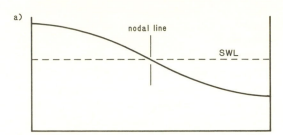

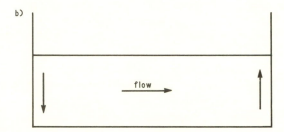

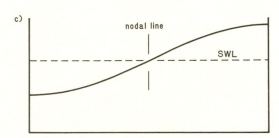

ocean. For instance, when it is high tide in the eastern end of the Pacific ocean (near the west coast of North America), the tide is low in the western end of the ocean (near the east coast of Asia).

However, there is one basic difference between tidal waves and the other shallow-water waves we have examined. The seas and swell we have dealt with so far are *free* waves. They are produced by the wind and are then free to travel under their own steam. In contrast, tidal waves are *forced*. They are created at fixed periods by the tide-generating forces and are continually acted upon by these forces as they propagate.

As a consequence, a standing tidal wave oscillates with the same period as the fluctuation in driving force. This period is not necessarily the same that the system would have if the standing wave were free. Consider, for example, a hypothetical equatorial canal connecting Gabon and Brazil. High tides occur at Gabon roughly every 12 hours and 25 minutes. This requires the crest of a tidal wave to travel from the west coast of Africa to Brazil and back (a distance of about 6700 km) in this period if the next high tide is to occur on time. Now, the average depth of the Atlantic Ocean is 3.9 km. Inserting this value into eq. 5.43, we see that a free wave would travel with a celerity of 196 m/s and require 19 hours to complete its circuit across the ocean. This time is known as the *natural period* for this system. In this case it is far longer than the period demanded by the tide-generating forces (12 hours, 25 minutes). Thus, the tractive forces must coerce the tidal wave to move faster than it otherwise would, with the result that the wave tends to lag behind the forces that produce it. A more precise examination (see Chapter 14) predicts that high tide would occur about six hours after the moon is overhead—the tidal wave would be 180° out of phase with the lunar tide-generating forces. The natural period of waves in the equatorial Pacific is even longer than in the Atlantic (42 hours), and again the tides will be out of phase with the moon. This explains the moonlit low tides in Central America.

Near the poles and in some small ocean basins the ocean is deep enough and the east-west distance between shores is small enough for the natural period for tidal waves to be shorter than the tidal period. In this case the tidal wave is in phase with the tide-generating forces—high tide occurs when the moon is overhead or directly underfoot. For an ocean with a depth of 3.9 km, the east-west dimension can be no more than 4380 km for the tides to be in phase with the tractive forces.

This switch from in-phase to out-of-phase tides is characteristic of forced oscillatory motion. We will deal with these motions in some detail when we discuss the dynamics of beam bending in Chapter 14. For the moment, a simple analogy may serve to illustrate how this phenomenon works. Tie a small weight to one end of a rubber band, and hold the other end in your hand. If you hold your hand still and tweak the weight, it will oscillate up and down with its natural period. Now, if you move your hand up and down slowly so that the period of your movement is longer than the natural period, the weight will move up and down in time with your hand. This is analogous to a standing tidal wave for which the natural period is less than the tidal period—the movement of the wave is able to keep up with the force that causes it. Moving your hand with a period shorter than the natural period, you find that the weight moves up when your hand moves down; the weight is now out of phase. This is analogous to tidal waves in the equatorial Atlantic or Pacific: the period of the generating force is less than the natural period of the system, and the wave moves out of phase with the force that generates it.

In these examples we have always confined our tidal waves to hypothetical canals. What happens if we knock down the canal walls? When we consider tidal waves that are not confined to canals, we immediately run into problems having to do with the centripetal force created by the earth as it rotates about its axis. To understand this situation we return to our experimental bathtub and move it onto a merry-go-round—the merry-go-round providing a handy analogy to the earth's rotation (Fig. 8.a). Before making any waves, we observe what happens to the water in the tub as the merry-go-round rotates. Because the water is not held in place, it tends to travel in a straight line until it is restrained by the walls of the tub. As a consequence, water piles up on

the outer wall, the wall farthest from the center of the merry-go-round. Water continues to pile up until the gradient in hydrostatic pressure is sufficient to provide the centripetal force. Thus at equilibrium the water in the tub is not level, but slopes upward from the inner wall of the tub to its outer wall. This equilibrium water level is the level that occurs in the absence of standing waves, and in future experiments we will regard this tilted—but—equilibrium—level as the still water level.

The same phenomenon happens to water in the oceans. As water on the earth's surface rotates about the earth's axis, it feels the same sort of centripetal acceleration as a rock on a string. This acceleration varies with latitude. Water at the equator is farthest from the earth's axis, moves at the greatest velocity, and therefore feels the greatest centripetal acceleration. Water at the poles rotates in place and experiences no centripetal acceleration at all. As a result of this latitudinally varying centripetal acceleration, water tends to move from the poles toward the equator, forming a bulge. This bulge continues to grow until the increased hydrostatic pressure at the equator is sufficient to offset the tendency of the water to move away from the poles. The height of this bulge is approximately the same all along the equator. Unlike the tidal bulges, this centripetal bulge moves with the earth and therefore does not create a tidal fluctuation. Tidal fluctuations in water level are measured relative to this equilibrium level.

We now return to the bathtub on the merry-go-round. We stand at one end of the tub, facing the direction in which the merry-go-round turns. To maintain the analogy between tub and ocean, we call our end of the tub the west end and the opposite end the east (Fig. 8.8a). Thus the wall of the tub farthest from the center of the merry-go-round is the south wall, and the inner wall is north. Now we try to create a standing wave. Consider the situation starting when the water in our end of the tub (the west) is high, and the water in the opposite end is low (Fig. 8.8b). At this point the water is stationary everywhere in the tub (relative to the tub, that is), and the equilibrium slope of the water upward from the north wall to the south wall is maintained. By analogy, the tide is high in the western end of the ocean and low in the eastern. As the wave crest in the west end subsides and the water in the east rises, water flows in the same direction as the merry-go-round turns. The horizontal (west-to-east) speed of the water is greatest at the nodal line, halfway along the tub, and is maximal when the water's surface is at its equilibrium level. However, because the water near the nodal line has

been speeded up, it feels a greater centripetal acceleration. In response to the force caused by this acceleration, water moves laterally across the tub, piling up higher along the central part of the southern wall. As the water along the southern wall rises (Fig. 8.8c), the water along the northern wall falls below its equilibrium level. By analogy, when the "tide" is at midlevel in the east and the west, it is high in the south and low in the north. As the west-east standing wave continues its motion, the water along the east wall rises, and along the west wall it falls until the water motion again comes to a halt. At this point the "tide" is high in the east and low in the west. Because there is no water movement at this time, the water at the nodal

Figure 8.8. Kelvin waves. (a) A bathtub on a merry-go-round is analogous to an ocean on a rotating earth. Imagine you are looking down on the earth from above the North Pole. (b) Surface elevations in the tub when the "tide" is high in the "west." (c) As the tide falls in the "west" and rises in the "east," it is high in the "south." (d) Surface elevations when the "tide" is high in the "east." (e) As the tide falls in the "east," it is high in the "north."

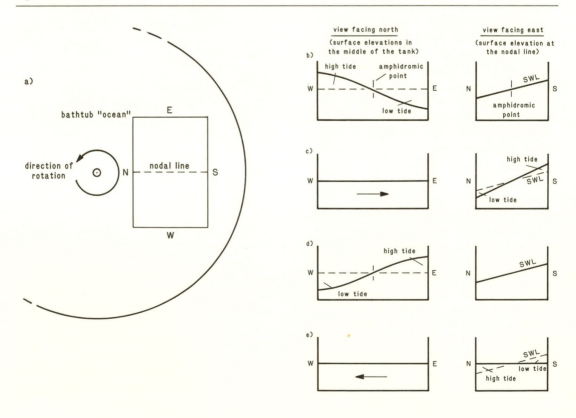

line is stationary and therefore is at its equilibrium level. This entire process is reversed as the wave crest falls in the east. Water moves from east to west, opposite the direction of the merry-go-round, and is therefore slowed down. As the water slows, it feels less centripetal acceleration and tends to return to the level it would maintain if the merry-go-round were not rotating. The water along the north wall rises above its equilibrium level, and along the south wall it falls. Thus, when the tide is falling in the east and rising in the west, the tide is high in the north. Through this entire process, the water at the center of the tub has not changed its level. This point where the water level remains constant is called the *amphidromic point*. Recalling what has happened during one wave period we see that the crest of the tidal wave moves in a counterclockwise fashion—from west to south to east to north. This sort of circularly rotating standing wave is called a *Kelvin wave*.

The analogous process occurs in the ocean. In the Northern Hemisphere the crest of the tidal wave travels in a counterclockwise path around the ocean basin. This helps explain why high tide occurs at Monterey before it occurs at the Golden Gate farther north. In the Southern Hemisphere the same physics applies, but the tidal Kelvin wave travels in a clockwise circle. The amplitude of the tidal Kelvin wave is greatest at the periphery of the ocean and is nearly zero in the center of the ocean, explaining at least in part why Hawaii has such a low tidal amplitude.

As if this were not complicated enough, there are a whole host of interactions between the tidal wave and the topography of the coastline. Just as shallow-water wind waves shoal, refract, and diffract as they approach shore, tidal waves are altered as they move toward a coast. These alterations can be spectacular. For instance, the tidal wave is funneled into the Bay of Fundy (between New Brunswick and Nova Scotia), producing tides that can exceed

12 m in amplitude. The behavior of shoaling tidal waves is strongly dependent on the shape of the ocean basin. For example, the Isthmus of Panama separates the Pacific Ocean (specifically the Bay of Panama) from the Caribbean. The difference in shape of these two basins leads to vastly different tidal amplitudes—about 0.3 m on the Caribbean side of the isthmus, and up to 7 m on the Pacific side. The interaction of forced tidal waves with the shape and size of the basin can be noticed even on a small scale. Because the tides within San Francisco Bay, for example, can vary by as much as an hour, we can stand on Nob Hill and simultaneously see expanses of water whose levels differ by a foot or more.

Predictive Methods: Harmonic Analysis

As interesting and as informative as dynamic theories of the tides can be, they are not yet developed to the point where they can be used to predict accurate tidal levels at arbitrary points. This does not mean, however, that tides cannot be predicted; accurate tidal predictions have been possible for well over a century. These predictions are based on a form of spectral analysis known as harmonic analysis.

We have seen how the equilibrium model for the tides provided an accurate explanation for the influence of the moon's orbital motion on the period between succeeding high tides. The same is true for other factors affecting the tides—the revolution of the earth around the sun, the interaction between the tractive forces of the sun and moon, the elliptical nature of the moon's and earth's orbits, the change in the plane of the moon's orbit with time, and so forth. Thus the equilibrium model provides an accurate method for predicting the *periods* of various tidal fluctuations. However, it does not accurately predict the amplitude of these fluctuations or their relative phases.

Figure 8.9. A mechanical device for predicting the tides. A_1, A_2, A_3—amplitudes of the component waves. ω_1, ω_2, ω_3—circular frequencies of the component waves (period = $2\pi/\omega$).

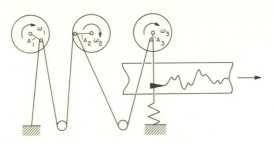

In practice, this poses no real problem. By applying the techniques of spectral analysis to the records of tidal fluctuations at a particular site, these phases and amplitudes can readily be determined. Once the phase, amplitude, and frequency are known for each factor influencing the tides, the Fourier series that sums all these effects can be run out into the future to predict the tides. It is on the basis of this sort of harmonic analysis that tide tables are calculated.

Before the advent of digital computers, a series of wonderful mechanical tide predictors were constructed (e.g., Fig. 8.9). These machines consisted of a group of wheels, each turning at a rate proportional to the frequency of the factor it was intended to monitor. A pulley attached to each wheel pulled on a string. By adjusting the distance from the center of the wheel to the pulley, and by starting all wheels at the appropriate phase, the distance moved by the end of the string reflected the summation of the various sine waves affecting the tide. A pen attached to the end of the string recorded the deflection on a moving chart, and from this the tidal predictions were read. Defant (1958) tells of a German machine that could take sixty-two different sine waves into account, requiring perhaps a day to predict a year's tides for a particular harbor.

The Reliability of Tide Tables

None of the approaches to tide prediction take into account the local weather, and a change in barometric pressure can directly affect the local sea level. For instance, a change from 29 to 30 inches of mercury can result in a change of about 0.14 m in the equilibrium sea level. Furthermore, storm winds acting via waves and the set-up can substantially raise the sea level in the surf zone. As a result of these unpredictable local effects, tide tables must be regarded as providing nothing more than a rough estimate of the actual fluctuations in sea level. If a precise measurement of the tides is essential, it must be made *in situ*. Methods for measuring the tidal fluctuation of sea level are discussed in Chapter 19.

CHAPTER 9

~ ~ ~ ~ ~ ~ ~ ~ ~ ~ ~

Benthic Boundary Layers

In our examination of linear wave theory (Chapter 5), we worked with a model of wave motion that, for simplicity, treated water as an "ideal," inviscid fluid. By ignoring the existence of viscosity we predicted that under waves in shallow water sizable horizontal velocities extend all the way to the sea bed (Fig. 4.6). For a viscous fluid constrained to satisfy the no-slip condition, this conclusion is clearly false. But, fortunately for this model, the actual consequences of viscosity are confined to areas very close to the substratum, and they have only minor effects on overall wave motion. Linear wave theory is still a useful predicter of wave-induced flow. There is one hitch, however. The goal in this exploration of wave motion is an understanding of the water movement around wave-swept organisms, and most of these organisms live on or very near the substratum where the effects of the viscosity, if any, will be most evident. As we will see, viscosity and the no-slip condition can be important on this fine spatial scale, and in this chapter we explore the nature of water motion near the substratum in what is known as the *benthic boundary layer*.

Laminar Boundary Layers

A Laminar Boundary Layer Model

Although the motion near the substratum under waves can be highly complex, the general idea of a boundary layer is simple: the no-slip condition and the water's viscosity act to slow the water as it moves over the seabed. The water nearest the bottom is slowed the most, and that far-

thest from the bottom is slowed the least, establishing a velocity gradient. The shape of this gradient affects the forces exerted on the bottom, and it changes with both the time and distance through which water moves. Our task here is to quantify the terms in this simple, qualitative concept.

The mathematics used to describe the velocity gradient near the bottom are not straightforward, and therefore do not serve well as a heuristic tool. It will be more edifying to attempt an understanding of the process through a mechanistic approach. We begin by taking a simple model and conjure up a hypothetical ocean whose bottom is flat and smooth (Fig. 9.1). For the purposes of this model, it is convenient to think of the ocean itself as being comprised of a stack of planar layers (or laminae) of water, a sort of fluid-dynamic deck of cards. The orderly movement of these layers is the laminar flow we referred to in Chapter 3. As in Chapter 5, we use the variable s to measure distance from the bottom. Each fluid layer has an infinitesimal thickness ds and extends to infinity in a horizontal plane. To keep track of the layers we number them starting at the bottom; the layer right next to the bottom is 1, the next layer up is 2, and so on (Fig. 9.1). The following explanation is mostly a matter of bookkeeping—what each layer is doing at each step in the evolution of a velocity gradient—and is best understood by continued reference to Figure 9.1.

Initially the water is everywhere stationary (Fig. 9.1a). In this condition the bottom layer of water has the same velocity as all other layers ($u = 0$). In other words,

Figure 9.1. A sequential model of boundary-layer generation (see text).

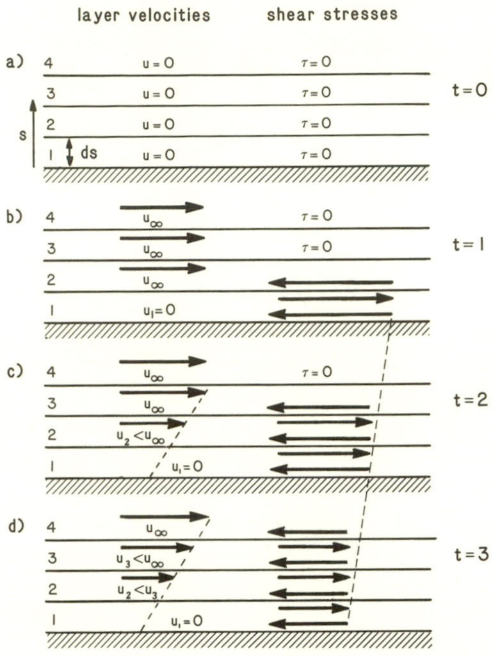

right. The no-slip condition dictates that the force on the bottom surface must win this tug-of-war, and layer 1 remains stationary.

The shearing force on the top surface of layer 1 is not without effect, however. From Newton's third law we know that the force tending to push the top of layer 1 to the right is accompanied by (and in this case caused by) an equal and opposite force tending to hold the bottom of layer 2 in place (Fig. 9.1b). As layer 2 slides to the right, it must overcome this opposing force acting on its bottom. As a result, work is done by the fluid as layers 1 and 2 slide past each other. This is the frictional work briefly mentioned in Chapter 3.

Aside from contributing to the frictional loss of mechanical energy from the fluid, the shear force acting on the bottom of layer 2 (which initially was moving with velocity u_∞ to the right) slows the layer down by an amount proportional to the shear force (Fig. 9.1c).

In turn, there are two consequences of layer 2 slowing down. First, the velocity gradient across layer 1 depends on the rate at which layer 2 slides. Because layer 2 has slowed, there is less of a velocity gradient across layer 1 and consequently less shear stress exerted on the bottom (examine the right-hand dashed line in Fig. 9.1). Second, because layer 2 has slowed down, it now travels at a different velocity from layer 3, and a velocity gradient is established across layer 2 (Fig. 9.1c). This new velocity gradient results in a shear force just as the velocity gradient across layer 1 did, with the same results: frictional work is done by the fluid as layer 3 slides past layer 2, and layer 2 causes layer 3 to slow down by an amount proportional to the shear gradient across layer 2 (Fig. 9.1d). In this manner the velocity gradient is extended up into the bulk of the fluid.

What is the profile of the velocity gradient at any time? As the velocity gradient extends layer by layer away from the bot-

no velocity gradient is present. We now give the ocean a velocity u_∞ to the right. When this happens, the fluid layer next to the bottom is caught in a bind. The seabed beneath it is not moving, while the fluid above it moves with velocity u_∞. As a result, the velocity of the fluid changes dramatically across the small thickness of this layer, creating a very steep velocity gradient, u_∞/ds (Fig. 9.1b). Recall from Chapter 3 that the product of a velocity gradient and the fluid's viscosity is a shear stress:

$$\tau = \frac{\text{force}}{\text{area}} = \mu\, du/ds. \qquad (3.8)$$

Because the velocity gradient across layer 1 is large, the consequent shear stress is large. Now, this shear stress acts over the area of layer 1, resulting in a pair of shear forces (Fig. 9.1b). At the bottom the shear force is directed to the left, tending to hold the layer in place, while the force on the top of the layer tends to slide it to the tom, the gradient any layer induces across the next layer out is equal to the difference in speed between the layer itself and the mainstream velocity (Fig. 9.1). For the slowly moving layers very near the substratum, this difference is nearly constant. Thus at any point in time the velocity near the bottom increases in a nearly linear fashion with distance away from the bottom. However, because each layer out from the substratum moves at a slightly faster velocity than the layer below, the velocity gradient induced in the next layer up is somewhat less than that in the layer itself. Beginning near the point where the velocity of a layer is half that of the mainstream, the velocity gradient (du/ds) begins to decrease substantially with an increase in the distance from the bottom, and the velocity gradient asymptotically approaches zero as the velocity of a layer approaches that of the mainstream.

From this simple model one can conclude that the results of fluid layers sliding past each other are threefold: (1) As time passes, layers of water farther and farther from the substratum are slowed down more and more; (2) As water farther from the bottom is slowed, the magnitude of the velocity gradient decreases at every level in the fluid, and the shear force on the bottom decreases; (3) Frictional work is done by the fluid at the expense of the fluid's mechanical energy.

These conclusions lead to two analogies that may be useful in understanding the behavior of velocity gradients. First, the velocity gradient may be thought of as a hypothetical "substance" that we can loosely refer to as *vorticity*.[1] A high value of du/ds is thus equivalent to a high concentration of vorticity. Immediately after flow is initiated, the vorticity is concentrated right at the bottom. However, as flow continues, the vorticity can be

thought of as diffusing outward from the substratum. The manner in which the velocity gradient extends away from the bottom is thus directly analogous to the manner in which heat or a chemical diffuses from a substratum into a stationary fluid, and the mathematical treatment of the velocity gradient involves the standard differential equations for diffusion (see Sabersky et al. 1971).

Alternatively, we can think of the growth of the velocity gradient in terms of the momentum of the fluid. As the velocity gradient extends into the mainstream, fluid slows down and loses momentum. Due to the no-slip condition, the velocity (and hence the momentum) of the fluid is zero at the substratum, and the substratum acts as a "momentum sink." Ultimately, of course, the mechanical energy from the change in momentum is degraded to heat via the action of viscosity, but one can think of the velocity gradient as being caused by the diffusion of momentum out of the mainstream and into the substratum. This analogy will be extremely useful in Chapter 11 when we examine the forces that moving fluids place on organisms.

In an attempt to convey an understanding of the mechanism of velocity-gradient formation, the process has been described here as being sequential: layer 1 affects layer 2, layer 2 then affects layer 3, etc. However, in a real fluid all layers affect each other simultaneously. While layer 1 is affecting layer 2, layer 2 is affecting layer 3, and so on through the entire fluid. Even when the fluid has just begun to move and the gradient is very steep near the substratum, layers far from the substratum are slowed by viscous forces, even if only infinitesimally so. This leads to a certain difficulty in defining the mainstream velocity. Only infinitely far from the bottom is the theoretical, true mainstream velocity maintained, this being the reason for the ∞ subscript on u_∞.

Now, the notion of a velocity gradient

---

[1] This is a crude use of a standard (and precisely defined) fluid-dynamics term. The cautious reader should consult a fluid-dynamics text for careful explanation of the concept of vorticity.

that extends to infinity, asymptotically approaching zero, may be mathematically and logically "nice," but it leads to practical problems. For example, when dealing with the pattern of flow past objects, it is desirable to be able to divide the fluid into discrete regions, one where viscosity is important, and one where the fluid behaves as if it were inviscid (Chapter 11). With a velocity gradient that smoothly approaches zero, there are no clear landmarks (or watermarks) as to where this dividing line should be drawn, and one must resort to a practical solution. Even though the velocity gradient never reaches zero, at some distance from the substratum the gradient becomes so small that for all pratical purposes the effects of viscosity are negligible. The exact distance is somewhat arbitrary and depends on the practical purpose one has in mind. Several such distances have been used (see Schlichting 1979 for a thorough discussion), but the one most commonly used for unidirectional flows in the biological literature is a distance from the substratum at which the local velocity is equal to 99% of u_∞. Farther away from the substratum the gradient is so small that shear forces are deemed to be negligible, and the fluid behaves essentially as if it were inviscid. This leads to an important definition: the fluid contained in the layers between a stationary substratum (a boundary) and our arbitrarily chosen distance is defined to be the *boundary layer* (Fig. 9.2). The distance from the substratum to the point where $u = 0.99u_\infty$ is the *boundary layer thickness*, δ. Obviously, a different cutoff distance yields a different boundary-layer thickness.

The boundary-layer thickness is a convenient, one-number means of characterizing the boundary layer, but we must be careful not to let the concept draw attention away from the fluid motions. Remember that the boundary-layer thickness provides information about flow at only one point in the velocity gradient and says

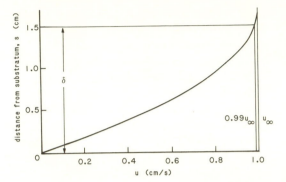

Figure 9.2. The laminar boundary layer. Water velocity as a function of distance from the substratum is calculated for a point 10 cm from the leading edge ($u_\infty = 1$ cm/s). δ is the boundary layer thickness.

nothing about the shape of the gradient itself.

Growth of a Laminar Boundary Layer

Consider fluid moving past a solid object such as a flat plate. Fluid upstream of the plate is not affected by a solid surface and therefore has no velocity gradient. However, as soon as the water arrives at the leading edge of the plate, it must satisfy the no-slip condition, and thereby a velocity gradient is established. Because it takes time for the gradient to diffuse out from the solid surface (and momentum to diffuse in), the thickness of the boundary layer grows as the water moves along the plate. For a *laminar boundary layer* (the sort described so far), the thickness, δ, is given as a function of distance, x, from the leading edge of a plate by the formula (Schlichting 1979)

$$\delta = 5\sqrt{\frac{x\mu}{\rho u_\infty}}. \qquad (9.1)$$

This expression is graphed in Figure 9.3 for one particular value of mainstream velocity.

Eq. 9.1 seems to imply that a laminar boundary layer can grow forever, but in any real-world situation there are limits to boundary-layer thickness. The boundary layers on discrete objects are limited by the size of the object itself—x can be no larger than the length of the object. At the downstream margin of the object (if not before), the boundary layer *separates* from the surface and moves downstream as a *wake*. The process of boundary-layer separation will be discussed in more detail in Chapter 11 when we examine the hydrodynamic forces exerted on plants and animals. For continuous boundaries such as the sea bed, the ultimate thickness of the boundary layer is limited by the water's depth. However, before a laminar boundary layer can fill the entire water column, the flow in the velocity gradient is likely to become turbulent, a phenomenon we will return to shortly.

Figure 9.3. Growth of a laminar boundary layer—δ as a function of distance from the leading edge. Calculated using eq. 9.1.

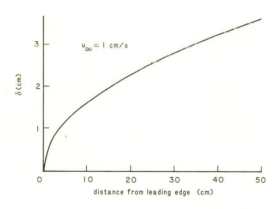

Boundary Layers in Oscillatory Laminar Flow

So far we have only considered boundary layers in unidirectional, steady flow, a situation that is hardly typical of the wave-swept environment. In water of intermediate and shallow depths ($d < L/2$), linear wave theory predicts that the horizontal velocity at the bottom is a harmonic oscillation along the direction of wave propagation:

$$u = \frac{\pi H}{T} \cos(kx - \omega t) \frac{1}{\sinh(kd)}. \quad (5.31)$$

Rather than moving at the constant velocity assumed by our simple boundary-layer model, the water gradually accelerates and then decelerates. The same principles apply to boundary layers in oscillating flow, but our simple approach does not readily yield a quantitative answer. In dealing with this problem, we rely on a solution devised by the same G. G. Stokes responsible for finite amplitude wave theory (Schlichting 1979, pp. 93–94).

Stokes considered the problem of a solid, flat boundary oscillating in an infinite sea of stationary fluid. The boundary has a velocity

$$u = u_0 \cos(\omega t). \quad (9.2)$$

The problem is to calculate the velocity of fluid at a distance s away from the boundary. It is assumed that the no-slip condition holds, so that the velocity of the water at $s = 0$ is the same as that of the boundary.

Stokes's solution to this problem shows that the velocity at time t and distance s from the plate is

$$u(s, t) = u_0 \exp\left\{-s \sqrt{\frac{\pi}{vT}}\right\} \cos\left\{\omega t - s \sqrt{\frac{\pi}{vT}}\right\}, \quad (9.3)$$

where v is the kinematic viscosity of water.

This solution is in terms of a moving boundary and a stationary sea, whereas we would like to know the velocity gradient for a stationary boundary (the sea bed) and an oscillating sea. However, by moving our frame of reference back and forth in time with the oscillating boundary, the boundary can be made to appear stationary. When we move in this fashion, the inertially stationary sea seems to move with a velocity equal in magnitude to our motion but opposite in direction:

apparent ocean velocity, $u = -u_0 \cos(\omega t)$. (9.4)

By adding this apparent velocity to the velocity caused by the presence of the boundary, we arrive at the velocity for an oscillating sea with a stationary bottom. In this new frame of reference, eq. 9.3 becomes

$$u(s, t) = \underbrace{u_0 \exp\left\{-s\sqrt{\frac{\pi}{vT}}\right\} \cos\left\{\omega t - s\sqrt{\frac{\pi}{vT}}\right\}}_{\text{difference from mainstream}}$$

$$- u_0 \cos(\omega t). \qquad (9.5)$$

When $s = 0$, eq. 9.5 simplifies to

$$u(0, t) = u_0 \exp(-0) \cos(\omega t) - u_0 \cos(\omega t) = 0.$$

Thus the velocity at the bottom is zero in our reference frame, as required for the no-slip condition. When s is large, $\exp -(s\sqrt{\pi/vT})$ approaches zero. In this case, eq. 9.5 reduces to

$$u(s, t) = -u_0 \cos(\omega t), \quad s \text{ large.}$$

Far from the plate the water oscillates, as expected.

So far these results are exactly what might be expected based on the reasoning given earlier. There is, however, one strange fillip. Note the term $s\sqrt{\pi/vT}$ in the cosine expression for velocity in eq. 9.5. The presence of this term introduces a phase shift in the velocity at distance s from the bottom relative to that of the oscillating ocean. For instance, where $s\sqrt{\pi/vT} = \pi$, the velocity in the boundary layer is exactly out of phase with the oscillations due to wave motion alone. This happens when

$$s = \pi\sqrt{\frac{vT}{\pi}}.$$

For a wave with 10 s period,

$$s = \frac{\pi}{560} = 5.6 \text{ mm.}$$

Thus, at a distance of 5.6 mm away from the bottom, water moves out of phase with the wave-induced motion.

The reduction in velocity caused by the plate decreases exponentially with increas-

ing distance from the plate due to the term $\exp -(s\sqrt{\pi/vT})$. However, the phase-shift term $s\sqrt{\pi/vT}$ in the argument of the cosine function makes it difficult to calculate a boundary-layer thickness. Due to the phase shift, the velocity near the plate can actually be higher than u_∞. For instance, when $\omega t = \pi/2$, $u_\infty = 0$, but water in the boundary layer has a nonzero velocity. Jonsson (1980) suggests an alternative definition for the thickness of this type of oscillating boundary layer:

$$\delta = \frac{\pi}{2}\sqrt{\frac{vT}{\pi}}. \qquad (9.6)$$

At this distance from the plate the velocity in the boundary layer is $\pi/2$ out of phase with the wave-induced velocity. Using this definition, $\delta = 2.8$ mm for a wave with a 10 s period, and smaller still for waves with shorter periods. It is clear that the effects of viscosity in wave-induced laminar flows are confined to a region very near the seabed.

The velocity gradient at any point can be calculated from the derivative of u with respect to s:

$$\frac{du}{ds} = u_0 \exp\left\{-s\sqrt{\frac{\pi}{vT}}\right\}\sqrt{\frac{\pi}{vT}}\left[\sin\left\{\omega t - s\sqrt{\frac{\pi}{vT}}\right\}\right.$$

$$\left. - \cos\left\{\omega t - s\sqrt{\frac{\pi}{vT}}\right\}\right]. \qquad (9.7)$$

The gradient in velocity is greatest at the bottom ($s = 0$) and decreases exponentially away from the sea bed.

When $\omega t - s\sqrt{\pi/vT} = \pi/4$, $5\pi/4$, etc., $\sin(\omega t - s\sqrt{\pi/vT}) - \cos(\omega t - s\sqrt{\pi/vT}) = 0$ and the gradient is zero. Now, $du/ds = 0$ at the point of maximum velocity. So eq. 9.7 allows us to locate the distance from the plate at which the maximum velocity in the boundary layer occurs. For instance, when $t = 2.5$ s, the maximum velocity occurs at a distance of 1.40 mm from the sea bed for a wave with a 10 s period (Fig. 9.4; eq. 9.5).

The primary biological consequence of these calculations results from the small thickness of the boundary layer. An organism extending only a centimeter or two into the water column under a wave sees virtually the same sort of flow it would see if the bottom were not there at all. However, this velocity is out of phase with the mainstream flow. The possible biological consequences of this phase difference have not been explored.

As with all theoretical predictions, these should be used with care when applied to the real world. Stokes's derivation assumes a smooth, flat bottom, a condition not typical of the ocean floor. The effect of bottom roughness on these findings is difficult to predict. Furthermore, Stokes's derivation assumes that the plane of the plate is parallel to the direction of water movement. If the plate is angled up or down by as little as a few degrees, the resulting pressure gradient may substantially effect the boundary layer.

Longuet-Higgins (1953) carried Stokes's approach to wave motion to a second approximation and concluded that water movement in the benthic boundary layer is not strictly oscillatory. Surprisingly, there is net transport in the direction of wave motion. The speed of transport varies with distance from the substratum:

$$\bar{u} = \frac{2\pi H^2}{16T} \frac{k}{\sinh^2(kd)} \left[5 - 8 \exp\left\{\frac{-s}{\delta}\right\} \cos\left\{\frac{s}{\delta}\right\} \right.$$
$$\left. + 3 \exp\left\{\frac{2s}{\delta}\right\} \right], \qquad (9.8)$$

where $\delta = \sqrt{\nu T/\pi}$, an alternative estimate of the boundary-layer thickness. This transport reaches a maximum velocity when $s = 2.3\delta$:

$$\bar{u} = 0.344 \frac{2\pi H^2}{T} \frac{k}{\sinh^2(kd)}. \qquad (9.9)$$

For example, when $H = 1$ m, $T = 10$ s ($L = 150$ m), and $d = 10$ m, $\bar{u}_{max} = 0.05$ m/s in the direction of wave propagation.

Although Longuet-Higgins's derivation assumes a constant depth, it is robust enough to apply to gently sloping beaches

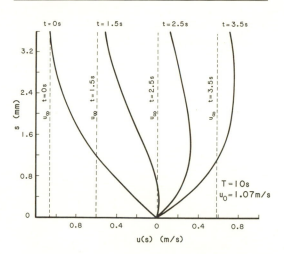

Figure 9.4. The velocity gradient in laminar, oscillatory flow (eq. 9.5).

as well. This shoreward flow in the benthic boundary layer has been observed in wave tanks (Longuet-Higgins 1953) and may be important in transporting planktonic larvae to the surf zone.

Turbulent Boundary Layers

Reynolds Stresses

In our discussion so far we have assumed that viscosity is the only mechanism by which a moving fluid interacts with a solid boundary to produce shear stresses. However, in many, if not most, wave-induced flows viscosity accounts for only a small fraction of the total shear stress. To understand the origin of the remainder, we examine the effects of turbulent velocity fluctuations.

Consider the flow in a small area of the fluid having dimensions dy and dz (Fig. 9.5a). Recall from Chapters 3 and 6 that we can examine the force acting on this area by examining the flux of momentum across it. The flux of momentum through face $dy\,dz$ is

$$\text{momentum flux} = (\rho u\,dy\,dz)u = \rho u^2\,dy\,dz. \qquad (9.10)$$

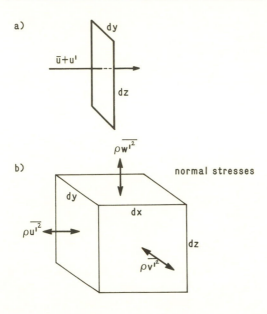

fact that $\overline{u'}$ is zero. Thus the presence of fluctuations in the fluid velocity lead to a stress $(\rho\overline{u'^2})$ in addition to the one we find in steady flow $(\rho\overline{u}^2)$. This additional stress is one example of a *Reynolds stress*.

The particular Reynolds stress we have calculated here is but one of three normal, compressive stresses that act on the fluid in turbulent flow. If there are fluctuations in velocity along the y and z-axes, the same procedure can be applied and the Reynolds stresses for these directions can be calculated (Fig. 9.5b).

In addition to these compressive stresses there are three Reynolds shear stresses. These are calculated as follows. Consider a flow in the x-direction and a small area lying parallel to the flow, $dx\,dy$ (Fig. 9.6a). The shear stress acting on this area is equal to the rate at which momentum in the direction of flow passes through the area parallel to flow. To convince yourself of this, recall the laminar boundary-layer model discussed earlier in this chapter in which the movement of each fluid lamina is retarded by the viscous interaction with the next lamina nearer the solid substra-

Dividing by the area of the face ($dy\,dz$), we end up with a term for the momentum flux per area, ρu^2. To this point we have tacitly assumed that u is constant. If instead u fluctuates about a constant mean velocity, we need to reevaluate the momentum flux. We separate the overall instantaneous value of u into the mean velocity, $\overline{u}$, and the velocity fluctuation, u'. Since the average of $(\overline{u} + u')$ equals $\overline{u}$, the average velocity fluctuation must be zero. Given this new form for u, we rewrite the equation for momentum flux:

$$\text{momentum flux} = \rho(\overline{u} + u')^2\,dy\,dz$$
$$= \rho(\overline{u}^2 + 2\overline{u}u' + u'^2)\,dy\,dz.$$

$$(9.11)$$

If we average this instantaneous momentum flux over a period of time, we see that the average stress (force/area) acting on the face $dy\,dz$ is

$$\text{stress} = \rho(\overline{u}^2 + \overline{u'^2}) = \rho\overline{u}^2 + \rho\overline{u'^2}, \quad (9.12)$$

where the bar designates the time average, and we have taken into account the

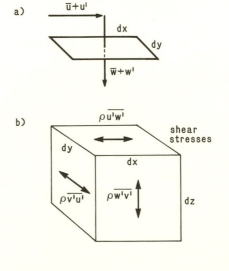

tum, thereby causing a shear stress. In the process, each layer of fluid gives up some of its momentum to layers nearer the substratum, the net effect being that the substratum acts as a sink for momentum. The ultimate sink in this case is the transduction of momentum into heat via the mechanism of viscosity, but it is easy to think of the momentum flux as being directed toward the substratum, perpendicular to the direction of flow. By analogy, the flux of x-component momentum across a surface lying in the x-y plane results in a shear stress. This momentum flux is simply the product of the rate at which mass flows in the z-direction (due to velocity w) and the velocity in the x-direction. If the velocities in both the x and z-directions fluctuate, this expression can be written as

$$\text{momentum flux} = \rho(\bar{w} + w')\, dx\, dy\, (\bar{u} + u')$$
$$= \rho(\overline{uw} + u'\bar{w} + \bar{u}w'$$
$$+\, u'w')\, dx\, dy$$
$$\frac{\text{momentum flux}}{\text{area}} = \rho(\overline{uw} + u'\bar{w} + \bar{u}w' + u'w').$$

$$(9.13)$$

Taking the temporal average of this expression, and again noting that $\bar{u'} = \bar{w'} = 0$, we arrive at an expression for the shear stress:

$$\text{shear stress} = \rho(\overline{uw}) + \rho(\overline{u'w'}). \quad (9.14)$$

If the flow were strictly in the x-direction ($\bar{w} = 0$) and not fluctuating ($w' = 0$), this shear stress would be zero and the only shear stress present would be that due to viscosity alone (not accounted for in this derivation). However, if the temporal average of $u'w'$ were nonzero, there would be a shear stress in addition to that due to viscosity. To obtain the temporal average of $u'w'$, we would measure u' and w' at N discrete times and average their cross-products, yielding a term describing the covariance of these two velocities:

$$\overline{u'w'} = \frac{1}{N}\sum_{i=1}^{N} u'_i w'_i. \quad (9.15)$$

If u' and w' vary randomly and independently, these two terms are of opposite sign half the time and $u'w'$ is negative. For the remaining time the two have similar sign and $u'w'$ is positive. The average of these half-negative, half-positive cross-products is zero. Thus, if u' and w' vary independently, $\overline{u'w'}$ does not contribute to the shear stress. Similarly, if every time $u' \neq 0$, $w' = 0$ (or vice versa), the average equals zero. However, this sort of very tight coupling between u' and w' does not seem likely. Finally, if the fluctuations in u and w are correlated such that, on average, u' is greater than zero when w' is also greater than zero (and vice versa), the temporal average has a positive value and the shear stress is augmented. This additional shear stress due to velocity fluctuations ($\rho\overline{u'w'}$) is a Reynolds shear stress, τ_R:

$$\text{Reynolds stress, } \tau_R = \rho\overline{u'w'}. \quad (9.16)$$

In a turbulent boundary layer the presence of a nearby solid surface insures that $\bar{w} = 0$, but the correlated fluctuations in velocity due to turbulent eddies (w' and u') result in a Reynolds stress. For analogous reasons, there are two other Reynolds shear stresses, $\rho\overline{u'v'}$ and $\rho\overline{v'w'}$ (Fig. 9.6b), but in the following discussion we deal only with $\rho\overline{u'w'}$.

The Reynolds shear stress is generally much larger than the viscous shear stress. Take, for example, a mainstream flow of 1 m/s over a flat plate with a boundary layer 1 cm thick. If the flow is laminar, the average gradient in velocity (du/dz) across the boundary layer is 100/s, resulting in a viscous shear stress ($\mu\, du/dz$) of about 0.1 Pa. If the flow is turbulent with velocity fluctuations in each direction equal to about 10% of the mean mainstream velocity, the Reynolds shear stress is of the order of 0.001 ρu^2 = 1 Pa, ten times that due to viscosity.[2] In other words, turbulent

---

[2] Note that a maximum fluctuation $u'_{max} = w'_{max} = 0.1\,\bar{u}$ does not mean that $u'w' = 0.01\,\bar{u}$. For every instance when $u' = w' = 0.1\,\bar{u}$, there is likely to be an instance where $w' = 0$ when $u' = u'_{max}$. Thus the cross-product $u'w'$ is on the average much smaller than $u'_{max}w'_{max}$.

mixing is much better than molecular mixing (diffusion) in transporting momentum. Because it is the transport of momentum that determines the velocity profile in a boundary layer, we can say at the outset that turbulent boundary layers are likely to behave differently than laminar ones.

The Reynolds shear stress can be used to define a characteristic velocity of a turbulent flow. This is the *friction velocity* or *shear velocity* defined as

$$u_* = \sqrt{\frac{\tau_{R,\text{bottom}}}{\rho}}, \qquad (9.17)$$

where $\tau_{R,\text{bottom}}$ is the Reynolds shear stress acting on the bottom. Thus, the term u_* (pronounced u-star) is a measure of the magnitude and correlation of turbulent fluctuations in velocity near the substratum. The utility of this expression as an index for turbulent flow will soon become apparent.

The Turbulent Transition

Under what circumstances is the boundary layer flow turbulent? The orderly flow in a laminar boundary layer breaks down into turbulence when the damping effect of viscosity can no longer restrain the tendency for the fluid's inertia to form eddies. This transition occurs either when the mainstream flow is sufficiently fast or when the laminar boundary layer is thick enough so that fluid along its outer edge is not sufficiently under the steadying influence of the solid boundary. This interplay between inertial and viscous forces suggests that Reynolds numbers (Chapter 3) are useful tools for describing the transition from a laminar to a turbulent boundary layer.

First, we can characterize the transition to turbulence in terms of the thickness of the boundary layer, which is a function of the distance from a leading edge. In this case we characterize the flow in terms of the *local Reynolds number*, Re_x:

$$\text{Re}_x = u_\infty x/\nu, \qquad (9.18)$$

where x is the distance from the leading edge of a flat plate. This is useful when examining flow past an object; but for the ocean bottom, which cannot be said to have a leading edge, it is often useful to express x as the distance traveled in time t by fluid moving with average velocity u_∞.

$$\text{Re}_x = u_\infty^2 t/\nu. \qquad (9.19)$$

Empirical experiments have shown that a laminar boundary layer on a smooth plate becomes turbulent when $\text{Re}_x = 3.5 \cdot 10^5$ to 10^6. Thus for flows that have continued at a constant rate for 5 s, a situation roughly analogous to the flow in one direction associated with surface waves of period 10 s, the velocity need only exceed 0.26–0.45 m/s before the boundary layer becomes turbulent.

The presence of roughness on the substratum lowers the critical velocity by an amount dependent on the size of the roughness elements, and for rough bottoms an alternative rule of thumb can be drawn for the presence of turbulence in the boundary layer. This rule is based on the *roughness Reynolds number*, Re_*:

$$\text{Re}_* = \frac{u_* D}{\nu}, \qquad (9.20)$$

where u_* is the friction velocity defined above and D is the height of the roughness elements (the diameter of pebbles forming the bottom, for instance). In many boundary layer flows, u_* is on the order of $0.1 u_\infty$. Thus,

$$\text{Re}_* \simeq \frac{0.1 u_\infty D}{\nu}.$$

The onset of turbulence in the outer reaches of the boundary layer occurs at a roughness Reynolds number of about 6. For roughness elements 0.5–1 cm high (barnacles, for instance), this corresponds to a mainstream velocity of 6–12 mm/s. Thus, unless the bottom is very smooth and the wave-induced flow near the sea floor is very slow (in other words, unless the wave height is small or the depth approaches one-half wave length), the benthic boundary layer is at least beginning to be turbulent. The boundary layer flow is fully turbulent (the turbulence extends all the way to the substratum) when $Re_* > 75-100$ (Schlichting 1979; Nowell and Jumars 1984). For roughness elements 0.5–1 cm high, this corresponds to velocities above about 0.1–0.2 m/s.

The conditions required for laminar boundary layers are clearly violated once a wave has moved far enough inshore and has broken. In moving inshore, the horizontal velocity increases to the point where, at breaking, it is close to the wave celerity, generally several m/s. The process of breaking, as we have seen, also introduces turbulence into the flow, and the flow patterns near the bottom can be expected to be highly chaotic. What is the velocity profile under these circumstances?

Turbulent Boundary Layers

At first the concept of a turbulent boundary layer seems to be a contradiction in terms. As we have encountered it so far, the idea of a boundary layer implies an orderly gradient of velocities extending away from a solid surface. In contrast, turbulence implies a chaotic distribution of flow speeds and directions. How can the two be reconciled? At any instant in time they cannot. Imagine that we could somehow take a snapshot of the velocities near a flat sea bed as a turbulent mass of water flowed over it. If velocity is measured as a function of distance away from the bottom, no simple pattern is found. The water immediately in contact with the bottom must be stationary due to the no-slip condition, but otherwise the flow changes frequently in a seemingly unpredictable fashion. However, it is possible to perceive some mean order in turbulent flows by averaging the velocities at one point in space over a period of time. If we take a large number of snapshots and average the velocities at distances away from the sea bed, on average the velocity is found

to increase with distance away from the substratum. It is this *time-averaged velocity gradient* that constitutes a turbulent boundary layer. The turbulent boundary layer is defined in exactly the same manner as a laminar boundary layer, except that all the measurements are made on the time-averaged gradient. Thus, the thickness of a turbulent boundary layer is the distance away from a solid surface where the average velocity is 99% of the average mainstream velocity.

As with laminar boundary layers, there are two questions that need to be answered regarding turbulent boundary layers: How does the gradient change through time, and at one time how does the gradient change with distance from the sea floor?

Growth of a Turbulent Boundary Layer

In unidirectional flow a turbulent boundary layer grows in thickness in a manner analogous to that of a laminar boundary layer. Schlichting (1979) gives a formula for the growth of a turbulent boundary layer as fluid moves along a flat plate:

$$\delta = 0.37x \left\{ \frac{\bar{u}x}{v} \right\}^{-0.2} \quad (9.21)$$

This gives δ in terms of x, the distance from the leading edge (Fig. 9.7). For continuous boundaries (such as the bottom) this expression can be restated in terms of time after the initiation of flow. Substituting $\bar{u}t$ for x in eq. 9.12 and working through the algebra, we see that

$$\delta = 0.37\bar{u}^{0.6}t^{0.8}v^{0.2}. \quad (9.22)$$

For example, a turbulent bore with a total depth of water $(d + H) = 2.5$ m beneath its crest is accompanied by a velocity of about 1.5 m/s at the bottom. Such a bore is a good model for a large broken wave running inshore. At one spot on the bottom, the shoreward flow caused by this bore continues for at most half the period of the waves; for waves with a 10 s period, no more than 5 s. Thus,

$$\delta = 0.37(1.5 \text{ m/s})^{0.6}(5 \text{ s})^{0.8}(10^{-6} \text{ m}^2/\text{s})^{0.2}$$
$$= 0.11 \text{ m}.$$

The water near the bottom is predicted not to reach 99% of the mainstream flow until the distance from the bottom is greater than 11 cm. This is much thicker than the equivalent laminar boundary layer, 0.01 m (from eq. 9.1). But before we start thinking that organisms up to 11 cm high can live under broken waves without experiencing appreciable flow, we must take a close look at the shape of the velocity gradient.

The Turbulent Velocity Profile

The velocity gradient in a turbulent boundary layer on a smooth substratum can be divided into three intergrading areas. As we might expect, these areas are delineated according to a Reynolds number that serves as an index of whether fluid motion is, or is not, controlled primarily by viscosity. This particular Reynolds number is usually expressed as the "dimensionless distance" s^+.[3]

$$s^+ = u_* s/\nu. \qquad (9.23)$$

The notion of a "dimensionless distance" may seem like a contradiction in terms, but thinking of this Reynolds number in this way makes it a useful tool for measuring distance from the substratum relative to dimensions of the boundary layer.

If the bottom is smooth, the fluid very near the substratum ($s^+ < 6$) is sufficiently under the steadying influence of the solid

[3] This Reynolds number is used primarily by fluid dynamicists (for whom "up" is in the positive y-direction) rather than physical oceanographers (who use s to denote vertical distance). Consequently, the value we are likely to see in the literature is symbolized by y^+. I thought it best here to continue the use of the oceanographical axes, and assign s^+ as an unconventional but easily transposed symbol.

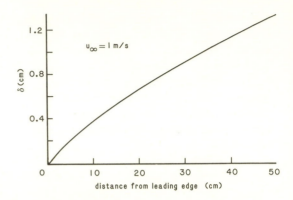

Figure 9.7. Growth of a turbulent boundary layer—boundary layer thickness as a function of distance from the leading edge. Calculated using eq. 9.21.

boundary and remains laminar. This *laminar* or *viscous sublayer* is usually very thin. For example, where $\bar{u}$ is 1 m/s, u_* is typically 0.1 m/s, implying that the laminar sublayer is only 60 μm thick. As we can see from Figure 9.8, the velocity gradient is very steep across the laminar sublayer, resulting in the imposition of a large shear stress on the substratum. Within the laminar sublayer the transport of momentum from the moving fluid is due solely to viscosity.

Figure 9.8. The turbulent boundary layer on a smooth plate. Although the turbulent boundary layer may be thicker than its laminar counterpart, high velocity is reached much nearer the substratum.

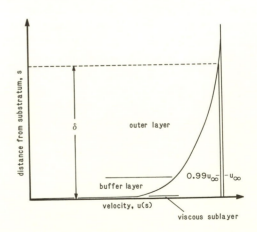

The area where $s^+ > 60$ is termed the *outer layer*. Here the flow is essentially independent of viscous effects and is characterized by the presence of relatively large turbulent eddies. As these eddies move about, they tend to uniformly distribute the fluid's momentum. Consequently, the time-averaged fluid velocity is maintained fairly evenly across the outer layer. This is just another way of saying that the change in velocity from the mainstream to the substratum must be confined largely to the viscous sublayer.

Between these two layers ($6 < s^+ < 60$) both viscous and inertial forces are important in the pattern of flow, resulting in what is called the *buffer layer* or *turbulence-generation layer*. It is here that small, energetic turbulent eddies are generated when kinetic energy associated with the mean velocity of the fluid is converted to kinetic energy associated with eddies. The Reynolds shear stresses are maximal in the buffer layer.

The turbulent boundary layer over a rough surface is qualitatively the same as that just described for a smooth surface, except that the viscous sublayer is missing. Each roughness element may have its own laminar boundary layer, but the combined effect of the wakes from multitudinous roughness elements disrupts any semblance of a general layer of laminar flow. This disruption of the viscous sublayer provides a mechanistic explanation for the division of surfaces into rough and smooth on the basis of Re_*. If the tops of the roughness elements are at $s^+ < 6$ (i.e., $Re_* < 6$), they do not poke out of the viscous sublayer, and the surface is considered smooth. If the tops of the roughness elements are at $s^+ > 6$, they extend into the buffer layer where they can affect turbulence generation, and the surface is effectively rough.

A dimensional analysis of turbulent boundary layers predicts that near a rough substratum the average velocity increases as the logarithm of distance from the substratum (Schlichting 1979; Middleton and Southard 1984). A typical velocity profile is shown in Figure 9.9. The universality of this type of profile has led to the widespread use of a particular equation for describing turbulent velocity gradients:

$$\bar{u}(s) = \frac{u_*}{\kappa} \ln \left\{ \frac{s - d_0}{s_0} \right\}, \qquad (9.24)$$

where $\bar{u}(s)$ is the time-averaged velocity parallel to the substratum at a distance s above the solid surface. The constant κ is known as von Karman's constant and has an empirically determined value of about 0.4. The term $\ln(s - d_0/s_0)$ adjust the equation for the presence of roughness elements. The *roughness parameter* or *roughness height*, s_0, is determined by plotting $\bar{u}(s)$ as a function of $\ln(s - d_0)$ (Fig. 9.10). An extrapolation of the regression line gives the height at which $\bar{u}(s)$ would go to zero if the logarithmic profile held true for heights less than D, where D is the height of the "peaks" of the roughness elements. This hypothetical height is s_0, usually equal to about 0.033 D (Schlichting 1979). Here d_0 is the *zero plane displacement*, in a sense the distance that the effective substratum has been lifted due to the presence of the roughness elements (Fig. 9.9). The zero plane displacement varies from one surface to the next and is measured through trial and error in the production of a graph like Figure 9.10; d_0 is chosen to give the best straight-line fit to the experimental data. For many rough surfaces d_0 is approximately 0.6 D. Note that though d_0 and s_0 have the dimensions of height, they are computational fictions with no physical significance.

The logarithmic profile cannot be expected to predict accurately $\bar{u}(s)$ for $s < D$.

In situations where d_0 is small compared to the thickness of the boundary layer (e.g., for sandy or muddy bottoms), eq. 9.24 reduces to

$$\bar{u}(s) = \frac{u_*}{\kappa} \ln \frac{s}{s_0}. \qquad (9.25)$$

The is the form used by Jumars and Nowell (1984), Eckman (1982), and other researchers of soft bottoms.

Eq. 9.24 is theoretically justifiable only in the portion of the boundary layer nearest the substratum ($s < 0.2\,\delta$), but often gives a good approximation to the true profile all the way to the outer edge of the boundary layer. More precise (and theoretically "nice") expressions for the velocity profile are available for the outer layer (see Schlichting 1979; Middleton and Southard 1984), but we will not deal with them here. Eqs. 9.24 and 9.25 have been used with success by meteorologists in describing the flow gradient over a variety of surfaces and flow conditions ranging from forests during hurricanes to corn fields during gentle breezes. These equations may be particularly useful in characterizing the velocity gradient over an ocean floor subjected to tidal currents or the motion at depth induced by long-period waves. For instance, Eckman et al. (1981) used eq. 9.25 to characterize the flow around worm tubes on a muddy-bottomed tidal flat. The flow patterns around these tubes has a substantial effect on the local bottom topography, and thereby on the community structure of the organisms living in the mud. Nowell and Jumars (1984) and Jumars and Nowell (1984) review the existing literature on this subject.

There are potential problems, however, in using a logarithmic profile to characterize velocity gradients in the shallow, wave-swept environment. First, most studies dealing with turbulent-boundary-layer characteristics have been conducted on "fully developed," "equilibrium" boundary layers. These are boundary layers that are no longer increasing in thickness. However, a boundary layer is fully developed only after water has flowed a considerable distance over a more-or-less uniformly rough surface. A good rule of thumb is

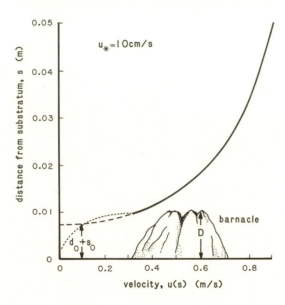

Figure 9.9. The turbulent boundary layer on a rough surface. Velocity increases logarithmically with distance from the substratum (eq. 9.24). The dashed line shows the predicted velocity gradient (which is not expected to hold true for $s < D$). The dotted line approximates the actual gradient for $s < D$. An acorn barnacle is shown as an example of a biological roughness element.

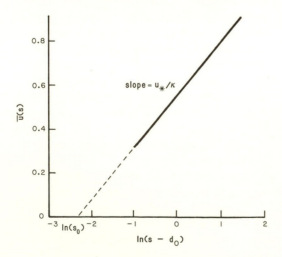

Figure 9.10. By extrapolating the line of $ln(s - d_0)$ vs. mean velocity ($\bar{u}(s)$), $ln(s_0)$ can be calculated.

that equilibrium is reached 50 δ from a change in surface roughness or a leading edge, where δ is the equilibrium boundary-layer thickness (Nowell and Jumars 1984). If the boundary-layer depth is limited by the water depth d (as might happen in a bore during run-up), δ in this rule of thumb is replaced by d. Thus, water must flow for at least 5 m over a uniform substratum to produce a 10 cm-thick equilibrium turbulent boundary layer. On many wave-swept shores it is unlikely that areas of uniform bottom rugosity extend for 5 m in any direction, and fully developed boundary layers may not exist in this environment.

We must be careful when we apply information regarding equilibrium boundary layers to nonequilibrium situations. For instance, Antonia and Luxton (1971, 1972) found that nonequilibrium boundary layers consist of two parts: an outer layer with a gentle gradient in average velocity, and an inner layer next to the substratum with a very steep velocity gradient. This inner layer should not be confused with a viscous sublayer; it is intensely turbulent. The magnitude of these gradients rapidly changes with distance away from a change in roughness. Even in some cases where the boundary layer is at equilibrium, the logarithmic profile is not an adequate description of the velocity gradient. Nowell and Church (1979) showed that two logarithmic profiles were needed to describe flow over densely packed roughness elements. One profile describes the gradient in the vicinity of the peaks of roughness where the flow is strongly affected by the wakes of individual roughness elements. Farther from the substratum, the standard logarithmic profile takes over.

The potential problems involved in characterizing boundary layers in the wave-swept environment are actually more intricate. Here we have examined only turbulent boundary layers in unidirectional flow. These findings are qualitatively correct for oscillatory turbulent boundary layers, but their quantitative predictions must be used with care. A quantitative treatment of oscillatory, turbulent boundary layers and their interaction with currents is complex, and we will not explore it here. Sleath (1983) and Grant and Madsen (1986) provide reviews of the recent findings on this subject.

The complexities involved in characterizing the benthic velocity gradient in the surf zone should be placed in perspective. By concentrating on the average gradient (as presented by the logarithmic profile, for instance) it is easy to lose sight of the instantaneous aspects of flow. At any one time the fluid motion near the wall can be quite different than the average predicted by the logarithmic (or any other) profile. For instance, recent advances in flow-visualization techniques have shown that on smooth surfaces large, energetic eddies (called *sweeps*) periodically dip down to touch the substratum, with two important results. First, the sweep causes a *burst* of turbulence production, and the resulting eddies are ejected outward into the flow. Second, and more important for benthic organisms, the sweep imposes velocities momentarily equal to or even exceeding the mean mainstream velocity at distances only a fraction of a millimeter away from the substratum. If an organism responds to the instantaneous fluid velocity in the force it feels, these periodic high velocities may have drastic consequences. For example, it has been shown under a variety of conditions that T_{sweep}, the period between the imposition of sweeps at a point on the substratum, is

$$T_{sweep} = 6\delta/u_\infty, \qquad (9.26)$$

although the value of 6 may vary between 2.5 and 10 (Cantwell 1981). For values of δ and u_∞ typical of the wave-swept environment ($\delta = 0.1$ m, $u_\infty = 1$ m/s), the time between sweeps is likely to be less than a second. Now, a planktonic larva attempting to settle on the substratum requires some time in which to adhere firmly. If this adhesion time is greater than the period between sweeps, the larva may not be able to settle. There is circumstantial evidence that this is the case on wave-swept shores. For instance, barnacles cannot (or at least do not) settle on smooth plates (for a review, see Crisp 1974). In a case such as this where it is the instantaneous rather than the average property of the velocity gradient that is biologically important, the concept of a turbulent boundary layer is likely to have little mechanical significance.

Consequently, in the turbulent flow present under breaking and broken waves, the water velocities caused by the wave can (for many mechanical purposes) be considered to extend right down to the bottom, leaving essentially no room above the substratum ($s > D$) in which organisms can hide. Perhaps this is the reason that many suspension feeders (e.g., bivalves and tunicates) have evolved internal filter apparatuses; external filters can be subjected to dangerously rapid flows (see Chapters 11, 16, and 17).

This is not to say that organisms cannot hide *in* the substratum. Below the peaks of the roughness elements the local flow is strongly controlled by the presence of individual wakes. In a sense, each roughness element sits in the wake of other elements and is protected from the flow. Nowell and Church (1979) showed that when roughness elements (in their studies, standard Lego blocks) covered one-twelfth of the substratum or more, the flow tended to "skim" over the tops of the roughness rather than extend down into the interstices. Most intertidal and shallow subtidal rocky shores are densely covered with rugose plants and animals, and it is likely that these substrata are characterised by this sort of skimming flow. The roughness of the substratum may thus serve as a refuge from high water velocities. For instance, Emson and Faller-Fritsch (1976) found that littorine snails tend to be limited to a size smaller than the roughness elements of the substratum on which they live. Presumably, by maintaining this small size and hiding in the pits and interstices of the substratum, these snails avoid the full brunt of the water flow. Sweeps can probably reach even between roughness elements, but the depth of their penetration and the period between the imposition of "deep sweeps" have not been examined for the rugose substrata typical of the wave-swept environment.

CHAPTER 10

~ ~ ~ ~ ~ ~ ~ ~ ~ ~ ~

Turbulence and Mixing

"Without turbulence there would be no life on earth."
—H. J. Lugt (1983)

Although moving fluids and organisms often form an adversary relationship, many aspects of biology actually require the movement of water. For instance, passive filter feeders rely on water currents to bring food within their grasp. Planktonic larvae depend on water flow to disperse any substantial distance, and, once dispersed, require flow to return to the substratum. The swash of waves prevents organisms from desiccating in the high intertidal zones, and all aquatic organisms need some water movement to carry away their waste and to bring in new supplies of oxygen or carbon dioxide.

Although water movement can be a "good thing," not all forms suffice. Consider the case of an algal frond in a unidirectional flow. If the surface of the frond is smooth and planar and the water velocity is sufficiently slow, the movement of water past the frond is laminar. As we have seen (Chapter 9), this sort of laminar flow results in the formation of a boundary layer. Each infinitesimal lamina of fluid slides relative to the layers above and below it, but does not move up or down. Apart from molecular diffusion there is no transfer of fluid between laminae. This lack of bulk mixing poses a problem for the frond that requires a constant influx of carbon dioxide (in the form of bicarbonate) and inorganic nutrients if it is to carry out photosynthesis at a rate sufficient for growth. Here we assume that the availability of bicarbonate is the limiting factor in photosynthesis, but the same arguments apply to inorganic nutrients. Bicarbonate moves into the frond by diffusion from the fluid layers adjacent to the frond's surface. The surface at the leading edge, where the boundary layer is very thin, is continuously exposed to new volumes of water and can be adequately supplied with bicarbonate. However, as fluid moves away from the leading edge its supply of bicarbonate is absorbed by the frond and is soon depleted. The only way in which bicarbonate can reach the surface of the frond at points away from the leading edge is to diffuse across the entire boundary layer from the bicarbonate-rich mainstream flow. Because the boundary layer increases in thickness with distance away from the leading edge, this diffusion distance increases, and the rate at which bicarbonate can be supplied is curtailed. Molecular diffusion is an extremely inefficient transport mechanism. The time required for a molecule to diffuse a distance x is $x^2/2D_m$, where D_m is the diffusion coefficient. For a bicarbonate ion in water, D_m is on the order of 10^{-9} m^2/s. Thus it would take a bicarbonate ion roughly 500 s to diffuse across a boundary layer only 1 mm thick. The outcome of this scenario is an algal frond seriously lacking in its ability to remove sufficient carbon dioxide from the water.

How then do algae survive? The answer lies in the fact that laminar flow is seldom found in nature. Once the flow reaches a critical local Reynolds number (approximately $3.5 \cdot 10^5$), the boundary layer becomes turbulent (Chapter 9). For instance, at a mainstream flow of 1 m/s in laminar mainstream flow, the boundary layer on a smooth, planar frond becomes turbulent about 28 cm from the leading edge. At 10 m/s the transition would occur at less than 3 cm from the leading edge, and if the mainstream flow is itself turbulent, or the frond is at all rough, the transition to a turbulent boundary layer occurs virtually at the leading edge of the frond. Thus many (if not most) algal fronds are primarily encased in turbulent rather than laminar boundary layers. The transport of bicarbonate (or any other particle in the fluid) by a turbulent bounday layer is entirely different than that by its laminar analog. As the water swirls in a turbulent eddy, particles a large distance away from the frond are advectively brought near it, and the converse is true for particles starting out near the frond's surface. The bulk mixing due to turbulence vastly increases the rate at which a substance can be transported to or away from a solid surface (for a review, see Okubo 1980; for algal examples, see Wheeler 1980, Gerard 1982, and Koehl and Alberte 1987).

The implications of turbulent mixing are important for every aspect of biology that relies on transport by fluids. This does not apply only at the molecular level. Turbulent mixing is just as important for the transport of macroscopic particles as it is for molecules. Consider two pertinent examples: (1) A planktonic larva attempting to return to the rock substratum from a position several hundred meters out at sea would have an extremely difficult time if it had to rely on its own feeble swimming abilities. As we will see, the process is aided by the turbulence present in breaking waves. (2) Many aquatic organisms sexually reproduce using external fertilization. Here turbulent mixing may serve to bring sperm and egg together. However, the consequences of turbulent mixing may not be entirely advantageous. The fertilization process is effective only if sperm and eggs are present in sufficient concentra-

tion. If the water into which gametes are shed is mixed too thoroughly, sperm and eggs may become too widely dispersed and never meet.

Because turbulent mixing is so important, it is well worth to explore its mechanism in some detail.

Transport by Fluctuations

We begin with an abstract example. Consider a fluid in which a number of particles are suspended (Fig. 10.1a). If we take a small cube of this fluid with dimensions dx, dy, and dz, we can at any time count how many particles are in the volume. The concentration of particles is

$$c = \frac{N}{dx\,dy\,dz},\qquad(10.1)$$

where N is the number of particles in the box. If the particles move around at random, N varies through time. If we keep track of N for a long time we can calculate the mean concentration, $\bar{c}$, and express the concentration at any time as

$$c(t) = \bar{c} + c',\qquad(10.2)$$

where c' is the fluctuation from the mean. Since the average of $(\bar{c} + c')$ must be equal to $\bar{c}$, the average of the fluctuations must be zero.

Now let the fluid move in the x-direction with a velocity u that varies through time. As in our treatment of Reynolds stress (Chapter 9) we express the instantaneous velocity as an average plus the fluctuation from this average:

$$u(t) = \bar{u} + u'.\qquad(10.3)$$

Given these conditions, at what rate are particles transported in the direction of the mean flow? The number of particles passsing through the downstream wall of the cube in a given time is simply the product of the concentration of particles in the cube, $N/(dx\,dy\,dz)$, and the velocity in the x-direction, dx/dt. When we carry out this

Figure 10.1. Transport associated with random fluctuations in velocity. (a) Correlated fluctuations in velocity and concentration can lead to transport of particles (see text). (b) Due to the gradient in concentration, there is a net movement of dye away from the center of the cloud (see text).

a)

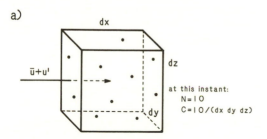

b)

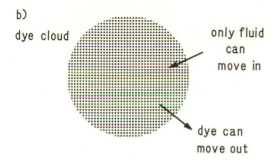

dye cloud

only fluid can move in

dye can move out

multiplication using our expressions for fluctuating concentration and velocity, we arrive at the following conclusion:

$$\text{flux} = (\bar{c} + c')(\bar{u} + u')$$
$$= \overline{uc} + \bar{c}u' + c'\bar{u} + u'c'. \quad (10.4)$$

We can average this instantaneous flux over time, and remembering that $\overline{u'} = 0$ and $\overline{c'} = 0$, we find that the average flux of particles in the x-direction is

$$\text{average flux} = \overline{uc} + \overline{u'c'}. \quad (10.5)$$

This is a surprising result. The average flux of particles in a fluctuating system is different from the average flux in a non-fluctuating system ($\overline{uc}$) by an amount $\overline{u'c'}$. Even if the average velocity is zero, there can still be transport of mass (as particles) in the x-direction simply due to the fluctuation of velocity around zero.

Surprising as this conclusion may be, it makes mechanistic sense given the appro-

priate correlation between velocity and concentration. Consider a case where the average velocity is zero, but the velocity fluctuates randomly. If velocity and concentration are correlated so that every time the velocity is in the positive x-direction the concentration is high, many particles are carried forward. Conversely, when the velocity is negative, the concentration is low and few particles are carried back. In this manner a correlation between u' and c' results in a net transport of mass. This type of mechanism can result in net transport in the negative x-direction if, on average, u' is positive when c' is negative (and vice versa).

This finding may seem nonintuitive, but it is manifested in everyday experience. If you release a small volume of dye into the middle of a large volume of still water, the dye particles gradually mix through the fluid. The dye is transported because, on the molecular scale, there are correlated fluctuations in the velocity and concentration of dye particles as they are moved about by thermal agitation.

Thermal agitation jostles molecules randomly. Why then should the fluctuations in velocity and concentration be correlated? The correlation depends on the presence of a concentration gradient. Consider, for example, the outer edge of a cloud of dye particles soon after its release (Fig. 10.1b). Inside the cloud the dye concentration is high, so that any motion away from the center of the cloud results in the outward transport of a large number of dye particles. Just outside the dye cloud there are no dye particles at all, so any movement of fluid from outside the cloud toward its center cannot transport dye particles inward. Therefore, because there is a gradient in concentration between the cloud and the surrounding fluid (in this case an unrealistically steep gradient), fluctuations in velocity and concentration are necessarily correlated, and the result is a net transport of dye away from the center of the cloud—the dye diffuses outward.

This explanation has its disadvantages. In practice, it is difficult to measure separately fluctuations in velocity and concentration on a molecular scale. Instead, one measures their combined effect by determining the flux between areas of different concentration. A molecular diffusion coefficient, D_m, can then be defined by the following equation:

$$\text{flux} = -D_m \frac{dc}{dx}, \qquad (10.6)$$

where the negative sign reflects the fact that flux proceeds from an area of high concentration to one of lower concentration. This is Fick's equation for diffusion, a readable derivation of which is presented by Berg (1983). The magnitude of the molecular diffusion coefficient is governed by how fast a molecule moves and by how far it moves in a particular direction before encountering another molecule, impacting, and flying off in a new, random direction. In water at room temperature the speed of a bicarbonate molecule is about 200 m/s, but the step length between molecular collisions (the *mean free path*) is only on the order of 10^{-11}m. This helps explain the long time it takes for a molecule to travel even a small distance in any particular direction. Every time it moves 0.01 nm it must change direction, and as often as not it moves backward instead of forward.

Although the mass transport mechanism explained here can be used to describe molecular diffusion, it is not intrinsically tied to any particular scale of movement. If local bulk water velocities and the concentration of particles vary through time, the same sort of transport phenomenon can operate on a macroscopic level. This is the basic idea behind the turbulent transport of biological particles.

Although we now have some clues as to how turbulent mass transport might operate, we are still a long way from a practical understanding of it. Before we can explore its consequences, we need to define a turbulence equivalent of the molec-ular diffusion coefficient and we must have a method for measuring correlated fluctuations in turbulent flow.

The Eddy Diffusivity

In defining the turbulence equivalent of a molecular diffusion coefficient, we rely on an analogy between the transport of mass—for example, the rate at which particles are diffusively mixed—and the transport of momentum, a process we examined in our discussion of boundary layers (Chapter 9). Recall that the viscous shear stress, τ, can be thought of as the rate at which momentum that is directed parallel to a solid substratum (that is, the momentum due to velocity u) is carried toward the substratum. Thus the equation relating viscous stress to the velocity gradient,

$$\text{viscous shear stress, } \tau = \mu \frac{du}{dz}, \quad (3.11)$$

tells us that the rate at which momentum due to velocity u is transported to the substratum depends on the velocity gradient and the fluid's viscosity. By analogy, we can write an equation relating the Reynolds shear stress (Chapter 9) and a velocity gradient:

$$\tau_R = \rho \overline{u'w'} = \gamma' \frac{du}{dz}. \qquad (10.7)$$

Or, by absorbing the density into the coefficient γ',

$$\overline{u'w'} = u_*^2 = \gamma \frac{du}{dz}. \qquad (10.8)$$

In this form the coefficient γ is called the *eddy viscosity*.

We now make a leap of faith and assume that in turbulent flow, mass is transported in the same way as momentum. Thus the flux of mass can be described by an equation similar to eq. 10.8. By analogy to Fick's equation for molecular diffusion (10.6), we write an equation for the mass flux due to turbulence:

$$\text{flux} = \overline{w'c'} = -K_z \frac{dc}{dz}. \quad (10.9a)$$

In this case K_z is the *eddy diffusivity* in the z-direction, a turbulence equivalent of the molecular diffusion coefficient. The negative sign reflects the fact that transport is directed away from areas of high concentration. Similar expressions can be written for diffusion along the y and z-axis:

$$\overline{u'c'} = -K_x \frac{dc}{dx}$$

$$\overline{v'c'} = -K_y \frac{dc}{dy}. \quad (10.9b,c)$$

Here we encounter a major difference between molecular and turbulent diffusion. D_m, the molecular diffusion coefficient, is usually independent of direction. In contrast K_x, K_y, and K_z are seldom equal, and thus the diffusivity can vary with direction.

There are two common methods by which turbulent diffusivities are quantified. Reasonably enough, the first of these is based on observations of the rate at which particles are dispersed. Consider the case shown in Figure 10.2. A large number of particles are released at one point in a turbulent flow. At the time of release the concentration of particles near the point of release is very high, but with passing time the cloud of particles spreads out. At any time we could, in theory, measure the distance from each particle to the center of mass of the cloud. We could then plot the distribution of the x, y, or z, component of these distances. If we did this several times, we would see that the distributions become "flatter" the longer the time elapsed after the particles were released; with passing time fewer and fewer particles are located near the center of the cloud. The faster the rate of turbulent diffusion, the faster the distributions flatten out. Thus if we can quantify the rate at which the distributions of the x, y, and z components of distance change we will have a measure of the directional diffusivities.

The simplest measure of the "flatness" of a distribution is its variance, σ^2. In this case the variance is the mean square of the directional component of distance from the center of mass of the cloud. For example,

$$\sigma_x^2 = \frac{1}{N} \sum_{i=1}^{N} x_i^2, \quad (10.10)$$

where x_i is the x component of the distance from the ith particle to the center of mass of the particle group. The directional diffusivity is directly related to the rate of change of the variance (Csanady 1973; Okubo 1980).

$$K_x = \frac{1}{2} \frac{d\sigma_x^2}{dt}$$

$$K_y = \frac{1}{2} \frac{d\sigma_y^2}{dt}$$

$$K_z = \frac{1}{2} \frac{d\sigma_z^2}{dt}. \quad (10.11)$$

In a turbulent boundary layer the square root of the variance (i.e., the standard deviation σ) can be related to the friction velocity:

$$\sigma_x = \alpha_x \frac{u_*}{\bar{u}} x^{\beta_x}$$

$$\sigma_y = \alpha_y \frac{u_*}{\bar{u}} x^{\beta_y}$$

$$\sigma_z = \alpha_z \frac{u_*}{\bar{u}} x^{\beta_z} \quad (10.12)$$

These equations provide a connection between diffusivity, a measure of the rate of mass transport, and friction velocity, a measure of shear stress and therefore of momentum transport. They quantify the analogy we proposed above: that in turbulent flow mass is transported in the same manner as momentum. This is usually true (at least as a first approximation) and thus allows us to use the friction velocity as an experimental tool for quantifying turbulent transport. Battjes and Sakai (1981) have shown in laboratory experiments that u_* can be as much as 9% of $\bar{u}$ (the water

Figure 10.2. As particles disperse, the distribution of their distances to the group's center of gravity changes. (a) Particles soon after release. They are closely packed and their distance distribution has a low variance. (b) Later the particles have dispersed and their distance distribution has a larger variance.

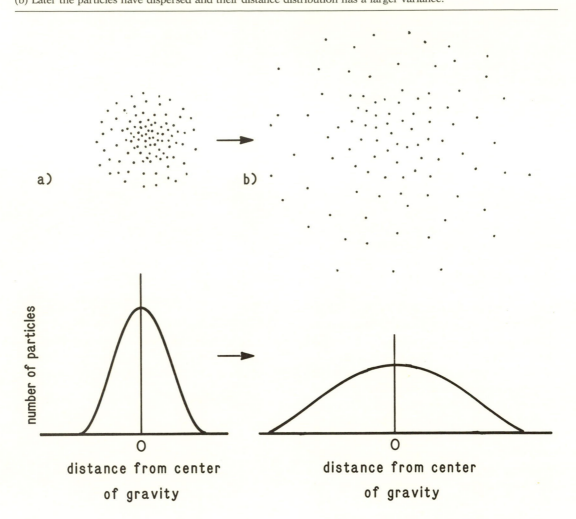

velocity at the crest) close behind the leading edge of a turbulent bore and can have a value of 5% $\bar{u}$ for a considerable distance behind the leading edge.

Although simple in theory, this approach to quantifying the turbulent diffusivities can be difficult in practice. The coefficients α and β must be determined experimentally and can vary as a function of both time and position. For example, β is approximately 1 during the early stages of dispersal where the group is small, and is typically between 0.9 and 1.5 as dispersal continues. α_x and α_y are typically greater than α_z. Panofsky (1967) cites data from atmospheric experiments suggesting that $\alpha_x = \alpha_y = 2.2$ and $\alpha_z = 1.25$. For the purposes of this introduction, we will assume that α_x, α_y, and α_z are constant through time, and that $\beta_x = \beta_y = \beta_z = 1$. Thus

$$\sigma_x \simeq \frac{\alpha_x u_* x}{\bar{u}}$$

$$\sigma_y \simeq \frac{\alpha_y u_* x}{\bar{u}}$$

$$\sigma_z \simeq \frac{\alpha_z u_* x}{\bar{u}}. \qquad (10.13)$$

Or, assuming that $x = \bar{u}t$ at any level in the boundary layer,

$$\sigma_x \simeq \alpha_x u_* t$$
$$\sigma_y \simeq \alpha_y u_* t$$
$$\sigma_z \simeq \alpha_z u_* t. \qquad (10.14)$$

Given these assumptions, we can specify approximations for the directional diffusivities in a turbulent benthic boundary layer (eq. 10.11):

$$K_x \simeq \alpha_x^2 u_*^2 t$$
$$K_y \simeq \alpha_y^2 u_*^2 t$$
$$K_z \simeq \alpha_z^2 u_*^2 t. \qquad (10.15)$$

Note that the diffusivity measured in this fashion increases with time. In other words, the longer a group of particles is allowed to disperse, the faster they will diffuse. This is because the rate of dispersal depends on the size of a group of particles relative to the size of turbulent eddies. When particles are closely grouped (i.e., shortly after they are released), any eddy with a diameter much larger than that of the group merely advects the group as a whole without changing the distances among particles within the group. Therefore, initially only the smallest eddies in the flow are effective at increasing the variance of particles about the group's center of mass. However, as time passes, these small eddies gradually disperse the particles, and larger and larger eddies become effective at increasing the group's variance. As a larger percentage of the eddies present in flow becomes effective at dispersing the group, the rate of diffusion increases. Okubo (1971, 1980) has shown that horizontal diffusivities (K_x, K_y) vary as $L^{4/3}$, where L is a length scale of the group size, for convenience taken to be $3\sigma_x$ or $3\sigma_y$.

Remember that the diffusivities shown in eq. 10.15 are approximations based on a simplifying set of assumptions that may or may not apply to a particular flow situation.

In most real flows these expressions will require modification.

An alternative approach is available for specifying K_z in a turbulent boundary layer. This approach relies on a theory of turbulent mixing developed by Ludwig Prandtl. We will not delve into the basis of this theory here (the interested reader should consult Schlichting 1979 or Sutton 1953) and will treat the result of theory as an established fact:

$$K_z = \frac{l_* u_*}{2}. \qquad (10.16)$$

In this expression, u_* is used as a turbulence analog to the velocity at which a molecule moves, and l_* is the *mixing length*, a convenient abstraction hypothesized by Prandtl. The mixing length is a turbulence equivalent to the mean free path for molecules and represents in a vague way the distance that an individual turbulent eddy moves before it loses its identity and either splits or merges with another eddy. In a turbulent boundary layer

$$l_* = 2\kappa s, \qquad (10.17)$$

where κ is von Karman's constant ($\kappa = 0.4$). As before, we use s to measure the distance from the substratum. Inserting this value for l_* into the expression for K_z (eq. 10.16), we see that

$$K_z = \kappa u_* s \qquad (10.18)$$

This definition of K_z using l_* leads to two problems. First, consider the magnitude of l_*. If, for example, we want to know something about the settlement of larvae in the presence of a turbulent boundary layer, we ask questions about the transport of the larva over distance δ, where δ is the boundary layer thickness. At $s = \delta$, $l_* = 0.8\,\delta$; the two are nearly the same. This makes it difficult to reconcile any tidy analogy between molecular diffusion and turbulent diffusion. Because mean free paths of molecules in water are on the order of 10^{-11} m, each molecule changes direction many times in diffusing any sensible distance. In such a situation the assumption of random motion used in the derivation of transport-by-fluctuation is clearly valid. However, because l_* is nearly the same size as the distance over which we measure transport, the assumption of random movement may not be justified. Despite its lack of theoretical justification, the use of l_* is a mainstay of practical research in turbulence processes: it seems to work.

The second problem with l_* is the presence of s in its definition. The dependence of the mixing length on the distance from the substratum makes intuitive sense. An eddy very close to the substratum is as likely to impact the substratum as it is to move away. Thus its average "free path" is quite small. The farther the eddy is from the substratum, the farther it can move, on average, before it hits the bottom. The "free path" is also determined by the interactions among eddies, but presumably this is constant with distance away from the substratum. The problem with the variable l_* lies not in its understanding but in its use. It again leads to a spatially variable eddy diffusivity, in distinct contrast to the constant diffusion coefficient found for molecular diffusion. As a consequence, some of the results for turbulent diffusion using this expression for K_z are more complicated than the equivalent results for molecular diffusion.

There are many situations in the wave-swept environment where one might want to know how the chaotic mixing of the fluid affects the organisms present. Here we deal with two simple situations—the probability of a larva encountering the substratum, and the spread of gametes away from a spawning organism.

Larval Settlement

Consider the plight of a planktonic larva such as a barnacle cyprid or the pluteus of a sea urchin or star fish. After spending the requisite time at sea, the larva must return to the rock substratum of the coast before it can metamorphose and grow into a reproductive adult. Large-scale oceanic phenomena (e.g., mass transport by surface waves) may serve to bring the larva into the surf zone, but once in the surf zone the larva has the problem of traveling that last short distance and actually coming into contact with the substratum. The water velocities in the surf zone are typically on the order of 1–10 m/s, vastly greater than the swimming velocity of the average planktonic larva (on the order of 1–10 mm/s). This difference in velocity would not be a problem if the larva could continuously swim toward the substratum, but the chaotic, turbulent nature of the flow would seem to preclude directed swimming, and the settling velocity for most larvae is very low (< 1 mm/s) and therefore may not be of much practical consequence in surf-zone flows. One would guess that larvae in the surf zone move at the whim of the local flow regime. What, then, are the chances that an individual will actually contact the rock?

We can arrive at an answer to this question by following two separate lines of reasoning. First, we will examine the biological evidence using barnacles as an example.

The density of larval settlement is notorious for being a fickle phenomenon. In any given year certain stretches of coastline are inundated with settling larva while adjacent stretches are not. The density of settlement varies widely from year to year. However, given the proper conditions (and no one is precisely sure what these are), larval settlement can be a surprisingly effective process. For example, in 1984 there was a dense settlement of the barnacle *Balanus glandula* in Monterey Bay in California. In the course of one week, barnacles settled at the rate of 10^4 per square meter on the rocks near Hopkins Marine Station. During this period

Gaines et al. (1985) measured the concentration of cyprids in the water arriving at the shore, which amounted to barely fifty cyprids per cubic meter. For this low concentration to have produced the observed settlement, they calculate that at least 25% of all cyprids coming within a meter of the shore managed to settle.

Additional biological evidence supports this conclusion. For example, it has been shown that barnacle cyprids, once they have arrived at the rock, can be very selective in searching for an appropriate space on which to settle. The cyprid attaches to the rock by its antennules and stumps around searching for the appropriate clues that may include the rugosity and inclination of the surface and the presence of adult barnacles. This selectivity is understandable in light of the fact that, once metamorphosed, the barnacle is permanently glued in place. If the site does not seem appropriate, the cyprid releases itself from the surface, presumably to land elsewhere and try again. This releasing behavior can be adaptive only if the larva has a reasonable chance of returning to the rock. If there is only one chance in a million that the cyprid will reach the rock in the first place, and a similarly small chance of returning once it releases, this behavior would not make sense. The behavior of the larva thus provides further evidence that the mechanism bringing larvae to the rock must be reasonably effective.

This biological evidence leaves us with the task of finding a flow mechanism that accounts for the apparent effectiveness of settlement. As one might guess, turbulent mixing is the likely candidate, and we make use of the idea of turbulent diffusivity in estimating the probability that a larva will encounter the rock surface.

In attempting to calculate this probability, we follow a line of reasoning similar to that presented by Berg (1983) in his discussion of molecular diffusion. Thus we are once again appealing to the analogy

Figure 10.3. (a) A model for the calculation of time-to-settle for a larva in the surf zone (see text). (b) Time to settle, $W(s)$, as a function of distance from the substratum (eq. 10.26).

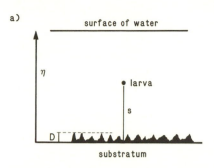

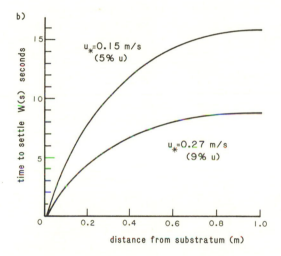

between random motion on a microscopic scale and the chaotic motion in turbulent flow.

Consider a larva in a turbulent bore as it travels over a rugose beach (Fig. 10.3a). The water has a depth η, and is turbulent throughout this depth. Larvae are moved freely by the flow through the water column, but when they encounter the water's surface they are "reflected" back, and when they encounter the rock substratum they stick and settle. We view the local movements of the fluid from a frame of reference moving at the mean fluid velocity. In this frame of reference the larva can be seen to move around chaotically in the x, y, and z directions, but it is only the movements in the z-direction—toward or away from the rock surface—that affect settlement, and these are the motions that interest us here. Given these assumptions, how long, on average, does it take a larva to encounter the rock surface?

We start with a larva at a distance s from the rock and define the mean time it takes this larva to reach the rock surface ($s = 0$) to be $W(s)$. In a short period of time, t, the larva travels a distance Δ along the z-axis; half the time this distance is toward the rock, and half the time away. If the larva moves away from the rock, at time t it is at $s + \Delta$, and the mean time to settlement from this distance is $W(s + \Delta)$. If the larva moves toward the rock, at time t, it is at $s - \Delta$ and has a mean time to settlement of $W(s - \Delta)$. We can thus redefine $W(s)$ as the average of the time it would take for the larva to settle from these two distances, plus the time it takes to arrive at the two distances:

$$W(s) = t + \left\{ \frac{W(s + \Delta) + W(s - \Delta)}{2} \right\}. \tag{10.19}$$

If we multiply both sides of this equation by 2, divide by Δ, and rearrange, we arrive at the conclusion that

$$\left\{ \frac{W(s + \Delta) - W(s)}{\Delta} \right\} - \left\{ \frac{W(s) - W(s - \Delta)}{\Delta} \right\}$$
$$+ \frac{2t}{\Delta} = 0. \tag{10.20}$$

The usefulness of this rearrangement may not be immediately apparent. However, the form of this equation may seem familiar to those recalling introductory calculus. The terms in curly brackets are in the form used to define the derivative. In other words, in the limit as Δ approaches zero, eq. 10.20 becomes

$$\left. \frac{dW}{ds} \right|_{s + ds/2} - \left. \frac{dW}{ds} \right|_{s - ds/2} + \frac{2t}{\Delta} = 0. \tag{10.21}$$

Dividing through by Δ once more and again applying the definition of the derivative, we see that

$$\frac{d^2 W}{ds^2} + \frac{2t}{\Delta^2} = 0. \tag{10.22}$$

Now, in the molecular analogy of this situation, $2t/\Delta^2$ is $1/D_m$ (Berg 1983), and we dutifully substitute our value for the vertical turbulent diffusivity (eq. 10.18):

$$\frac{d^2 W}{ds^2} + \frac{1}{\kappa u_* s} = 0. \tag{10.23}$$

This second-order differential equation can be solved provided we can supply two boundary conditions. Easily done. First, we can show that if larvae are reflected at the surface ($s = \eta$), $dW(\eta)/ds$ must equal zero. Consider a larva right at the water's surface. In a short period of time t it can move a distance Δ up or down. Movement down places the larva at $s = \eta - \Delta$, from which a time $W(\eta - \Delta)$ is required to reach the substratum. If the movement is up, the larva is reflected from the surface and the step Δ actually results in a step $-\Delta$, again placing the larva at $s = \eta - \Delta$. The time to reach the substratum from this position must be the same as when the step was initially down. Since varying s by $\pm \Delta$ results in the same time to settlement, the change in W with respect to s must be zero—in other words, $dW/ds = 0$ at η. This is our first boundary condition. The second boundary condition is established by noting that when the larva is at the rock surface, the time required to reach the substratum is, by definition, zero. Because we are dealing with a rough surface, however, we need to be careful as to what height we consider to be "the" surface. In a practical sense, the larva can be considered to have arrived when it reaches the tops of the roughness elements, a height D above the actual bottom. We thus establish as one of our boundary conditions that $W(D) = 0$.

To solve Eq. 10.23 we integrate with respect to s and see that

$$\frac{dW}{ds} = -\frac{1}{\kappa u_*}(\ln s + a), \quad (10.24)$$

where a is the constant of integration. Using our first boundary condition ($dW/ds = 0$ at $s = \eta$), we see that $\ln \eta + a = 0$, so that $a = -\ln \eta$. Inserting this constant into eq. 10.24 and integrating again, we have

$$W = -\frac{1}{\kappa u_*}(s \ln s - s - s \ln \eta + b). \quad (10.25)$$

From our second boundary condition we know that $W(D) = 0$. This can be true only if $b = -D \ln D + D + D \ln \eta$. Thus

$$W(s) = -\frac{1}{\kappa u_*}(s \ln s - s - s \ln \eta$$
$$- D \ln D + D + D \ln \eta) \quad (10.26)$$

This is the mean time to settlement from distance s. If we want to know the overall mean time to settlement, we average this equation over the range D to η, with the result that

$$\overline{W} = \frac{1}{\kappa u_*(\eta - D)} \left\{ \frac{3\eta^2}{4} - D\eta - D\eta \ln \eta \right.$$
$$\left. + D\eta \ln D - \frac{D^2}{2} \ln D + \frac{D^2}{2} \ln n + \frac{D^2}{4} \right\}. \quad (10.27)$$

This unwieldy equation can be simplified considerably in situations where the water depth η is much greater than the roughness height D. In this case D^2 is small compared to η, and terms in D^2 can be neglected. For example, if η is 1 m and D is 1 cm, D^2 is only one ten-thousandth η. Thus, to an accuracy of a few parts in ten thousand,

$$\overline{W} = \frac{1}{\kappa u_*(\eta - D)}$$
$$\cdot \left\{ \frac{3\eta^2}{4} - D\eta - D\eta \ln \eta + D\eta \ln D \right\}. \quad (10.28)$$

If we are willing to accept an error on the order of D/η (a few percent for the example cited above) the equation can be simplified

even further by setting D equal to zero, in which case

$$\overline{W} = \frac{3\eta}{4\kappa u_*}. \quad (10.29)$$

Thus, to a first estimate the mean time to settlement for a larva in the surf zone is directly proportional to the depth of the water, and inversely proportional to u_*. If we take a value of 5% $\bar{u}$ as a conservative estimate for u_* in the surf zone and assume that $\bar{u} = 3$ m/s (appropriate for a bore with 1 m of water below the crest), eq. 10.29 becomes

$$\overline{W} = \frac{3\eta}{4\kappa(0.15)} = 12.5 \text{ s.}$$

In other words, for a turbulent bore with a total depth ($H + d$) of 1 m, it takes a larva on average 12.5 s to get to the rock surface. If u_* is 9% of $\bar{u}$, $\overline{W} = 6.9$ s.

These calculations assume that both the water depth and turbulence structure are constant through time, while in the real world they change as the bore travels up the beach. Any precise treatment of the effects of these time-varying parameters is problematic and we will not attempt it here. Instead, we will use our estimates of mean time to settlement in making a simple ball-park estimate of the probability that a larva will encounter the rock. The upsurge of a wave on a beach typically lasts from 2 to 7 s. Thus if the mean time to settlement is anywhere near our calculated estimate of 12.5 s, it seems quite probable that a larva in the surf zone will encounter the rock surface at least once in the course of a few waves. Furthermore, we can use Eq. 10.26 to give us an estimate of the probability that, once on the rock, a larva will get back to the rock if it releases (Fig. 10.3b). We can easily see that once the larva is in the vicinity of the rock surface (s small), the mean time to settlement is quite small. Thus, even if a released larva is kicked up into the water column a distance of 10 cm, the mean time to settlement is only 3–5 s. It seems reasonable that larvae that release from the rock can expect to reencounter the rock on the same or next wave.

These estimates of mean time to settlement should be regarded as educated guesses. We have made a number of gross assumptions in performing these calculations and therefore should not expect them to be exact. A more exact calculation, however, requires detailed knowledge of the turbulence structure in the surf zone, especially the spatial and temporal pattern of u_*. This kind of information is not easy to come by, and for the present we must be content with the approximations used here.

External Fertilization

Many wave-swept organisms reproduce using external fertilization. Eggs and sperm are released into the water, usually with some synchronization between spawning males and females, and the motility of the sperm and the bulk mixing of the water brings gametes together. There is a central problem inherent in this form of reproduction. Gametes shed by one individual may have to travel centimeters or even meters before encountering another member of the same species. If it is uncertain in which direction the nearest neighbor lies, there are advantages to broadcasting gametes so that at least some of them come close to another spawning individual. However, the advantage of dispersal must be weighed against the probability of fertilization once gametes arrive near a neighbor. If in dispersing gametes over a wide area the concentration at any point is too dilute, the probability of fertilization is decreased.

In the wave-swept environment, both the processes of dispersal and dilution are largely under the control of turbulent mixing. The swimming speed of sperm (on the order of 0.1 mm/s) is so much lower than the mixing speed due to turbulence (on the order of 0.1 to 1 m/s) that to a first approxi-

mation it can be neglected. If we wish to explore the mechanics of external fertilization, we must employ the concepts of turbulent diffusivity.

We will explore external fertilization in three steps. First we will examine an expression that tells us how particles are dispersed after they are released into turbulent flow. We will then use the sea urchin as an example in calculating the concentration of sperm as a function of distance away from a spawning organism. Finally, we will examine how sperm concentration effects fertilization efficiency.

Particle Dispersion in Turbulent Flow

Again we begin with an analogy to molecular diffusion. Consider a hypodermic needle extending into an infinite volume of water with its tip located at the origin of our coordinate system (Fig. 10.4a). Dye is pumped out of the hypodermic needle at a constant rate of Q particles per second, and the water moves past the needle with constant velocity u in the x-direction. How are these particles dispersed?

We approach the problem as follows. In a short period dt, $N = Q \, dt$ particles are released. During this time the water has traveled a distance $u \, dt = dx$. To simplify matters, we assume that each particle is confined to the thin layer of water into which it is released, a layer that extends to infinity in the y and z-directions but is only dx thick in the x-direction. This reduces the situation to a two-dimensional diffusion.

As soon as each particle is released, it is advected downstream with velocity u, and it begins to wander randomly due to its thermal agitation. As each particle bounces around, its chance movement in any one direction is likely to be offset by an equal and opposite movement. Consequently, most particles remain in the vicinity of the x-axis and we expect the concentration of particles downstream from the origin to be highest along this axis.

This clustering, however, is strictly a mat-

Figure 10.4. The turbulent diffusion of particles from a point source. (a) The coordinate system used. (b) Particles are assumed to be confined to the infinitesimally thin layer (dx) into which they are released (see text).

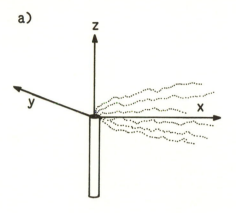

a)

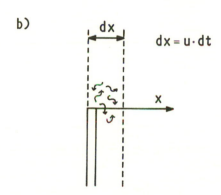

b)

$$dx = u \cdot dt$$

ter of chance. There is a probability, albeit a low one, that every time a particle moves it moves farther away from the x-axis. The more movements the particle makes, the farther it could conceivably travel. However, the more movements made, the less the probability that every movement is away from the axis. Now, the number of movements depends on how long the particle has been allowed to wander, and therefore on how far its layer of water has advected downstream of the release point. At any point downstream (i.e., for a given number of movements), the farther from the axis we look, the less the chance that a particle will have wandered that far. The farther downstream we look, the more

movements particles have made and the farther, on average, they have wandered.

From the statistics of a two-dimensional random walk (Crank 1975) it can be shown that the spatial distribution of concentrations in an infinitesimally thin layer is

$$c(R) = \frac{Q}{4\pi D_m x} \exp - \left\{ \frac{R^2 \bar{u}}{4Dx} \right\}, \quad (10.30)$$

where R is the radial distance from the x-axis ($= \sqrt{y^2 + z^2}$).[1]

Let us examine eq. 10.30 (Fig. 10.5).

Along the x-axis ($R = 0$) the exponential term is zero and eq. 10.30 reduces to

$$c(0) = \frac{Q}{4\pi D_m x}. \quad (10.31)$$

Thus the concentration decreases hyperbolically with distance from the release point, and at any point depends linearly on the rate at which particles are released (Fig. 10.5a). The larger the diffusion coefficient, the more rapid the decline in concentration. At any position along the x-axis, the concentration decreases exponentially with increasing R (Fig. 10.5b). The faster the velocity, the faster the concentration decreases with R at any x. This is because as u increases, particles have had less time to wander by the time they arrive at x. The distribution of concentrations is a flame-shaped plume, as shown in Fig. 10.5c.

The concentration distribution predicted by this simple calculation resembles the plume of smoke that trails from a smokestack in a steady wind, and we propose that it is analogous to the plume of gametes downstream from a spawning organism. However, at the large scale of a

[1] This expression is only an approximation to real diffusion because it assumes that there is no diffusion along the x-axis. However, the exact solution

$$c(r) = \frac{Q}{4\pi D_m r} \exp - \left\{ \frac{(r - x)\bar{u}}{2D_m} \right\},$$

where $r = \sqrt{x^2 + y^2 + z^2}$ (Bird et al. 1960) is less easily manipulated. The two solutions converge for all but the smallest values of x, and we use the simpler equation here.

Figure 10.5. Particle concentrations in turbulent diffusion. (a) Concentration along the *x*-axis (the axis of mean flow). (b) Concentration as a function of distance from the *x*-axis. (c) The turbulent plume—lines of equal relative concentration downstream from the point of release.

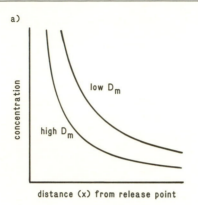

a)

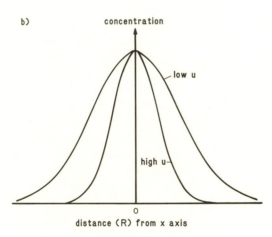

b)

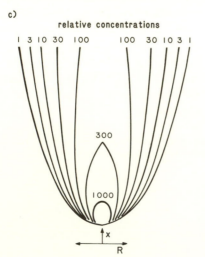

c)

smoke or gamete plume it is not molecular diffusion that causes the distribution of particles; it is diffusion due to turbulence. As before, we could arrive at an analogous equation for turbulent diffusion by substituting a turbulent diffusivity K for D_m.

$$c(R) = \frac{Q}{4\pi Kx} \exp - \left\{ \frac{R^2 \bar{u}}{4Kx} \right\}. \quad (10.32)$$

But there is a problem with this analogy: Which eddy diffusivity do we use? In this example we are forced to acknowledge the fact that because eddy diffusivities vary with direction, it is impossible in some cases to use them in direct analogy to the molecular diffusion coefficient. Instead we must use an expression that incorporates the directionality of turbulent transport (Csanady 1973):[2]

$$c(x, y, z) = \frac{Q}{2\pi \bar{u} \sigma_y \sigma_z} \exp - \left\{ \frac{y^2}{2\sigma_y^2} + \frac{z^2}{2\sigma_z^2} \right\}. \quad (10.33)$$

Although this equation looks quite different from its molecular analog, it is functionally equivalent.

Using eqs. 10.13 and 10.14 for σ_y and σ_z, we can rewrite eq. 10.33 to describe concentration either as a function of distance from the point of release,

---

[2] For the sake of simplicity, we have tacitly assumed that the release point is far enough away from the substratum so that particles reflecting from the bottom do not affect the pattern of concentration. Close to the substratum a modified expression should be used (Csanady 1973):

$$c(x,y,z) = \frac{Q}{2\pi \bar{u} \sigma_y \sigma_z} \left\{ \exp - \left\{ \frac{y^2}{2\sigma_y^2} + \frac{(z-s)^2}{2\sigma_z^2} \right\} \right.$$

$$\left. + \exp - \left\{ \frac{y^2}{2\sigma_y^2} + \frac{(z+s)^2}{2\sigma_z^2} \right\} \right\},$$

where z is measured from the substratum rather than the point of release, and s is the distance above the substratum at which particles are released. This modified equation is derived by assuming that the reflection of particles from the substratum is equivalent to having a second, hypothetical source a distance s into the substratum. The method of using such "mirror image" sources to account for reflection is discussed by Csanady (1973).

$$c(x, y, z) = \frac{Q\bar{u}}{2\pi\alpha_y\alpha_z u_*^2 x^2}$$

$$\cdot \exp - \left\{ \frac{y^2\bar{u}^2}{2\alpha_y^2 u_*^2 x^2} + \frac{z^2\bar{u}^2}{2\alpha_z^2 u_*^2 x^2} \right\}.$$

(10.34)

or as a function of time after release,

$$c(t, y, z) = \frac{Q}{2\pi\bar{u}\alpha_y\alpha_z u_*^2 t^2}$$

$$\cdot \exp - \left\{ \frac{y^2}{2\alpha_y^2 u_*^2 t^2} + \frac{z^2}{2\alpha_z^2 u_*^2 t^2} \right\}.$$

(10.35)

This approach has been shown to be qualitatively correct for the dispersal of pollen by wind (see Grace 1977 for a review), and we hypothesize that it can be used to describe the pattern of dispersal and dilution of gametes in the wave-swept environment.

The Sperm Plume: Patterns of Sperm Concentration

We will now examine the hypothetical case of an organism that releases sperm from a single point source. In order to calculate the concentration of sperm downstream from the source, we need to specify $\bar{u}$ and u_* and measure Q. Water velocity varies widely in the wave-swept environment. In this case we examine velocities varying from 0.1 m/s, near the lower limit for turbulent flow, to 1 m/s. For reasons that will become apparent it is not necessary to examine higher velocities. As we have seen (Chapter 9), u_* can vary widely. Here we use values from 5% to 10% of the mainstream velocity.

Q has not been measured in the field for any marine organism, and we will rely on laboratory data for sea urchins to provide us with an estimate. An adult *Strongylocentrotus purpuratus* of a size commonly found along the coast of California (about 7 cm test diameter) can extrude from 1 to 2 ml of semen. It takes about an hour to force this volume out of the five gonopores after the animal has been injected with a strong potassium chloride solution. This in-

jection causes the muscles of the gonad to contract maximally, and the eflux of sperm measured in this fashion can reasonably be assumed to be the maximum rate the animal can manage. Thus a maximum of $1.1 \cdot 10^{-10}$ m^3 of semen are extruded per second per gonopore. Tyler et al. (1956) estimate the sperm concentration in sea urchin semen (*Lytechinus pictus*) as $2 \cdot 10^{16}$/m^3. Combining these facts, we arrive at an estimate for maximal Q of about $2 \cdot 10^6$ sperm/s per gonopore. We use this value as an estimate of Q for a point source of sperm.

These various values for Q, $\bar{u}$, and u_* are used to examine the spatial variation in gamete concentration. First we examine the concentration distribution directly downstream of the point source (i.e., on the x-axis, $z = y = 0$). The turbulent analogs of eq. 10.31 are

$$\frac{Q\bar{u}}{2\pi\alpha_y\alpha_z u_*^2 x^2} \quad \text{or} \quad \frac{Q}{2\pi\bar{u}\alpha_y\alpha_z u_*^2 t^2}. \quad (10.36)$$

The first of these equations is graphed in Figure 10.6a using various values of u_* and $\bar{u}$, and using $\alpha_y = 2.2$ and $\alpha_z = 1.25$. As expected, the higher the friction velocity, the greater the rate of mixing, and the more rapid the dispersal of gametes.

Figure 10.6b,c shows the shape of the sperm plume by giving the contours of equal concentration as a function of y and x. At low average velocities with relatively high values for u_* (Fig. 10.6b), the plume spreads rapidly after release, while at high velocities and relatively low u_* (Fig. 10.6c), the plume is much narrower.

Fertilization Efficiency

The rapid dilution of sperm in the plume raises a question as to whether sperm concentrations in turbulent flow are sufficient to effectively fertilize eggs. Again we rely on sea urchins to provide us with data to evaluate this problem. In a laboratory study, H. Vogel et al. (1982) determined the rate at which various concentrations of sea

Figure 10.6. Sperm concentrations in the turbulent plume. (a) Sperm concentration along the *x*-axis downstream from the point of release as a function of u_* (eq. 10.36). (b,c) Lines of equal concentration downstream of the point of release.

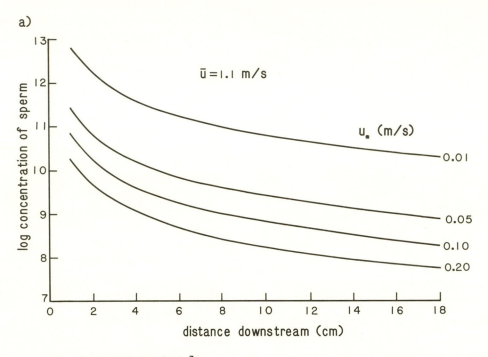

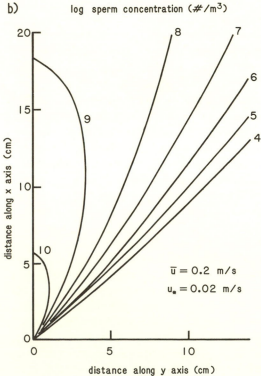

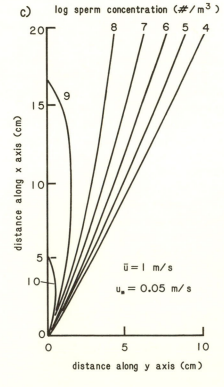

urchin (*Paracentrotus lividus*) sperm fertilized a standard concentration of eggs (6 · $10^9/m^3$). They assumed that the sperm swam at a constant speed but randomly changed direction, and found that a simple equation gave a reasonable fit to their data. The fertilization ratio (i.e., the fraction of eggs fertilized at time t after the introduction of sperm) is approximately

fraction of eggs fertilized, $F = 1 - \exp - (B[S]t)$,
(10.37)

where $[S]$ is the concentration of sperm in numbers/m^3, and B is a constant equal to the product of the fertilizable surface area of the egg and the swimming velocity of the sperm. For reasons that are not yet clear, only 1% of the surface area of a sea-urchin egg allows for fertilization when contacted by a sperm. This effective area is about 10^{-10} m^2. For present purposes we will replace the swimming speed of sperm with the speed at which sperm are carried by the turbulent flow; to a first approximation this is u_*. Thus 10.37 can be rewritten as

fraction of eggs fertilized
$$= 1 - \exp - (10^{-10} u_*[S]t). \quad (10.38)$$

Unless the concentration of eggs is so high that they compete for sperm (an unlikely situation in nature), the fraction fertilized is insensitive to egg concentration.

Now consider the sperm plume along its x-axis, where the concentration is highest. We assume that at some time after release the plume encounters eggs in the water. The concentration of sperm as a function of time after release is given by eq. 10.36. In a small period of time, Δt, the fraction of virgin (i.e., unfertilized) eggs fertilized is

$$F(t) = 1 - \exp - \left\{ \frac{10^{-10} Q}{2\pi \bar{u} \alpha_y \alpha_z u_* t^2} \Delta t \right\}. \quad (10.39)$$

In the first time interval, the fraction fertilized is $F(t_1)$. Note that this interval starts at time $t_1 = D\bar{u}$ where D is the distance downstream from the release point where sperm first encounter eggs. In the second

interval, a fraction $F(t_2)$ of the remaining virgin eggs is fertilized, or in terms of the overall fraction of eggs, $F(t_2) \times (1 - F(t_1))$. We can continue calculating the fraction of virgin eggs fertilized in each interval and sum them appropriately to give the overall fraction:

$F = F(t_1) + F(t_2)(1 - F(t_1))$
$\quad + F(t_3)[1 - (F(t_2)(1 - F(t_1)))] \ldots \quad (10.40)$

This summation can be continued until sperm lose their viability (about 30 min), yielding results such as those shown in Figure 10.7. The faster the $\bar{u}$, the lower the percentage of eggs fertilized (Fig. 10.7a). Even for a very low mean velocity only a few percent of eggs are fertilized in the sperm's viable period. The farther downstream sperm encounter eggs, the lower the F (Fig. 10.7b). Even when sperm encounter eggs only 10 cm downstream (about as close as two urchins could space themselves), the fertilization efficiency is very low. Fertilization efficiencies would be lower still at points away from the x-axis.

The situation is not much improved if the source of eggs lies next to rather than downstream of the source of sperm. A glance at Figure 10.6 shows that high sperm concentrations (above $10^{10}/m^3$) do not extend laterally farther than 2 cm from the x-axis. These calculations suggest that the only effective method for increasing the probability of fertilization is to increase Q, the rate at which sperm are extruded. In order to ensure that more than half of the eggs present at a distance 10 cm downstream from the release point are fertilized, sperm must be released at a rate ten to one hundred times that estimated here (Fig. 10.7c).

These calculations lead one to suppose that unless spawning organisms are very close together, external fertilization in the wave-swept environment may be a highly ineffective mode of sexual reproduction. Mixing due to turbulence is likely to dilute gametes so rapidly that most eggs are never fertilized. In this sense wave-swept organ-

Figure 10.7. Fertilization efficiencies calculated for the sea urchin. The percentage of eggs fertilized within the sperm's viable period (approximately 30 minutes) as a function of (a) mean velocity and (b) distance between the spawning individuals. (c) Fertilization efficiency as a function of the rate of sperm release. These figures were drawn using values of α_x, α_y, and α_z averaged over depth (see Csanady 1973, pg. 69–70, for details).

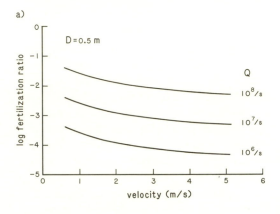

a)

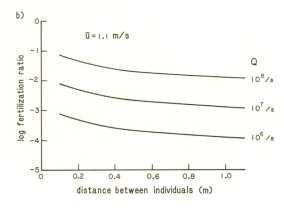

b)

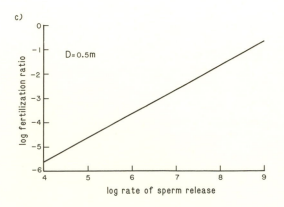

c)

isms may be similar to wind-pollinated plants. Before we accept this surprising speculation, we need to examine carefully the validity of our calculations.

Examining the Assumptions

There are several points in this calculation where we have made dubious assumptions. However, most of these have been conservative in that they are likely to lead one to underestimate the rate of mixing.

1. The value for Q used in these calculations is simply an educated guess. However, as noted, it is likely to be the maximal rate at which sea urchins can extrude semen. In the case of sea urchins and many other invertebrates, semen is extruded through a fine pore at a rate representing the balance between the pressure applied by gonadal muscles and the pipe resistance caused by the viscosity of the semen. Increasing the concentration of sperm in the semen is likely to increase its viscosity, possibly resulting in little increase in Q. As we have seen, the value of Q must be increased by one to two orders to magnitude to substantially affect the fertilization rate.

2. The equation used to estimate the fraction of fertilized eggs was validated in the laboratory in water that was stirred just sufficiently to keep the eggs in suspension. These conditions may bear little resemblance to those in the field, and therefore the quantitative predictions of this equation may be in error. However, it seems unlikely that sperm will be any more successful in the highly sheared, turbulent flow in the natural environment than they would be in the benign flow in a beaker. If anything, eq. 10.37 should overestimate the fraction of eggs fertilized under natural conditions.

3. The approach we have taken here assumes that gametes are released in a form that can readily be mixed with seawater. At least with sea-urchin semen, this seems to be the case. There is no problem involved in diluting sea-urchin semen in water; gentle stirring rapidly disperses the sperm.

4. Eq. 10.35 assumes that the mean flow is unidirectional. This is clearly not the case in most wave-swept regimes. If after a plume is formed the flow reverses direction, the plume is blown back upon itself. This is qualitatively the sort of plume one would expect in intermediate depths under steady waves, where in the absence of any long-shore current the flow oscillates along the x-axis. In the surf zone, when the direction of flow is much less regular, it becomes nearly impossible to predict the exact shape of the concentration distribution. However, in the surf zone the values we have used for $\bar{u}$ and u_* are very likely to be underestimates, and the calculated concentrations are therefore likely to be overestimates regardless of the exact shape of the plume.

In conclusion, although the calculations made here are extremely rough, they are probably robust enough to support the speculation that the turbulent mixing of water in the wave-swept environment could pose a real barrier to effective external fertilization. This speculation has been borne out in one experiment involving sea urchins (Pennington 1985). Here syringes containing a standard concentration of eggs were used to sample water at varying distances downstream of a spawning male. The sperm sucked up by each syringe were allowed several minutes to fertilize the eggs before the contents of the syringe were fixed and examined. Because of this sampling technique, the measured fertilization efficiency is likely to be much higher than that in nature, where sperm are continuously being diluted. However, even under the advantageous conditions of this experiment, the percentage of fertilization decreased rapidly with distance from the male, falling below 15% at a distance only 20 cm downstream.

There are, of course, strategies that would avoid the problem of sperm dilution. For instance, organisms in tidepools at low tide inhabit a confined volume of still water. Under these conditions the concentration of sperm and eggs may be substantial, and the time allowed for fertilization (the time between low and high tide) can be enhanced many thousandfold. Similarly, organisms small enough to live in the interstices of the substratum ($s < D$) could avoid the effects of turbulent mixing to some extent.

Organisms living intertidally but outside of tidepools could increase their fertilization effectiveness by spawning at low tide. Although water movement would not be available to disperse gametes, and the probability of finding another member of the same species is thereby reduced, gametes would not be unduly diluted. If gametes encountered an appropriate neighbor, fertilization could be effectively achieved. This strategy would seem to be most advantageous to mobile organisms that could congregate and thus virtually ensure that egg and sperm would meet.

Mixing and Chemical Cues

There are many other aspects of wave-swept life that are affected by turbulent mixing. For instance, many organisms have been shown to sense chemical cues in the water. A good example is the sensing at a distance of potential predators. For instance, limpets and abalone perform stereotypic escape behaviors when they are immersed in water that has flowed past a starfish (e.g., *Pisaster ochraceus*). This sort of "early warning system" is effective only if the chemical cue is swept from the starfish to the limpet at a distance great enough to allow the limpet to escape before the starfish arrives. Conversely, the cue can be sensed only if the chemical is present in the water at sufficient concentration. This situation is clearly open to the same sort of analysis we have applied to external fertilization. This approach has been used to examine problems of terrestrial chemoreception (e.g., the detection of pheromones by moths), and Okubo (1980) provides a good review.

Figure 10.8. Turbulence intensities over a congrega-
tion of limpets. The intensity $\overline{u'^2}$ peaks at the top of
the limpets. (Redrawn from Gallien, unpublished.)

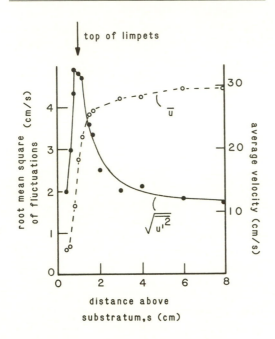

Turbulence and Suspension Feeding

Before we leave this discussion, we will
very briefly examine one final role of turbulent mixing. Passive suspension feeders rely
on water motion to bring food-bearing water to their vicinity. If the benthic boundary
layer were laminar, these organisms would
experience difficulties akin to that of the
algal frond examined at the beginning of
this chapter: an organism located well
away from a leading edge would be trying
to feed from water already sampled by
other suspension feeders. Even active suspension feeders might have problems under
these conditions— one mussel in the middle of a large bed might only be pumping
prefiltered water.

As with the example of an algal frond,
turbulent mixing probably alleviates this
problem. In fact, the mere presence of organisms on the substratum contributes to
the production of turbulence. Careful measurements of the spatial pattern of turbu-

lence (Nowell and Church 1979; Antonia
and Luxton 1971, 1972; Gallien, unpublished) have shown that the intensity of tur-
bulence ($\overline{u'^2}$) and $\overline{u'w'}$ reach a maximum
near the top of roughness elements in both
equilibrium and nonequilibrium turbulent
boundary layers (Fig. 10.8). Thus barnacles,
mussels, anemones, etc., may be increasing
their own feeding efficiency by acting as
roughness elements. This effect has been
noted for zooanthids (Koehl 1977d) and for
one species of octocoral (Patterson 1984).

Other Approaches
to Turbulent Diffusion

Here we have taken a simplified, traditional
approach in exploring the biological effects
of turbulent mixing, and in the process have
ignored at least two aspects of turbulent diffusion. The first of these deals with the dispersion of particles in a velocity gradient.
Consider the small group of particles shown
in Figure 10.9a. All the particles lie at the
same height in the boundary layer, and
therefore move at the same average velocity. However, if particles are mixed up or
down in the velocity gradient, they encounter fluid that, on average, travels faster
or slower. Particles that are transported
upward move faster than the center of
mass of the particle cloud, and particles
that are transported downward move
slower. As a result, the initially spherical

Figure 10.9. Shear dispersion. Particles mixed up
or down in a velocity gradient are dispersed relative
to the center of mass of the particle cloud (see text).

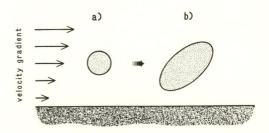

cloud is stretched in the direction of mean flow, and the variance of the cloud is increased (Fig. 10.9b). This mechanism of diffusion is known as *shear dispersion* (Csanady 1973; Okubo 1980). It is particularly important where the velocity gradient is steep—in other words, for turbulent boundary layers in the viscous sublayer or in the innermost portion of the buffer layer (Chapter 9). The biological examples we have examined here operate primarily in the outer regions of the boundary layer where velocity changes little with distance from the substratum, and in these cases we are probably justified in ignoring shear dispersion, at least for the level of accuracy we require. For processes operating nearer to the substratum, the neglect of shear dispersion can lead to a serious underestimate of the rate of turbulent mixing. The cautious reader should consult Csanady (1973) for an indepth discussion.

We have also ignored recent experimental results that bring into question the basic assumptions of our traditional approach to turbulent mixing. In photographic studies of boundary layers at high Reynolds numbers, various researchers have found large-scale coherent structures similar to those observed in "vortex streets" (see Chapter 11) at lower Reynolds numbers (e.g., Cantwell 1981). In this emerging view of turbulence structure, large masses of fluid maintain their identity for considerable distance as they are stirred about, and adjacent volumes of fluid interact only on a very fine spatial scale. These coherent motions cannot be reconciled with the assumption of random movement that is basic to the traditional approach to turbulent transport. At some point in the near future, it is likely that this new evidence will lead to a revised theory for the mechanism of turbulent diffusion. This does not necessarily mean that the usefulness of the traditional approach will be supplanted. Despite its theoretical shortcomings, the traditional approach used here has given accurate predictions in many circumstances. At present the limiting factor in our understanding of the biological effects of turbulence is not the limitations of theory, but rather it is a near total lack of accurate measurements of water flow. There is ample room for exploration on this important aspect of the wave-swept environment.

CHAPTER 11

~ ~ ~ ~ ~ ~ ~ ~ ~ ~ ~

Hydrodynamic Forces

In chapters 4 through 7 we saw how waves are accompanied by complex water flows. We now use this information in an attempt to understand and predict the forces that these water motions impose on wave-swept organisms.

Flow Forces

The Origin of Pressure Drag

Consider the situation shown in Figure 11.1a. An inviscid fluid flows past a horizontal circular cylinder, here seen end-on. The bulk of the fluid moves at a constant velocity, u_0, although, as we can see from the pattern of streamlines, the velocity varies in the vicinity of the cylinder. In particular, the velocity along the top and bottom of the cylinder is higher than mainstream, and the velocity at points directly upstream and downstream is lower. By applying Bernoulli's equation, we can immediately say that the pressure is high on the upstream and downstream sides of the cylinder, and that it is low laterally. The momentum the fluid gains as its velocity increases around the sides of the cylinder is just sufficient enough to be traded for an increase in pressure behind the cylinder. As a result, the pressures upstream and downstream are identical. Given the symmetry of the system, it is easy to see that the pressures are also identical on the top and bottom. Now, pressure imposes a force proportional to the area over which it acts. Thus the high pressure imposed on the upstream (left) side of the cylinder exerts a force that tends to push the cylinder to the right. This force is offset by the force exerted by the high pressure downstream

Figure 11.1. (a) Flow of an inviscid fluid past a cylinder. The downstream flow pattern and pressure distribution are mirror images of those upstream. (b) Flow of a viscid fluid past a cylinder. Because the flow separates, the pressure downstream is lower than the pressure upstream. (c) Boundary-layer flow and separation (see text).

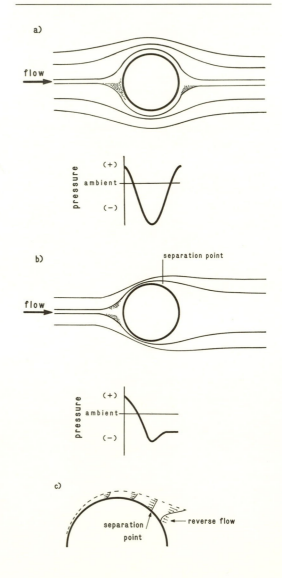

that pushes the cylinder to the left. The same argument can be applied to the forces acting laterally, and we can conclude that the cylinder feels no net force as an inviscid fluid moves by it at a constant rate.

We reach a different conclusion, however, when the effects of viscosity are taken into account (Fig. 11.1b). Here the flow upstream is much the same as that for an inviscid fluid: the fluid has a low velocity and a high pressure. However, in this case the fluid at the surface of the cylinder is stationary, due to the no-slip condition, and a velocity gradient (i.e., a boundary layer) is established. This boundary layer increases in thickness as the fluid moves along the cylinder's surface (Fig. 11.1c).

It is the behavior of the fluid in the boundary layer that results in a force being exerted on the cylinder. On the upstream portion of the cylinder, fluid in the boundary layer moves from an area of high pressure to an area of lower pressure along the sides. Because it is moving from high to low pressure along what is termed a favorable pressure gradient, the fluid speeds up. However, the fluid in the boundary layer attains less speed (and therefore less momentum) than it would in the absence of viscosity (see Chapter 9), and this loss of momentum has an important consequence. Along the side of the cylinder the pressure, although still low, begins to increase (Fig. 11.1c), forming an adverse pressure gradient. Because the fluid in the boundary layer has lost some momentum to viscosity, it cannot travel as far into this adverse gradient as it otherwise would and is brought to a halt. At this point flow *separates*, deviating from the path it would have taken if the fluid were inviscid. Beyond the separation point, flow in the boundary layer may actually be reversed. When this happens the boundary layer "rolls up," forming eddies that move downstream as a wake. The pressure in the wake is approximately equal to the pressure at the separation point, and it is considerably lower than the one acting on the upstream face of the

cylinder. The farther upstream the flow separates, the lower the wake pressure, and the larger the difference in pressure between front and back. This difference in pressure results in a net force—the *drag*—acting in the direction of flow. Because this drag is caused by a fore-aft difference in pressure, it is often referred to as *pressure drag* to differentiate it from skin friction (Chapter 3). The total drag on any object is the sum of pressure drag and skin-friction drag.

Due to the symmetry of the system, the top and bottom of the cylinder are at the same pressure, and there is no net lateral force.

Because drag is largely due to a difference in pressure, we can assume that it is proportional to S_p, the area of the object projected in the direction of flow (force = pressure × area), and to the average difference in dynamic pressure between the upstream surface of the object and the wake. Thus

$$f_d = \tfrac{1}{2}\rho u^2 S_p C_d, \qquad (11.1)$$

where f_d is the drag and C_d, the *drag coefficient*, is the appropriate coefficient of proportionality. This equation is used as a definition of the drag coefficient:

$$C_d \equiv \frac{2f_d}{\rho S_p u^2}. \qquad (11.2)$$

Because the wake pressure depends on the location of the separation point, the magnitude of the drag coefficient depends on the precise pattern in which fluid moves around the object in question.

For streamlined shapes (fish, for instance) flow is less likely to separate, and drag is entirely due to skin friction. In such a case it is more appropriate to use the entire wetted surface area of an object, S_w, in the definition of the drag coefficient. Because the wetted area is larger than the project area, this alternative definition leads to a smaller value for the drag coefficient than that calculated using S_p. For a more complete discussion of the many ways in which

C_d can be defined, see Vogel (1981). Unless otherwise noted all drag coefficients used here are based on the projected area.

Added Mass and the Acceleration Reaction

To this point we have assumed that fluid moves past the cylinder in a steady fashion. What happens if the fluid is accelerating? Consider a horizontal cylinder immersed in an infinite sea of inviscid fluid and moving relative to the fluid at velocity u_0. This is equivalent to having the object stationary and the fluid moving at velocity u_0; but in this particular case it is easier to follow the action if it is the object that moves. Because we have assumed that the fluid has no viscosity, no velocity gradient is set up around the cylinder, and no fluid is "dragged along" by the viscous nature of the fluid. However, this does not imply that the fluid remains stationary as the cylinder moves by. Obviously, the fluid in front of the cylinder must get out of the way, and other fluid will fill in the space left behind.

The precise motion of the fluid is complicated and varies with distance from the cylinder. A small volume (dV) of fluid near the upstream or downstream side of the cylinder moves at a speed approaching that of the cylinder itself, while fluid elsewhere moves more slowly. Each volume of moving fluid has a certain momentum ($\rho\,dV\,u$). In a theoretical treatment of the fluid motion accompanying a moving cylinder, Darwin (1952) summed all these individual momenta to calculate the total momentum of the fluid. Although the total momentum is distributed throughout the fluid, its magnitude is the same as that of a volume of fluid equal to the cylinder traveling at the same speed as the cylinder. In other words, even in the absence of viscosity, the fluid behaves *as if* a cylinder-sized, equivalent volume were dragged along with the cylinder. This equivalent volume is unaffected by the velocity of the cylinder.

We are now in a position to calculate the force required to accelerate a cylinder of mass m and volume V in an inviscid fluid. If we wish to accelerate the cylinder alone with an acceleration, a, we must apply a force $f = ma$. However, as the cylinder accelerates, the equivalent volume of fluid behaves as if it moves with the cylinder, and this volume has mass ρV. Thus an additional force must be applied to accelerate this *added mass*, the magnitude of the additional force being ρVa. To a first approximation, the presence of viscosity does not change this situation, and the concept of added mass is applicable to real as well as inviscid fluids.

It is a coincidence peculiar to cylinders that the equivalent volume of fluid carried along is exactly equal to the volume of the object itself. The ratio of equivalent fluid volume to object volume is different with different shapes. Thus the force required to accelerate an object in a fluid is, in general,

$$f_a = a(m + \rho C_a V),$$

moving object, stationary fluid,   (11.3)

where C_a is the ratio of the added volume of fluid to the volume of the object itself, and is termed the *added mass coefficient*.

The magnitude of the equivalent volume depends on the manner in which fluid flows around the object. Thus changes in the flow pattern due to viscous effects, the presence of local eddies in the fluid, or any other sort of imposed flow pattern can affect the added mass coefficient.

To this point we have considered the force necessary to accelerate an object in a stationary fluid. When dealing with wave-swept organisms, we are much more likely to encounter a situation where fluid moves past a stationary object. What is the force imposed on the object in this case? Because the object itself does not move, there is no force required to accelerate its mass. However, there is an added mass of fluid that is "carried along" by the object, remaining stationary while the fluid accelerates. We might be tempted to assume that the force due to the water's acceleration is simply that required to accelerate the added mass, $\rho C_a Va$, where a is the acceleration of the fluid relative to the stationary object. This force is indeed imposed, but it is not the only force.

To explain the origin of the remaining force, we carry out the following thought experiment. If the stationary object were not there, a volume of fluid equal to the object's volume would take its place. This fluid, having nothing to hold it back, would, like all the fluid around it, accelerate with rate a. Thus by simply being stationary the presence of the object has the same effect as accelerating a volume of fluid equal to its own volume in the direction opposite the fluid movement. This acceleration requires a force ρVa. This additional force is called the *virtual buoyancy*, by analogy to the buoyancy felt by objects in a stationary fluid when acted upon by the acceleration due to gravity. Thus the total accelerational force applied by a fluid to a stationary object is the *acceleration reaction*,

$$f_a = \rho(1 + C_a)Va = \rho C_m Va,$$

stationary object, moving fluid,   (11.4)

where C_m is the inertia coefficient.[1] Note that this is the force due solely to the fluid's acceleration. Except for the initial instant when a stationary fluid begins to move, this acceleration reaction is accompanied by a drag force. The total force along the direction of flow is the sum of these two forces:

total force in direction of flow
$$= \tfrac{1}{2}\rho u^2 S_p C_d + \rho C_m Va. \qquad (11.5)$$

This is known as the Morison equation (after J. R. Morison et al. 1950), and it

___

[1] Note that this definition is different from that of Batchelor (1967), who does not include the virtual buoyancy in the acceleration reaction. Because in this text we are primarily concerned with sessile organisms, on which a virtual buoyancy is exerted, it is handier to lump the two forces together under the term "acceleration reaction."

assumes that the drag and acceleration reaction act independently and therefore can legitimately be added. There is some question as to whether this is strictly true (Sarpkaya and Isaacson 1981; Sarpkaya and Storm 1985), but to date the Morison equation is the only approach that has gained widespread acceptance. We use the Morison equation here without further discussion, though we must keep in mind that it is a less-than-exact description of the real world.

We should also note that although the acceleration reaction acts along the line of flow, it need not act in the same direction as flow. For example, if the flow has a velocity to the right but this velocity is decreasing, the direction of acceleration is to the left. In this case the drag force and the acceleration reaction oppose each other.

Lift

In addition to drag and the acceleration reaction, a force known as *lift* acts at right angles to the direction of flow. It is the same sort of force that keeps birds and airplanes aloft, but we will not limit ourselves to defining lift as a force that acts in the direction opposite gravity. Instead, any force that acts normal to the direction of flow, be it down, up, or sideways, qualifies as a lift force.

Consider a horizontal circular cylinder oriented with its axis perpendicular to the direction of flow. As we have seen, water flowing around a cylinder speeds up as it passes along the sides, and then slows down (more or less) as it leaves the cylinder behind. Because the cylinder is symmetrical, the low pressure on one side is just offset by the low pressure on the other side, and there is no net force perpendicular to the flow.

Imagine now what would happen if the cylinder were to be sliced in half along the direction of flow (Fig. 11.2a). If we ignore the weight of the cylinder itself, in the absence of flow no force is required to keep

Figure 11.2. Lift. (a) The two lateral halves of a cylinder experience offsetting lateral forces. (b) A half-cylinder on the substratum experiences a lift due to the pressure difference between the internal pressure and the pressure of the moving fluid.

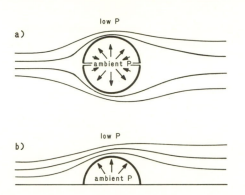

and the water outside leads to a *net* lift force on the half-cylinder.

This kind of lift force is responsible for lifting the roofs off buildings in tornadoes. If the windows of the building are closed, the air inside remains at approximately ambient pressure. The wind speeds up as it whistles over the roof, leading to a pressure difference between the inside and the outside. If the velocity of the wind is sufficient, the roof comes off, or, alternatively, the windows explode outward.

The same phenomenon can be observed under less stressful conditions. Semitractor trailers carrying grain or vegetables often have a fairly solid boxlike structure with an open top. During transport this top is covered with a canvas tarpaulin. As the truck cruises down the highway, the wind passing over the top of the trailer creates a pressure difference between the inside and the outside of the trailer and the tarp bulges upward.

It should be obvious from these examples that this sort of lift is not specific to cylindrical shapes. In fact, almost any shape will do. A limpet is an excellent example of how this principle operates for a wave-swept organism (Fig. 11.3). The edge of the conical shell is closely applied to the surface of the substratum. As fluid flows past the shell·the average pressure around the shell's edge is ambient or slightly higher, and this pressure is communicated to the foot, guts, and any free water trapped under the shell. Fluid flowing over the top of the shell is at a pressure considerably lower than ambient, and the stage is

the two halves together. The ambient pressure of the water pushing in on the cylinder's surface is just offset by the internal pressure of the solid cylinder, and the cut faces of the cylinder are thus at ambient pressure. However, in flow the water at the sides of the cylinder is at lower than ambient pressure due to its increased speed. Thus, in a flowing fluid there is a pressure difference between the faces of the cut and the fluid flowing past the cylinder. This pressure difference imposes a lift, and a force would be required to keep the cylinder halves together. It is only because the lift on the two halves are equal and opposite that this force is not noticed when the cylinder is intact. In the intact cylinder this adhesive force is provided by a slight deformation of the cylinder.

Consider now one of the half-cylinders situated with its cut face attached to a flat wall (Fig. 11.2b). Again, the internal pressure of the half-cylinder is assumed to be at ambient pressure. However, any water flowing past the half-cylinder behaves as it would if it flowed around a whole cylinder far away from a solid surface. The water speeds up as it moves around the cylinder and therefore is at less than ambient pressure. The resulting pressure difference between the material of the cylinder

Figure 11.3. Lift on organisms. A limpet is subject to lift.

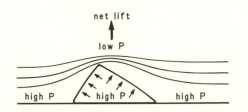

thus set for the imposition of a lift. This lift has been measured on models of limpets placed in waves and it can be quite substantial, often being larger than the drag force (Denny 1985a).

Many other wave-swept organisms should similarly be subject to lift forces. Starfish, chitons, and snails are but a few of the more obvious examples. Arnold and Weihs (1978) have described the lift imposed on a flatfish, the plaice. When the water velocity is too high, this sedentary fish tends to lift up into flow and be carried downstream. To avoid this problem, the fish employs an entire suite of behaviors, including using its fins to pump water under its body, thereby reducing the velocity difference between its upper surface (exposed to nearly mainstream flow) and its lower surface, which rests on the sandy substratum. One example that may not be so obvious is that of mussels in a mussel bed. Water flowing over the top of a tightly packed mussel bed has a lower pressure than the fluid trapped in the interstices of the bed. The result is a lift. Figure 11.4 shows the pressure difference measured across a mussel bed when the mainstream flow was approximately 4 m/s.

Lift forces arise from the same basic mechanism (a pressure difference) as the pressure drag, and by analogy the lift is proportional to the dynamic pressure and to the area over which the pressure difference acts. In this case the area is not that projected in the direction of flow, but rather it is the area projected in the direction of the lift (i.e., perpendicular to the flow). This area is often referred to as the *planform area*, S_{plan}. Thus

$$f_l = \tfrac{1}{2}\rho u^2 S_{\text{plan}} C_l, \qquad (11.6)$$

where C_l, the *lift coefficient*, is defined as

$$C_l \equiv \frac{2f_l}{\rho u^2 S_{\text{plan}}}. \qquad (11.7)$$

The magnitude of this lift coefficient depends on the precise flow pattern around the object. Although the form of the equa-

Figure 11.4. Pressure difference measured across the thickness of a mussel bed. Pressure was lower near the tops of the mussels than at the substratum, resulting in lift. Water velocity was approximately 4.4 m/s, resulting in a lift coefficient (pressure difference/$(\tfrac{1}{2})\rho u^2$) of about 0.8.

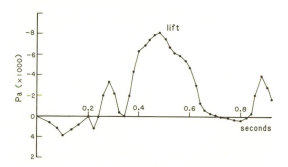

tion predicting the lift force is quite similar to that for pressure drag, it cannot be assumed that the variation in C_l will be similar to that of C_d.

The expression for lift given in eq. 11.6 allows us to complete the description of the overall force on an object in flowing fluid. Because lift acts at right angles to the drag and acceleration reaction, we must account for the direction as well as the magnitude of the overall resultant force. The Morison equation can be expanded to suit this purpose:

$$\text{force} = \sqrt{\{\tfrac{1}{2}\rho u^2 S_p C_d + \rho C_m Va\}^2 + \{\tfrac{1}{2}\rho u^2 S_{\text{plan}} C_l\}^2}$$

$$(11.8)$$

$$\text{direction} = \arctan\left\{\frac{\tfrac{1}{2}\rho u^2 S_{\text{plan}} C_l}{\tfrac{1}{2}\rho u^2 S_p C_d + \rho C_m Va}\right\}.$$

$$(11.9)$$

Figure 11.5 gives an example of the application of these equations, using a limpet as an example.

The lift we have examined so far is steady—as long as the velocity is constant, the lift is constant. In addition, a second, transient lift may act on wave-swept objects. As we have seen, the movement of water around an object is often accompanied by the shedding of eddies into the

Figure 11.5. The combined action of forces on a limpet.

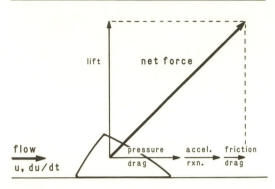

object's wake (Fig. 11.6). Often the eddies are not shed simultaneously from both sides of the object. In this case the shedding of eddies can result in a transient difference in velocity between the two sides, causing a difference in pressure, or a lift. The direction of this lift reverses as eddy shedding alternates from one side of the object to the other. This cyclically varying lift can cause the object to vibrate; this is the mechanism behind the thrumming vibration of power lines and ships' halyards in high winds. For a thorough discussion of this phenomenon, consult Blevins (1977).

Figure 11.6. As vortices are shed from a cylinder (alternating side to side), a transient lift is imposed.

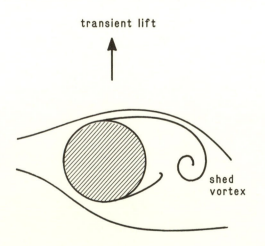

Because the direction of the transient lift reverses periodically, its temporal average is zero. The frequency with which vortices are shed is a function of the size and shape of the object and the water velocity. These factors will be examined later in the discussion of the Strouhal number.

Reynolds Number and Period Parameter

The magnitude of C_d, C_l and C_m depends on the precise flow patterns past an object. Unfortunately, the pattern of flow depends on a substantial number of factors, the salient ones being the size and shape of the object, the velocity of the fluid, the magnitude and duration of velocity oscillations, and the density and viscosity of the fluid. It would be possible to measure the drag, lift, and inertia coefficients separately as a function of each of these variables, but the work involved would be immense, and the presentation of results awkward. Instead, it is better to measure coefficients as functions of certain combinations of these variables. There are any number of ways in which the variables listed above can be combined, but two are particularly useful. Both combine variables in such a fashion that the result is a dimensionless number.

The first of these dimensionless numbers is the Reynolds number, Re. As we discussed in Chapters 3 and 9, the beauty of the Reynolds number lies in its ability to predict similarity of flow: in steady flow, equality of Re implies similarity in flow pattern. Thus a large cylinder in slow flow can have the same Re as a small cylinder in fast flow, and the flow patterns around the two are the same. As we will see, unsteady flow can throw a wrench into the works.

To calculate a Reynolds number, we need to know what value to use for *L*, the length scale. The convention when dealing with hydrodynamic forces is to use a characteristic length of the object, usually its length in the direction of flow.

An example will serve to illustrate the calculation of a particular Reynolds number. A limpet has a shell length of 0.02 m. As a wave rushes over the animal, a water velocity of 10 m/s is reached. The density of seawater is roughly 1000 kg/m^3, and its dynamic viscosity is 0.001 Pa s. Thus

$$\text{Re} = \frac{\rho u L}{\mu} = \frac{(1000 \text{ kg/m}^3)(10 \text{ m/s})(0.02 \text{ m})}{0.001 \text{ N·s/m}^2}$$
$$= 2 \cdot 10^5.$$

Reynolds numbers are not something one needs to calculate to any great degree of accuracy. It is just as useful to know that Re is approximately $2 \cdot 10^5$ as it is to know that (using exact values for ρ and μ) Re exactly equals $1.9123 \cdot 10^5$.

What range of Reynolds numbers are likely to be encountered in the wave-swept environment? We are dealing with seawater, so the density will always be approximately 1025 kg/m^3. The viscosity of seawater varies with temperature, from 0.9 to $1.4 \cdot 10^{-3}$ Ns/m^2 over the temperature range found in the oceans. The plants and animals we will be examining have sizes that in their maximum dimension vary from 1 mm up to about 10 m for some macroalgae. Flow velocities are generally fast, ranging from 0.5 m/s to 20 m/s. Thus the lowest Re we could expect to encounter (the smallest organism at the lowest velocity with the highest viscosity) would be on the order of 10^2. The highest Re would be on the order of 10^8, but a Re this high will be seen only infrequently. The vast majority of organisms are less than 0.1 m in maximum dimension, and for them the maximum Re is on the order of 10^6.

Few organisms in the wave-swept environment operate at Reynolds numbers around 1, and consequently we will not explore the mechanics of fluids where viscous forces are of a magnitude equal to inertial forces. Those readers interested in the many fascinations of life at low Re should consult Purcell (1977), Vogel (1981), or Berg (1983).

In this chapter we use Re primarily as a means of expressing the variations in drag, lift, and added mass coefficients. For example, we might be working with a species of snail, the individuals of which all have the same shape but whose length can vary by an order of magnitude from 5 mm to 5 cm. This species lives in an area exposed to tidal currents where water velocities vary from less than 1 m/s up to 10 m/s. We would like to be able to calculate the drag imposed on any member of the species, but to do this we need to know the drag coefficient, and the drag coefficient varies with both the size of the organism and the water velocity. A brute force approach to this problem would be to measure the drag coefficient for each size of snail at a series of water velocities between 1 and 10 m/s. The work involved, however, would be far more than necessary. A large snail (length 5 cm) at 1 m/s has the same Reynolds number ($5 \cdot 10^4$) as a small snail (length 5 mm) at 10 m/s. The equality of Reynolds number implies that the flow patterns around the two organisms are equivalent, and that the drag coefficients should be the same. Thus, rather than measure each size of individual at each velocity, we need only measure the drag coefficients over the range of Reynolds numbers experienced by the species.

To get the right Re we can fiddle with either the size of the snail or the water velocity.[2] The result of these measurements is a graph of the sort shown in Figure 11.7. Given this graph one need only know Re to be able to ascertain the drag coefficient.

The Reynolds number does not incorporate any factors relating to oscillatory flow. Thus, by specifying Re we have not completely specified the flow pattern, and for

[2] We could also vary the viscosity or density, but this would involve changing temperature, salinity, or even the type of fluid itself (to glycerine or alcohol, say). These options are often not practical when using live animals.

Figure 11.7. The drag coefficient (C_d) of a circular cylinder as a function of Reynolds number. (Redrawn from S. Vogel, Life in Moving Fluids, 1981, Wadsworth Inc., by permission of Brooks/Cole Publishing Co., Monterey, Calif.)

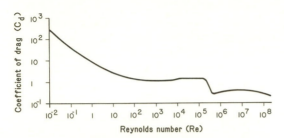

a more complete description we need another dimensionless number, the *period parameter K*. The idea behind the period parameter is quite simple. Consider an object with dimension L along the direction of flow. The flow past the object travels first in one direction, stops, and then flows back in the other direction. If the distance the water moves when flowing in one direction is large relative to the length of the object, the flow pattern is much as it would be if the flow were unidirectional. Any aberrations in flow due to the changing direction have sufficient time and distance to dissipate as the flow continues to move in one direction. If, however, the distance moved by the fluid in one direction is not much larger than the length of the object, any peculiarities due to a change of direction do not have sufficient time or distance to dissipate before the fluid again changes direction.

In a harmonically oscillatory flow, the distance traveled in one direction is proportional to $u_m T$, where u_m is the maximum velocity reached in each excursion and T is the overall period of oscillation. This distance can be compared to a characteristic length of an object to arrive at a definition of the period parameter:

$$K = \frac{u_m T}{L}. \qquad (11.10)$$

Because this is a ratio of lengths, it is di-

mensionless. When the period parameter is large, it implies that the fluid travels many times the length of the object in each oscillatory excursion, and when K is greater than about 30, the flow patterns may be expected to mimic those in steady flow (Sarpkaya and Storm 1985). When the period parameter is small, the fluid travels a distance comparable to the size of the object, and the flow patterns may be substantially affected by changes in flow direction.

What range of period parameters are encountered in the wave-swept environment? The period of flow oscillation is generally the same as that of the incoming waves, on the order of 10 s, and the velocity ranges from 0.5 to perhaps 20 m/s. For animals between 1 mm and 10 cm in length, this implies a range of period parameters between 50 and $2 \cdot 10^5$. Thus the period parameter is large enough for these organisms so that the oscillatory nature of the flow does not substantially affect the flow pattern around them. As a result, the drag, lift, and inertia coefficients are nearly independent of K. This considerably simplifies the prediction of the hydrodynamic forces imposed on wave-swept plants and animals. For a thorough discussion of the problems that can arise at $K < 30$, see Sarpkaya and Isaacson (1981).

Force Coefficients and the Reynolds Number

The drag, lift, and inertia coefficients vary as a function of Reynolds number, and we now turn to the pattern of these variations. The following discussion assumes that the period parameter is greater than 30.

Drag

We begin our examination of drag by making a careful observation of the flow patterns around a cylinder at various Reynolds numbers.

At low Reynolds number ($10 < \text{Re} < 40$ or so) the flow around a smooth circular cylinder closely matches that shown in Figure 11.8a. Streamlines are smooth, the flow is laminar, and the only noticeable difference between this situation and that at a very low Re is the presence of a pair of eddies behind the cylinder. Fluid coming around the sides separates and veers off downstream. The "gap" left in the fluid by this change in direction is filled by a volume of water that remains with the cylinder, circulating around in two counterrotating, attached eddies or vortices (Fig. 11.8b).

For reasons that are not entirely understood, at a Reynolds number around 40 the steady nature of the flow gives way to an oscillatory behavior. The flow coming around one side of the cylinder gains the ascendancy, and its attached vortex is pushed medially, causing the other vortex to break loose from the cylinder and be shed downstream. In the process a new attached vortex is established that pushes

Figure 11.8. Flow around a circular cylinder (see text). (Redrawn from S. Vogel, *Life in Moving Fluids*, 1981, Wadsworth Inc., by permission of Brooks/Cole Publishing Co., Monterey, Calif.)

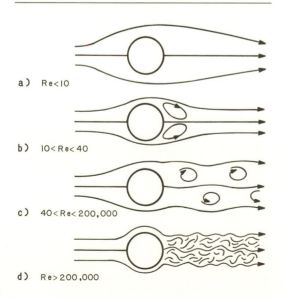

a) Re < 10

b) 10 < Re < 40

c) 40 < Re < 200,000

d) Re > 200,000

Figure 11.9. Strouhal number for a circular cylinder as a function of Reynolds number. (Redrawn from Lienhard 1966.)

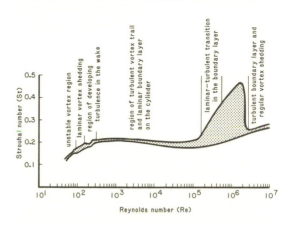

the remaining vortex laterally and causes it to be shed. The process continues, vortices being shed alternately from the two sides, forming a "vortex street" in the wake of the cylinder. If the velocity of flow is increased, the frequency with which these vortices are shed increases. Over a considerable range of Reynolds numbers, the frequency-to-velocity ratio is constant. This constancy is the impetus for defining another dimensionless number, the *Strouhal number*,

$$\text{St} = \frac{fL}{u}, \qquad (11.11)$$

where f is the frequency of vortex shedding, u is the fluid velocity, and L is a characteristic length of the object (i.e., the diameter of a cylinder). For smooth cylinders in the Re range of 10^2–10^5, St is approximately constant at 0.2 (Fig. 11.9). For example, a cylinder 1 cm in diameter in a flow of 10 m/s sheds vortices at a rate of about 200/s. At 1 m/s, vortices are shed at a rate of 20/s. For cylinders that are flexible enough to oscillate slightly, the range of constant Strouhal number can be extended out to Re of about 10^7.

As velocity increases, the rate of vortex shedding increases, and each vortex is shed in a less discrete manner. Above a Re of 500 or so, although the frequency of vortex shedding can be identified as the dominant frequency of velocity fluctuation in the wake, to casual observation the wake appears chaotic or turbulent. The general flow pattern presented by a cylinder with a turbulent wake has already been introduced. Water approaches the cylinder in an orderly, laminar fashion and flows around the sides of the cylinder. In the process, momentum is lost to viscosity, and at some point the flow separates from the cylinder and joins the turbulent wake. The separation point moves forward on the cylinder as the velocity increases. As long as this general flow pattern is maintained, the drag imposed on the cylinder increases in proportion to the square of the velocity, and the drag coefficient is constant (see eq. 11.2 and Fig. 11.7).

At Reynolds number below 10^5, the effects of turbulence are confined to the wake. However, at a critical Reynolds number the boundary layer itself becomes turbulent. As we have seen (Chapter 9), the shape of the velocity gradient in a turbulent boundary layer is such that relatively high velocities are present closer to a solid substratum than in a laminar boundary layer. For this reason turbulent boundary layers are less susceptible to separation, allowing the separation point to move downstream on the cylinder. The result is a smaller wake with a higher pressure. As a result there is a drastic decrease in the pressure drag, and the drag coefficient goes down. For a smooth circular cylinder, the transition from a laminar to a turbulent boundary layer can be quite abrupt (Fig. 11.7). Spheres, which have a C_d vs. Re curve that very nearly parallels that of cylinders, have a turbulent transition that is so abrupt and the drop in C_d so large that an increase in velocity from Re just below transition to Re above it actually results in a decrease in drag (not just a decrease in C_d). As has often been noted (Shapiro 1961; Vogel 1981), this is the reason golf balls have

dimples. A well-driven golf ball travels at a velocity corresponding to Re just a bit too low to reach the transition to a turbulent boundary layer. The irregularities introduced by the dimples are sufficient to initiate transition at Re lower than usual, below the Re of a driven ball. The drag is consequently reduced, and the travel of the ball is enhanced. (It should be noted that the decrease in drag afforded by surface irregularities such as dimples is effective only for Re's near the transition. At Re's that are either higher or lower, the presence of surface irregularities can substantially *increase* drag.)

After the turbulent transition has occurred, a further increase in Re does not result in any gross change in the flow pattern. The drag coefficient rises slowly, but no more drastic changes have been noted.

Because of the importance of smooth circular cylinders in engineering practice, their C_d vs. Re curve (Fig. 11.7) is the one most often reported in books on fluid mechanics. It is indeed instructive in that it dramatically demonstrates the changes in flow pattern as Re increases. However, from a biological standpoint this curve is much less useful and may actually be misleading. Very rarely does a wave-swept organism closely resemble a circular cylinder, and those that do are almost never smooth. What effects do roughness and changes in shape have on the nature of the C_d vs. Re curve?

We can begin to answer this question by examining the drag of rough cylinders. Figure 11.10 shows the C_d vs. Re behavior near the turbulent transition point for cylinders of varying roughness. The roughness is measured as the ratio of the mean diameter of the roughness elements k expressed as a fraction of the cylinder's diameter D. As the roughness increases, the Re at which the turbulent transition occurs decreases. More important, the change in C_d associated with the transition is drastically decreased. For cylinders with roughness elements one-fiftieth the diameter, the effects of the turbulent transition on the drag coefficient are barely noticeable. A ratio of $k/D = 1/50$ amounts to roughness elements of only 200 μm for a biological cylinder 1 cm in diameter. Most biologists would consider such a "rough" biological cylinder to be quite smooth. Thus the natural roughness of biological cylinders probably renders them more or less immune to the abrupt change in C_d accompanying turbulent transition, and their C_d vs. Re curves are expected to be more orderly.

Furthermore, the behavior exhibited by circular cylinders cannot necessarily be extrapolated to other shapes. As a pertinent example, we examine the behavior of flat plates. Because of a plate's sharp edges the flow around a thin flat plate oriented perpendicular to flow must be quite different from that around a cylinder. Even at quite low Re the flow moving from the front of the plate to the back cannot follow the contours of the plate. To do so would require making an abrupt 180° turn at the edge. The inertia of the fluid resists any such quick about face, and the flow separates at the edge. With the separation point effectively imposed on the flow, any transition to a turbulent boundary layer

Figure 11.10. The drag coefficient of circular cylinders as a function of Reynolds number. The larger the size of the roughness elements on the cylinder (k, expressed as a fraction of the cylinder's diameter, D), the less pronounced the turbulent transition. (Redrawn from Sarpkaya 1976 by permission of the author.)

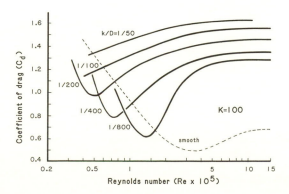

does not substantially affect the flow pattern, and consequently a flat plate does not experience the dip in C_d seen by a cylinder. The drag coefficient for a thin plate oriented perpendicular to a steady flow is virtually constant at 1.17 for all Re above 10^3 (Hoerner 1965; Vogel 1981).

In contrast, when oriented parallel to flow a thin plate causes very little disturbance of the flow pattern. If it were not for viscosity, a very thin plate would cause essentially no change in pattern at all regardless of the Re. Because of the no-slip condition, however, a real fluid flowing near the surface of the plate is retarded to form a boundary layer, and a force is required to establish this velocity gradient, as described earlier (Chapter 4). This friction drag does not depend on pressure differences at all—it consists solely of the force required to establish the velocity gradient. This sort of drag is present for all objects; but for objects that are not very well streamlined (and therefore do not resemble flat plates) the pressure drag is much larger than the friction drag.

Friction drag behaves in a much more manageable fashion than pressure drag. The drag coefficient (based on wetted area rather than projected area) of a flat plate parallel to flow decreases with Re. For flows with a laminar boundary layer (Schlichting 1979),

$$C_{d,\text{friction}} = \frac{1.33}{\sqrt{\text{Re}}}. \qquad (11.12)$$

For flows with a turbulent boundary layer (Re greater than about 10^6), the appropriate equation is (Schlichting 1979)

$$C_{d,\text{friction}} = \frac{0.072}{\text{Re}^{0.2}}. \qquad (11.13)$$

These drag coefficients are quite low. For instance, at Re of 10^5, $C_{d,\text{friction}}$ is 0.004 or 0.007, depending on whether the boundary layer is laminar or turbulent.

The drag coefficients cited here were measured using rigid flat plates. A few wave-swept organisms such as the hermatypic corals produce plates of sufficient rigidity to approximate those used in these measurements, and for these the drag coefficients cited here are applicable. However, the vast majority of biological flat plates are very flexible; obvious examples are the fronds of macroalgae. These flexible plates cannot effectively be oriented perpendicular to the flow; they simply bend around to a more or less parallel orientation. However, even when oriented parallel to the flow, their behavior differs from that of a rigid plate. Thin flexible plates have a tendency to flutter and flap in much the same manner as a flag, and the flapping substantially increases the drag coefficient. Hoerner (1965) cites values for the C_d of flags varying from 0.025 to 0.1, depending on the length-to-width ratio of the flag. Preliminary results obtained in my laboratory by J. Coombs suggest that the fronds of *Gigartina corambyfora*, an intertidal macroalga, have a drag coefficient of 0.05–0.1, well above the value for skin friction alone.

Figure 11.11 shows the C_d vs. Re rela-

tionship for two wave-swept organisms (a limpet and a barnacle). The decrease in C_d with increasing Re indicates that these organisms experience some transition in drag coefficient, but the measurements were not extended to Reynolds numbers high enough to see the complete transition amplitude. Table 11.1 lists a variety of empirically measured drag coefficients for wave-swept organisms. These values should be used with caution, for they apply only at the specified Reynolds numbers.

There are several invaluable sources for further data concerning drag coefficients. *Fluid-Dynamic Drag*, written and published by Sighard Hoerner (1965), is full of odd and interesting facts about an amazing variety of drag situations. Although very few of the examples cited directly concern organisms, one can almost always find some example that is a reasonable approximation to the biological problem at hand. Some of the more obvious of these are shown in Figure 11.12. Vogel (1981) is another gold mine for drag information.

Figure 11.11. Drag coefficients for wave-swept organisms and shapes resembling wave-swept organisms. (Cone data from Ullensvang, unpublished; all other data from Denny et al. 1985.)

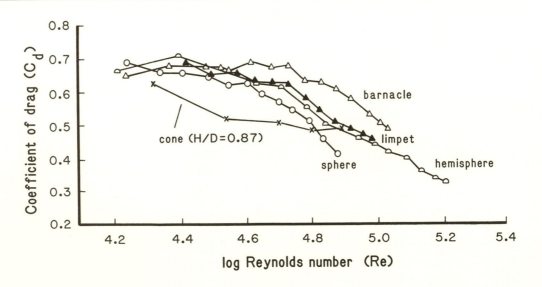

Table 11.1
Measured Force Coefficients for
Wave-swept Organisms

Organism	Re	C_d	C_l	C_m
Limpets				
Colissella				
digitalis[a]	10^5	0.52	0.25	1.84
C. pelta[a]	10^5	0.45	0.47	1.68
Patella				
cochlear[b]	$5 \cdot 10^4$	0.28		
P. argenvillei[b]	$7 \cdot 10^4$	0.30		
P. longicosta[b]	$4 \cdot 10^4$	0.67		
P. granularis[b]	$7 \cdot 10^4$	0.30		
P. granatina[b]	$7 \cdot 10^4$	0.42		
P. oculus[b]	$7 \cdot 10^4$	0.39		
Barnacles				
Semibalanus				
cariosus[a]	10^5	0.5		1.31
Balanus				
glandula[a]	10^5	0.5		1.73
Snail				
Thais				
canaliculata[a]	10^5	0.67		1.72
Coral				
Acropora				
reticulata[c]	$2 \cdot 10^6$	0.71–1.02		
Kelp				
Eisenia arborea[d]	$\sim 7 \cdot 10^4$	1.5		
Gigartina				
corymbifera[e]	$\sim 10^6$	0.15[f]		

[a] Denny et al. 1985.
[b] Branch and Marsh 1968.
[c] Vosburgh 1982.
[d] Charters et al. 1969.
[e] Coombs, unpublished.
[f] C_d based on wetted surface area rather than projected surface area.

The Lift Coefficient

Figure 11.13 shows the curve of steady C_l vs. Re for the limpet *C. pelta* and for cones of various shapes at Re's approaching those applicable to the wave-swept environment. The steady lift coefficient varies little over this range of Reynolds numbers. Although this relationship is encouragingly simple,

there is at present no way of knowing whether it is general. No other measurements of lift force vs. Re have been conducted for wave-swept organisms, nor has the relationship between lift and period parameter been examined.

Few reliable data are available in the engineering literature regarding the variation in C_l as a function of Re. For example, several studies (reviewed by Sarpkaya and Isaacson 1981) have examined the steady lift on a cylinder lying on the substratum. But the C_l values they report, which should be identical, instead vary over a large range—from nearly 0 to about 1. No clear correlation with Re is evident.

In contrast, there is a wealth of information concerning the transient lift forces on circular cylinders (for a review see Sarpkaya and Isaacson 1981), indicating that instantaneous lift coefficients as high as 6 are possible. However, it is unlikely that these findings have much bearing on the forces acting on wave-swept organisms. For example, experiments on cylinders are carefully set up so that effects of flow around the ends of the cylinder are minimized, allowing the cylinder to act as if it were infinitely long. By this stratagem, experimental results can most easily be compared to theoretical predictions for two-dimensional flow. However, most wave-swept organisms are blunt, consisting as much of end as they do of span. In this situation, any difference in pressure between the sides of an organism is likely to result in a flow around the end rather than a substantial lift. Furthermore, the presence of a large transient lift requires that a vortex be shed simultaneously by one entire side of the organism. This "spanwise coherence" seems unlikely given the complex shapes of most wave-swept organisms. Finally, note that the vortices creating transient lift forces are shed at frequencies that depend on the water's velocity and the organism's size. If one assumes a Strouhal number of 0.2, this means that vortices are shed at fre-

Figure 11.12. Drag coefficients for various shapes. (Data from Hoerner 1965 and Vogel 1981.)

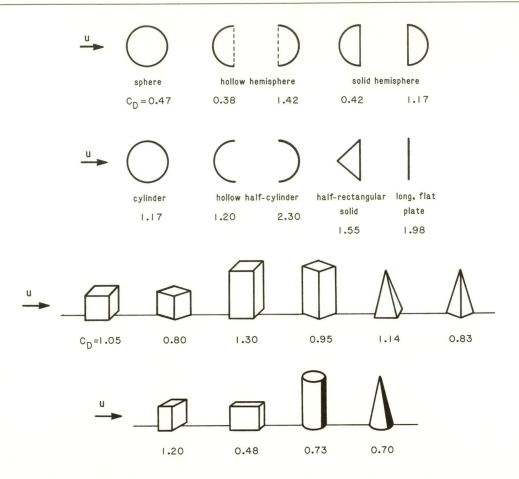

quencies from 2 to greater than 100 Hz for typical flows (1–10 m/s) and typical sizes of organisms (1–10 cm). As we will see in Chapter 14, because of the short duration of these forces they have little effect on the flexible skeletons of wave-swept organisms. Although we cannot rule out transient lift forces as a factor in the design of wave-swept organisms, there is no solid evidence to suggest that they are important. No direct measurements have been made of transient lift force in wave-swept plants and animals.

The Inertia Coefficient

Figure 11.14 shows the relationship between C_m and Re for a smooth cylinder.

The inertia coefficient increases with increasing Re, reaching a plateau at a value of about 1.9 at Re = 10^6. However, a much different picture is found for a roughened cylinder. Here the added mass coefficient goes through a peak and then decreases over the same range that C_m for the smooth cylinder steadily increases. The rougher the cylinder, the more nearly constant the added mass coefficient, and we may assume that the same holds true for biological objects.

Unfortunately, measurements of inertia coefficients for biological shapes are even more scarce than drag coefficients; the only measurements are those made by Denny et al. (1985) for limpets, snails, and barnacles (Table 11.1). There are, how-

Figure 11.13. Lift coefficients for a limpet and cones of various "peakedness." (Limpet data from Denny et al. 1985; cone data from Ullensvang, unpublished.)

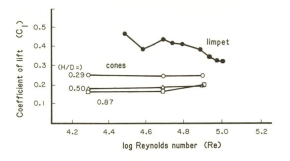

Figure 11.14. The inertia coefficient of a circular cylinder (C_m) varies with Reynolds number and period parameter. The variation is minimized for rough cylinders. (Redrawn from Sarpkaya 1976 by permission of the author.)

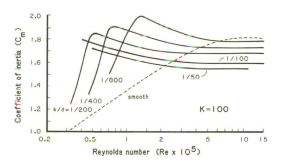

ever, quite extensive lists of theoretically derived or empirically measured values for a variety of shapes. A sampling of the more pertinent of these values is given in Figure 11.15. Note that these theoretical values are intended to apply to a body in an infinite fluid—in other words, far away from a solid surface. The inertia coefficient of a circular cylinder has been shown to be higher the nearer the cylinder is to a solid surface (Sarpkaya 1976b), reaching values in excess of 4 when the cylinder is very near the bottom. Exactly how the presence of the substratum affects other objects will depend on their shape, and there is no general rule to help guide us in this matter.

By comparing the information in Figures 11.12 and 11.15, it is clear that an object

with a high value for C_d also has a high value of C_a ($= C_m - 1$). When viewed in terms of the overall force that is exerted on an object, there is little in the way of possible trading off between pressure drag and acceleration reaction.

Rules of Thumb

We have now explored the various hydrodynamic forces, but to this point we have treated them all as if they were of equal importance. There are cases in which one or several of these forces are negligible

Figure 11.15. Added mass coeficients ($C_a = C_m - 1$) for various shapes. (Data from Sarpkaya and Isaacson 1981.)

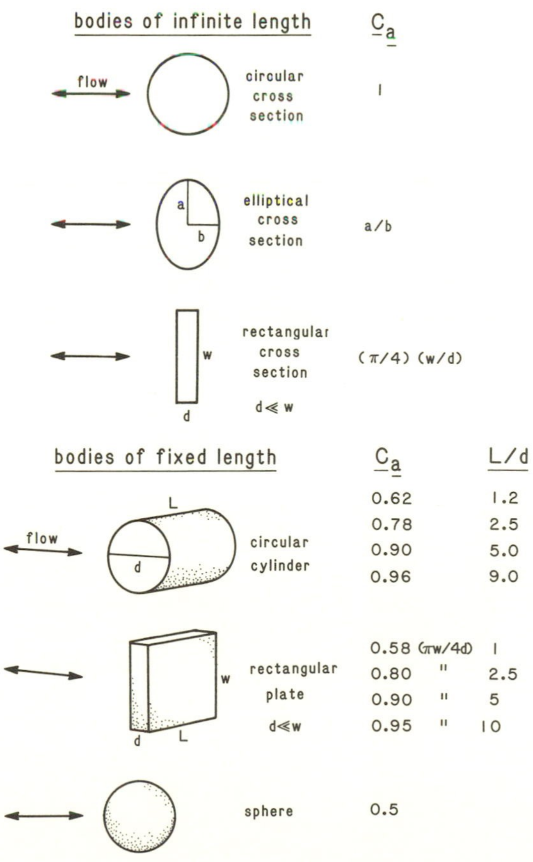

bodies of infinite length	C_a	
circular cross section	1	
elliptical cross section	a/b	
rectangular cross section, $d \ll w$	$(\pi/4)(w/d)$	

bodies of fixed length	C_a	L/d
circular cylinder	0.62	1.2
	0.78	2.5
	0.90	5.0
	0.96	9.0
rectangular plate, $d \ll w$	0.58 $(\pi w/4d)$	1
	0.80 "	2.5
	0.90 "	5
	0.95 "	10
sphere	0.5	

relative to the others, and we can draw up rules of thumb for the analysis of forces on wave-swept organisms.

In making comparisons among various forces, it is convenient to choose one particular force as a standard. The choice is arbitrary, but here we will use the force that is typically the largest—pressure drag.

Friction Drag

Combining the equations for friction drag coefficient (11.12 and 11.13) with the equation for pressure drag (11.1), we arrive at the conclusion that

$$\frac{\text{friction drag}}{\text{pressure drag}} = 1.33 \frac{S_w}{S_p} \frac{1}{C_d\sqrt{\text{Re}}}, \quad (11.14)$$

or

$$\frac{\text{friction drag}}{\text{pressure drag}} = 0.072 \frac{S_w}{S_p} \frac{1}{C_d\text{Re}^{0.2}}, \quad (11.15)$$

depending on whether the boundary layer flow is turbulent or laminar, respectively. In either case it is clear that the friction drag is of a size comparable to pressure drag in only two circumstances. If the Re is very low—in other words, if the flow is slow or the organism is very small—the friction drag can be important. However, the Reynolds number for all but the smallest wave-swept organisms is greater than 100, so this criterion is seldom realized. Alternatively, if the wetted area (S_w) is very large compared to the projected area, the friction drag can be important. Friction drag could thus be consequential for a kelp frond. For example, a planar kelp frond 1 m long, 10 cm wide, and 1 mm thick probably experiences a friction drag about seven times its pressure drag. But this assumes that the blade does not flutter—flutter, as we have seen, substantially increases the pressure drag. A more typical example is a limpet, where the wetted area is usually about four times the projected area, and C_d is about 0.4. At Re of 10^5, friction drag amounts to less than 0.1% of the pressure drag.

Acceleration Reaction

The appropriate expression for the acceleration reaction is given by eq. 11.4. Dividing by the pressure drag, we get

$$\frac{\text{acceleration reaction}}{\text{pressure drag}} = \frac{C_m}{C_d} \frac{2aL}{u^2}, \quad (11.16)$$

where the volume/projected area is represented as a length, L. Recalling that shapes with high drag coefficients usually have high inertia coefficients, we can safely say that the ratio C_m/C_d is likely to be in the range of 1–3 for most biological shapes. The primary exception to this rule would be a platelike organism oriented perpendicular to flow, in which case this ratio could become large. However, platelike organisms are rare. If we assume that $C_m/C_d = 2$, we are left with the conclusion that the acceleration reaction is of comparable magnitude to the pressure drag only when accelerations are high, the organism is large, or both. For example, velocities of 10 m/s are not uncommon on wave-swept shores, and a typical ratio of volume to projected area is 0.01 for organisms a few centimeters long. Using these values, we see that the ratio $2aL/u^2$ is near 1 only if the acceleration is near 2500 m/s^2. Such accelerations are possible (especially in the impact situations discussed below) but are not likely to be common. A more typical value of acceleration would be 100 m/s^2 in the surf zone, and considerably less in subtidal regions (see Table 5.2). The acceleration reaction is thus likely to be at most about 4% of the pressure drag (assuming an acceleration of 100 m/s^2), and will usually be much smaller. For a very large organism (length on the order of 1 m), the ratio of volume to projected area can be on the order of 0.1, and the acceleration reaction might be 40% of the pressure drag. Even though the acceleration reaction is comparatively small for most organisms, the manner in which it scales with size (being proportional to volume rather than area) is of importance in determining the physical limits to size of wave-swept organisms. This limitation is discussed in detail in Chapter 17.

Lift

The ratio of steady lift (eq. 11.6) to pressure drag depends only on the relative force coefficients and the ratio of the areas over which the force is applied:

$$\frac{\text{lift}}{\text{pressure drag}} = \frac{C_l}{C_d} \frac{S_{\text{plan}}}{S_p}. \quad (11.17)$$

Both of these ratios vary considerably for wave-swept organisms, and consequently no general rule of thumb can be drawn. Two examples serve to emphasize this point:

1. *High Lift/Drag.* The limpet *Lottia gigantea* has a low drag coefficient (0.4) and an area projected in the direction of flow only about 20% of the animal's basal area. The lift coefficient is relatively high (0.2). The net result is a lift force that is about 2.5 times the drag force.

2. *Low Lift/Drag.* The platelike coral *M. complanata* extends perpendicular to both the substratum and the prevailing direction of flow. It thus has a very large projected area, and a small planform area over which steady lift can act. Further, the steady lift coefficient is extremely low, while the drag coefficient is approximately 1.2. The net result is a drag that is many times larger than the steady lift.

What Shapes Give the Least Drag?

Using these rules of thumb, we seem justified in assuming that drag is often the largest force imposed on wave-swept organisms, and we have already seen that a flat plate lying parallel to flow has a particularly low drag coefficient. Why aren't all wave-swept plants and animals shaped like thin flat plates? Although there are a few platelike organisms—encrusting

sponges, coralline algae, bryozoans, etc.—the fact that most organisms have different shapes illustrates two of the more general rules of biological design with respect to drag.

First, we must remember that no matter what shape the animal or plant, the organism must live inside its body. Although a flat plate has little drag when it is oriented appropriately, it contains little volume in which an organism can reside. A more appropriate criterion for designing an organism than the minimization of drag per se may be the minimization of drag per enclosed volume. There are several ways in which this can be accomplished, depending on the flow environment.

As we have seen, the overall drag on a bluff body (such as a cylinder) is primarily due to pressure drag, and pressure drag is primarily due to the separation of flow on the downstream side of the object. Thus, to minimize pressure drag, we should pay particular attention to the shape of the downstream end. Shapes that do this best are the so-called "streamlined bodies," the sleek-looking forms that one associates with airplane fuselages and expensive cars. These shapes have a rounded upstream end and a tapering downstream end with a sharp trailing edge. The idea is to fit the shape to a pattern of flow that streamlines can easily follow, thereby minimizing the chance of separation. These shapes can be extremely effective in reducing pressure drag in unidirectional flow. Webb (1975) reports that a small fish with a streamlined shape (a mackerel) has a drag coefficient at Re = 10^6 of only 0.0043. This is essentially the same as a flat plate at the same Re, indicating that pressure drag has virtually been eliminated. Exactly what streamlined shape is the best depends on the design criterion chosen. Here we have decided to minimize the drag per body volume, and a shape such as that shown in Figure 11.16a is appropriate (Alexander 1968). This streamlined shape is about 4.5 times as long as it is wide, and it has a

Figure 11.16. Shapes with minimum drag per enclosed volume. (a) A streamlined shape for flows away from the substratum. (b) A streamlined shape for an object attached to the substratum. (c) A chiton.

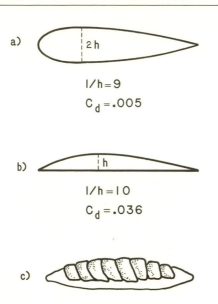

shape of Figure 11.16a, for instance, has a very low C_d when its long axis is parallel to flow, but it experiences a drag several hundred times that when it is perpendicular to flow. Even when flow is along the long axis of the shape, but directed from the rear, the drag is substantially increased (Vogel 1981).

These highly direction-dependent drag properties are disadvantageous to wave-swept organisms that experience flows from unpredictable directions. In fact, examples of streamlined shapes are difficult to come by among wave-swept organisms. Perhaps the best examples are small limpets (e.g., *Patina pellucida*, *Notoacmea insessa*), which spend their entire adult life attached to (and eating) the fronds or stipes of macroalgae. The shells of these limpets are reasonably streamlined, with a length two to three times their width. They can afford to have this shape because the blades to which they are attached are flexible and orient to the flow. Thus, even though the limpet is sessile, it is passively oriented, and streamlining is therefore appropriate. The same principle may hold for some snails (such as *Thais* [= *Nucella*] *emarginata*), whose shell is somewhat streamlined and conceivably could passively rotate to an appropriate orientation.

When the direction of flow is unspecifiable and the organism is rigid enough to preclude passive orientation, what shape gives the least drag per volume? We confine ourselves to the consideration of shapes with circular cross sections such as cylinders, cones, and sections of spheres. Data on these shapes are summarized in Figure 11.17, leading us to the tentative conclusion that a hemisphere attached to a wall has the lowest drag per enclosed volume, though cones with small height per diameter ratios are not far behind. This conclusion can be refined somewhat using data provided by Hoerner (1965) for round rivet heads. He cites data suggesting that at Re < 10^4 the drag of these

drag coefficient of approximately 0.005 for the Re range of 10^4–10^6. This shape, or one nearly like it, would give the best drag reduction characteristics for an organism surrounded by flow. Organisms attached to a wall are best off with a shape similar to that of Figure 11.16b. The length of the shape is about ten times its depth, and the upstream end is not as rounded as in Figure 11.16a to reduce the possibility of flow separation near the wall (Hoerner 1965). Hoerner reports a C_d value of 0.036 for a lump on a wall such as this, at a Reynolds number above the turbulent transition.

By deciding to minimize drag per volume, we have provided a reason why organisms should not look like flat plates, but we have really only pushed the question back one step. Most organisms in the wave-swept environment are not streamlined, either. This is because the drag-reduction characteristics of streamlined shapes depend on a precise orientation to the direction of flow. The streamlined shapes rises as the cube of the ratio of their height to diameter. At higher Re the drag goes up approximately as the square of the height/diameter. Assuming the rivet heads are sections of a sphere, we can calculate that the drag per enclosed volume is minimized for a height/diameter of about 0.4 ($Re < 10^4$), and for a height/diameter of about 0.2 at higher Re's. We will return to this subject in Chapter 16 when we examine the shapes of limpets.

On the basis solely of drag per volume, an upright cylinder is not a particularly desirable shape. However, many cylindrical organisms are found in the wave-swept environment (tube worms, gooseneck barnacles, algal stipes, etc.), suggesting that factors other than drag have controlled their design.

Impact Forces

The forces we have discussed so far have been examined under the implicit assumption that the object experiencing the force was completely submerged at all times. This is a valid assumption for subtidal organisms, but intertidal organisms are emersed for at least part of each tide. When in the surf zone, an organism is subject to the sudden imposition of force as the wall of water of an upsurging wave reaches the organism. Under certain circumstances, the force exerted by moving water hitting a solid surface can be immense.

Consider a cubic mass of water approaching a flat wall with velocity u. The water has a momentum mu, where m is the mass of the cube. One face of the cube smacks flat onto the wall. If the wall is unyielding, the water is in that instant brought to a complete halt. Once halted, the momentum of the water is zero. The force that was required to halt the water is equal to the time rate of change of its momentum—in other words, to mu divided by the length of the instant over

Figure 11.17. Radially symmetrical shapes for minimizing drag in flows in which the direction varies randomly. (Data for cones from Denny, unpublished; all other data from Hoerner 1965.)

	C_D	volume	$\dfrac{(drag/volume)}{0.5\rho u^2}$
h	0.73	$\pi r^2 h$	0.43/r
r	0.32	$(2/3)\pi r^3$	0.24/r
$h/r=.5$	0.30	$(1/3)\pi r^2 h$	0.29/r
$h/r=1$	0.46	$(1/3)\pi r^2 h$	0.44/r
$h/r=2$	0.60	$(1/3)\pi r^2 h$	0.57/r

which the velocity was brought to zero. For a very stiff wall this time is very short, and the force applied by the wall on the water (or by the water on the wall) is *very* high. In this hypothetical example, where we can make the wall as unyielding as we want, it is possible for the force to tend toward infinity.[3]

An infinite force is difficult to imagine, and certainly undesirable to deal with in nature. The real world differs from our hypothetical example in two important respects. First, no real wall is totally unyielding. If a cubic mass of water were to smack into an actual wall, the wall would deform (even if only slightly) under the impact. This deformation increases the time over which the water is brought to a halt and therefore decreases the impact

[3] If you care to acquire an empirical "feeling" for this concept, drop a brick on your toe.

force. The more *compliant* the wall (the more it gives on impact), the less the impact pressure.

Second, an actual cube of water (if one could be constructed) would not maintain its cubic form upon impacting the wall. As one side of the cube hit the wall, the pressure created by the impact would be transmitted into the rest of the fluid. This transmission is not instantaneous, however, and portions of fluid near the back of the cube may still be at ambient pressure while portions nearer the impact face are at high pressure. As a result, fluid at the back of the cube flows laterally around the high-pressure fluid in front of it, traveling a considerable distance sideways before impacting the wall, and different parts of the water mass are halted at different times. Again, the overall time-to-halt would be increased and the impact force decreased. How much the cube would deform laterally depends to a certain extent on how fast the initial pressure of impact can be transmitted to the far reaches of the cube. The faster the transmission, the more of the cube is at the same pressure, and the less the tendency to flow sideways. The speed at which a pressure wave is transmitted is simply the speed of sound. Thus, to a first estimate, the higher the speed of sound in the fluid hitting the wall, the greater the impact pressure. We can think of the speed of sound as being an inverse measure of the compressibility (= compliance) of the fluid. Thus a fluid with a low speed of sound is compressible and, by analogy to the compliance of the wall, results in a lowered impact pressure.

What factors affect the compliance of walls and the compressibility of water in nature? The most important factor is the presence of air. For example, the compressibility of pure seawater is very low, and the speed of sound in water is high, about 1530 m/s, but this speed goes down drastically if there is any entrained air (i.e., bubbles) present. The presence of 1% (by volume) air causes a tenfold decrease

in the speed of sound, equivalent to a tenfold increase in the compressibility. Longuet-Higgins and Turner (1974) report a value of greater than 10% entrained air in spilling waves, a value that would yield a speed of sound only 1% of that of pure water.

Carstens (1968) provides formulae for quantitatively evaluating the effects of water compliance on the impact pressure. The maximum pressure possible is

$$\text{maximum pressure} = \rho u c, \quad (11.18)$$

where u is the velocity of the fluid before hitting the wall, and c is the speed of sound in the pure fluid. The speed of sound in water with a volume fraction e of entrained air is

$$c(e) = \frac{c}{\sqrt{1 + e\left\{\dfrac{B_w}{B_a} - 1\right\}}}, \quad (11.19)$$

where B_w and B_a are the bulk moduli of water and air, respectively. The bulk modulus is a measure of the change in pressure required to bring about a change in volume, and is about 15,000 times greater for water ($2.18 \cdot 10^9$ N/m^2 at 20°C) than air (about $1.4 \cdot 10^5$ N/m^2 at 20°C).

Eq. 11.18 assumes an unyielding wall. However, this is unlikely for natural substrata where air can be trapped in the texture of the rock, between animals, or beneath algae, effectively increasing the compliance of the wall. The wall compliance can also be increased by the presence of a layer of water, such as the seaward ebb of a previously broken wave.

The highest shock pressure recorded in nature (as cited by Carstens 1968; Blackmore and Hewson 1984) is approximately $7 \cdot 10^5$ Pa. This occurred for a wave with a breaking height of 2.5 m. Assuming that the water velocity for this wave was about 7 m/s (as predicted by eq. 7.12), this pressure is less than 10% of that theoretically possible (eq. 11.18). The researchers who recorded this pressure (Rouville, Besson, and Petry 1938) reported that they measured substantial shock pressures on only 2% of the waves measured. Their pressure recorder was mounted in a flat sea wall where the presence of trapped air pockets were minimized. Most impact pressures measured on sea walls and breakwaters are considerably smaller than those reported by Rouville et al. (Blackmore and Hewson 1984), and the impact pressures measured on a natural rock face covered with intertidal organisms would presumably be lower still. In more than 100 hours of observation, Denny (1985) did not record a single shock pressure at an exposed intertidal site.

Although it seems unlikely that large impact pressures are a common occurrence on natural shores, it is entirely possible that they occur as rare events. What would the effect be of a locally applied impact pressure? The forces accompanying impact pressures are directed into the rock. Thus the primary effect would be to compress the organism impacted. Because most wave-swept organisms are soft-bodied with no internal airspaces, they can be assumed to behave under hydrostatic compression in a manner similar to water itself; applying a rapid pulse of hydrostatic pressure to an anemone or alga is likely to cause no particular damage. If an air pocket were trapped inside the shell of a barnacle, mussel, or limpet, the applied hydrostatic pressure could collapse the shell, but such a collapse seems unlikely. For instance, an applied pressure of $7 \cdot 10^5$ Pa results in a force of only 70 N when applied to a barnacle with an area of 10^{-4} m² (1 square cm). Such a barnacle can often withstand the 700 N force exerted when a 70 kg biologist walks across intertidal rocks.

Another possible result of a local impact pressure would occur in an animal located to one side of the center of impact. The pressure decreases away from the center of impact and thus will vary across the width of the animal. If the pressure difference is sufficient, the organism could be dislodged laterally (in shear). Consider the case of a hypothetical cubic animal with sides 1 cm long. The animal is attached to the substratum by one of its faces and has a strength in shear of 10^5 Pa, a value typical of barnacles (see Chapter 12). For an impact pressure to dislodge this organism, a pressure difference of 10^5 Pa must be applied across two of the cube's lateral faces. Because these faces are only 1 cm apart, the average gradient in pressure must be 10^7 Pa/m. For an impact pressure of $7 \cdot 10^5$ Pa, this means that the whole zone of impact is only 7 cm in radius, and the radius would need to be smaller for smaller impact pressures. Although such areas of local impact are conceivable, their small size would limit the amount of damage they could do. It thus seems unlikely that impact pressures often dislodge wave-swept organisms in shear.

The discussion here has assumed that it is water that impacts organisms on the shore. Shanks and Wright (1986) have shown that waves can "throw" small rocks, and that the impact of these noncompliant missiles can damage intertidal organisms. One example of biological design in response to point loading by impacting objects is discussed in Chapter 14.

Sources of Information

The information presented here on the hydrodynamics of oscillatory flow is only the tip of a very large and complex iceberg. Excellent sources for further information include Streeter and Wylie (1979), Sarpkaya and Isaacson (1981), Sabersky et al. (1971), Batchelor (1967), and Wiegel (1964). Current research results are published in a variety of periodicals, including the *Journal of Experimental Biology*, the *Journal of Fluid Mechanics*, the *Journal of Geophysical Research*, *Applied Ocean Research*, and the *Journal of Waterway, Port, Coastal and Ocean Engineering*.

CHAPTER 12

Properties of
Biological Materials

What happens when a force is applied to a solid material? Imagine yourself grabbing a rubber band, pulling on it, and noting the results: the more force you place on the rubber band, the farther it deforms. If you maintain a constant force, the rubber band maintains a constant length, and when you remove the force the rubber goes back to its original dimensions. If you stretch the rubber band beyond some limit, it breaks. These simple observations embody all the concepts required to describe accurately the properties of solid materials. They are, however, qualitative. In order to make use of these concepts we need a quantitative framework for describing the mechanical properties of materials.

We begin by examining materials deformed to equilibrium by systems of static forces. The term static implies that the applied force is constant. In response to this static force the material deforms until it pushes back in equal measure to the force placed on it. Once equilibrium is reached, all portions of the material are stationary.[1] From Newton's second law of motion (eq. 3.1), we can deduce that for any system to be stationary the *net* force on the system must be zero.

The simplest way to deform a material is to apply two equal but oppositely directed

forces. If coaxial forces are applied to a material (Fig. 12.1a,b) the material is placed in *compression* (forces directed into the material) or *tension* (forces directed out of the material). Compressive and tensile forces are both "direct" or "normal" in that they are applied perpendicular to the surface of the material. If forces are applied along different axes, tangential to the surface, the material is placed in shear, just as a fluid is sheared in a velocity gradient.

Shearing forces require a closer examination. Consider the sample shown in Figure 12.1c. Two equal and opposite forces are applied, so the net force on the sample is zero, but this alone is not sufficient to ensure static equilibrium. Instead, these tangential forces cause the sample to rotate, in this case in a clockwise direction about an axis through the center of the cube (as shown by the curved arrow).

Before examining this rotation further, we pause to define several important terms: a force that acts to rotate an object applies a *moment*, or *torque*, to the object. The magnitude of the moment is the force times the perpendicular distance between the line of action of the force and the point about which the object rotates (Fig. 12.1e). This distance is referred to as the *moment arm*, or *lever arm*. In the case at hand (Fig. 12.1c), two moments are applied; they have the same magnitude and tend to rotate the sample in the same direction. This is shown schematically in Fig. 12.1f. A pair of parallel forces acting in opposite directions is called a *couple*.

---

[1] In theory, you could reach an equilibrium if the material deformed at a constant, nonzero rate as long as it pulled (or pushed) back in equal measure to the force applied. By our definition in Chapter 3, such a material would be a fluid rather than a solid. We deal with this distinction again later in this chapter.

As we have seen, the couple applied to the sample in Figure 12.1c tends to rotate the sample. For equilibrium to exist, an equal but opposite couple must be applied to the sample to counteract this tendency. In the deformed material, this second couple acts at right angles to the first (Fig. 12.1d), placing shear forces on the material in planes perpendicular to those in which the first two forces are applied.

Stress

The force required to deform a material depends on the size of the piece being deformed. Given two rubber bands both the same length but one twice as thick as the other, the thicker band takes twice the force to be stretched to a given length. If we are to describe the properties of the *material* from which a certain organism is constructed rather than just the properties of a particular *sample*, we must normalize the applied force to the size of the sample. To do this, we divide the force by the particular cross-sectional area of the sample that resists the force.

The term given to the value force/area is *stress*, σ.[2] Stress is expressed with the units of pascals. Conversion constants to pascals from various alternative units are given in Table 12.1.

When stress is specified for tension or compression the relevant cross-sectional area is measured perpendicular to the axis along which the force is applied (Fig. 12.2a,b). This cross-sectional area is likely to decrease as the sample is stretched, or increase as the sample is compressed. In either case, *true stress*, σ_t, is calculated for each extension by dividing the applied force by the cross-sectional area present at

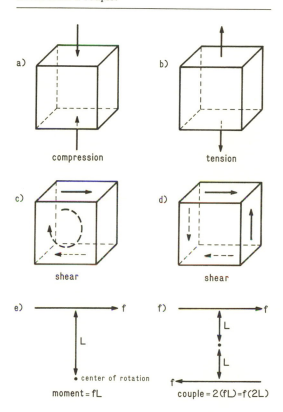

Figure 12.1. Forces on a sample. (a) Compression. (b) Tension. (c) A single couple applied to a sample causes it to rotate (dashed arrow). (d) A sheared sample is at equilibrium under the action of two orthogonal couples. The couples are shown here as they act on a cube in the middle of an already deformed sample. (e) A moment is the product of force and the perpendicular distance to the center of rotation. (f) Two oppositely directed, equal moments form a couple.

compression

tension

shear

shear

moment = fL

couple = 2(fL) = f(2L)

---

[2] It is unfortunate that "stress" has very different connotations in biological parlance (e.g., environmental stress, water stress, etc.). This often causes problems when biologists are faced with using the term in its engineering sense. The reader who is a biologist should pause to concentrate on the fact that stress = force/area.

that extension. Alternatively, we can refer the applied force to the undeformed area, in which case we obtain the *nominal stress*, σ_n.

If forces are applied in shear, the cross-sectional area is taken to lie in a plane parallel to both forces and perpendicular to the plane defined by the axes of the forces (Fig. 12.2c). As in fluids, shear stress is given the symbol, τ. Because there must be two, orthogonal couples for a sheared material to be at equilibrium, the shear stress on perpendicular planes must be equal.

Table 12.1
Conversion Factors by Which Various Pressures
Can Be Expressed As Pascals

To convert from: to Pa (N/m²)	Multiply by:
Atmospheres	$1.013 \cdot 10^5$
Pounds per square foot	47.88
Pounds per square inch	$6.895 \cdot 10^3$
Tons (short) per square foot	$9.576 \cdot 10^4$
Tons (short) per square inch	$1.379 \cdot 10^7$
Dynes per cm²	0.1000
Kg per cm²	$9.807 \cdot 10^4$
Kg per m²	9.807
Inches of water (4°C)	249.1
Feet of water (4°C)	$2.988 \cdot 10^3$
Cm of water (4°C)	98.06
Meters of water (4°C)	$9.806 \cdot 10^3$
Cm Hg	$1.333 \cdot 10^3$
Inches of Hg	$3.386 \cdot 10^3$

Strain

The amount that a sample deforms as the result of an applied force depends on the size of the sample. For example, consider two rubber bands with the same cross-sectional area, one twice the length of the other. A force is applied to each rubber band and is increased until each is stretched to twice its initial length. The force sufficient to stretch the shorter band to twice its length is exactly the same as that required to stretch the longer band to twice its length. However, the deformation of the longer sample is twice that of the shorter one. Therefore, to describe the deformation of a material independent of the size of the sample, deformation must be expressed relative to the sample's undeformed size. There are four definitions of normalized deformation in common use. Since each is a ratio of deformed to undeformed dimension, each is dimensionless.

Figure 12.2. Definitions of stress. (a) Tensile stress. (b) Compressive stress. (c) Shear stress.

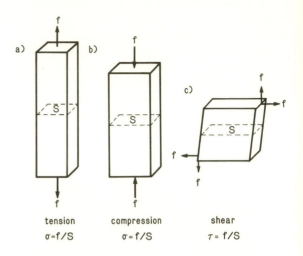

tension
$\sigma = f/S$

compression
$\sigma = f/S$

shear
$\tau = f/S$

1. *Extension ratio*, λ, is applicable to samples deformed in tension or compression and is equal to the deformed length divided by the undeformed length (Fig. 12.3a):

$$\lambda = x/x_0. \qquad (12.1)$$

Extension ratio is commonly used when describing rubbery materials but is seldom found in engineering literature that deals with structural mechanics.

2. *Engineers' strain*, ϵ, is the extension ratio minus one; it is used for deformations in tension and compression. Engineers' strain is equal to the *change* in length, Δx, divided by the undeformed length (Fig. 12.4a):

$$\epsilon = (x/x_0) - 1 = \Delta x/x_0. \qquad (12.2)$$

This is the strain definition most often used by engineers, and it is often used in the literature without explicit definition. Engineers' strain should be used only to describe small deformations ($\epsilon < 0.01$).

3. *"True" strain*, ϵ_t, is the natural logarithm of the extension ratio:

$$\epsilon_t = \ln \lambda = \ln(x/x_0). \qquad (12.3)$$

As the name implies, this is the definition of choice when describing deformations in

Figure 12.3. Definitions for normalized deformations. (a) Deformation in tension (see text). (b) Deformation in shear (see text).

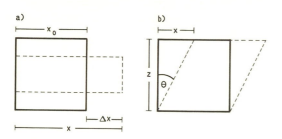

tension and compression. The idea behind what may appear to be a nonintuitive definition is quite simple: extension of a sample from an undeformed length of 10 cm to a length of 11 cm results in a 10% change in length. For the same sample, a 1 cm increment in length from 20 to 21 cm is only a 5% change in length *relative to the length immediately preceding the increment*. Now, the properties of the sample stretched to a length of 20 cm are, in all likelihood, different from those exhibited by the material at its undeformed length. This is due at least in part to the fact that stress applied in tension or compression has a tendency to change the volume of the sample (increase or decrease it, respectively). Consequently, the increment in stress required for deformation from 20 to 21 cm is best predicted by viewing this deformation as a 5% increment in length of the material with properties exhibited at 20 cm. This reference of each increment in length to the preceding length is handily accomplished by the logarithmic form of the definition of true strain. For small deformations, true strain and engineers' strain are nearly identical.

4. *Shear strain*, γ, is the gradient of deformation:

$$\gamma = dx/dz, \qquad (12.4)$$

where x is the deformation of the sample along the axis of the applied force and z is the dimension perpendicular to this axis.

In a discrete sample such as that shown in Figure 12.3b, this is equal to the deformation of the sample as a whole, divided by the thickness of the sample, or to the tangent of the angle θ. Shearing forces do not result in a sample's tendency to change volume. Consequently, the complications inherent in defining strain for tension or compression do not apply, and shear strain as defined here is a "true" strain.

Modulus

In terms of everyday experience, a stiff material is one that does not deform much when a force is applied. The stiffer the

Figure 12.4. Quantifying material properties. (a) The force-deformation curve. (b) The stress-strain curve calculated from (a), assuming the material is isovolumetric. In this calculation true stress is used. As the sample is strained, its cross-sectional area decreases, resulting in the increasing slope of the stress-strain curve. If nominal stress had been used in the calculation, a straight stress-strain line would have resulted.

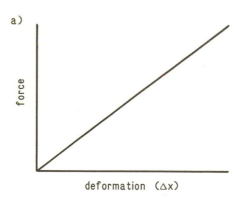

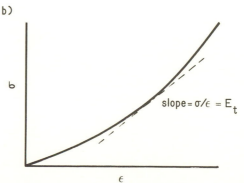

material, the more the force needed for a given deformation. Translating this idea into the terms just defined, we see that the ratio of stress to strain is a measure of the stiffness of a material. This ratio is called a *modulus*. When measured at infinitesimal strain, the modulus in tension and compression is referred to as *Young's modulus, E.*

$$E = \sigma/\epsilon \quad \text{as } \epsilon \to 0. \qquad (12.5)$$

The Young's modulus is a constant of the material. The tensile and compressive moduli for a given material are often (but not necessarily) the same.

Stiffness in shear is measured as the shear modulus, *G.*

$$G = \tau/\gamma. \qquad (12.6)$$

For a given material, G has a different magnitude than E. In both cases the units of the modulus are Pa.

The mechanical properties of any solid material can be described by measuring the stress required to deform the material to a range of strains. This is usually done by imposing an increasing deformation on a sample and recording the force corresponding to each deformation (Fig. 12.4a). Force and deformation are then converted to stress and strain by normalizing to the sample's dimensions, and the result is graphed as shown in Figure 12.4b. The slope of the stress-strain curve at any point is the tangent modulus E_t or G_t.

Materials need not have a linear relationship between stress and strain (Fig. 12.4b). Indeed, it is highly unusual for a particular biological material to have a linear (often referred to as a *Hookean*) stress-strain curve for more than the first few percent of its overall strain range. For materials with curvilinear stress-strain relationships, the modulus is a function of strain, and when a value for E_t or G_t is cited the strain at which it was measured must be noted.

It is often necessary as a means of simplification to assume that materials behave in a Hookean manner. This assumption will be made in numerous places throughout the remainder of this book, usually when modeling the behavior of a structure. It is important to keep in mind, however, that the assumption of a linear stress-strain curve is at best an approximation.

Failure

At a sufficiently large strain, every material loses its usefulness or fails. Failure can occur in a number of different fashions. The simplest is breakage—the sample breaks into two pieces at a certain strain (the *breaking strain*) and at a corresponding stress (the *breaking stress*; Fig. 12.5a). Breaking stress is also referred to as *strength*. Alternatively, failure can occur as the sample *yields*—in other words, when it shows an abrupt reduction in modulus (Fig. 12.5b). For example, if a skeletal element must have a certain stiffness to function properly (a leg bone, for instance), the reduced modulus associated with yielding can prevent the skeleton from working even though it has not broken. Any precise definition of failure depends on the task expected of the material, and for biological applications should be defined for each specific case.

Strain Energy

Figure 12.6 demonstrates a further use of the force-deformation curve obtained from a material sample. In applying a force to a material and having the material deform, energy is expended. The energy used in extending a sample, the *strain energy*, is defined to be the integral of force times extension, and is therefore equal to the area under a force-extension curve. As with force and deformation, the energy of extension depends on the size of the sample and is normalized by dividing by the

Figure 12.5. Failure. (a) Failure by breaking. (b) Failure by yielding.

a)

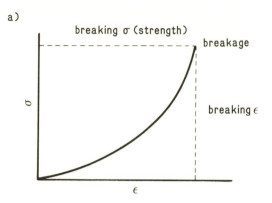

b)

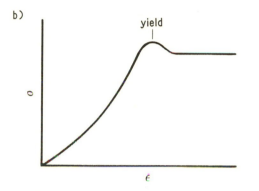

Figure 12.6. Strain energy is measured as the area under a force-deformation curve.

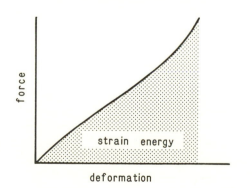

sample's volume. The result is a value with units of joules/cubic meter, or the *strain energy density*. Note that because strain is the ratio of lengths, the product of stress and strain ($N/m^2 \cdot m/m$) has the units J/m^3. Thus the area under a stress-strain curve is a measure of the strain energy density. The strain energy required to break a material is its *breaking energy*. Materials that have a high breaking-energy density are said to be *tough*; those with a low breaking-energy density are *brittle*.

Poisson's Ratio

As mentioned earlier, when a material is deformed in tension or compression, its cross-sectional area tends to decrease or increase, respectively. In order for this to happen, the dimensions of the sample must change along axes perpendicular to the primary stress axis (Fig. 12.7). The ratios of strain along these axes to that along the primary stress axis are *Poisson's ratios*, v. If the primary stress axis is the x-axis, the strain in the x-direction caused by the applied force can be denoted as ϵ_x, and the lateral strains as ϵ_y and ϵ_z. Poisson's ratios are

$$v_{yx} = -\frac{\epsilon_y}{\epsilon_x} \quad \text{and} \quad v_{zx} = -\frac{\epsilon_z}{\epsilon_x}. \quad (12.7)$$

Figure 12.7. Terms used in defining Poisson's ratio (see text).

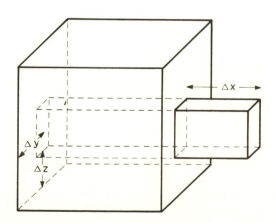

The two ratios are usually, but not necessarily, the same. For our purposes here, the usefulness of Poisson's ratio lies in its ability to describe the relationship between a material's modulus in tension or compression and its modulus in shear. Provided that $v = v_{yx} = v_{zx}$,

$$G = \frac{E}{2(1 + v)}. \qquad (12.8)$$

Very stiff materials such as mollusc shells and bone have low Poisson's ratios (0.1 or less), and their stiffness in shear is approximately one-half that in tension. Softer materials, especially those swollen in water such as protein rubbers and mucus, have a Poisson's ratio very near to 0.5 and are therefore proportionally less stiff in shear. Unfortunately, Poisson's ratio has not been measured for most biological materials. If $v_{yx} \neq v_{zx}$ (as occurs for skin and other fibrous, fabriclike materials), eq. 12.8 cannot be used.

At this point we digress briefly to illustrate the usefulness of true strain. Another way of looking at Poisson's ratio is as a measure of the change in volume of a material when it is deformed in tension or compression. The unit volume change in a material is (Gere and Timoshenko 1984)

$$\Delta V/V = \epsilon_t(1 - 2v), \qquad (12.9)$$

again assuming that $v = v_{yx} = v_{zx}$. This implies that when $v = 0.5$, the volume of a material does not change. This relationship is strictly true *only* if true strain is used to measure the material's deformation. If we were to use engineers' strain to measure Poisson's ratio for an isovolumetric material, we would find that v decreases as strain increases and equals 0.5 only at infinitesimally small strains (see Appendix 12.1).

Viscoelasticity

In the preceding discussion, the materials in question have been carefully referred to as "solids." This precision in language was not without purpose, because the described relationships between stress and strain are strictly true only if the material is a perfect solid. But all real materials behave under certain circumstances as if they have the properties of fluids. The distinction between solids and fluids rests in the relationship between stress and strain.

As described above, an ideal solid gives a stress proportional to strain, and we can define a tangent modulus at each strain that is independent of time and the same whether the strain is increasing or decreasing. Such materials are said to be *elastic*.[3] In contrast, the stress required to deform an ideal fluid does not depend at all on the *amount* of deformation; rather, it is solely determined by the *rate* of deformation. The proportionality constant between stress and the rate of strain is the fluid's dynamic viscosity, μ (Chapter 3):

$$\mu = \frac{\sigma}{d\epsilon_t/dt}, \qquad \mu = \frac{\tau}{d\gamma/dt}. \quad (12.10)$$

For biological materials, the properties of fluids and solids are mixed together in a wide range of proportions, and these materials are best described as being *viscoelastic*. Mollusc shell, for instance, behaves very nearly as an ideal solid; some mucins behave nearly as ideal fluids.

The fluid and solid natures of viscoelastic materials are best illustrated by an examination of the effects of viscosity on the stiffness and strain energy. The stiffness of a viscoelastic material depends on the strain rate—the faster the material is deformed, the stiffer it is. An elegant example of this effect is afforded by the sea anemone *Metridium senile*. These graceful creatures live in protected waters, and when undisturbed they assume an inflated posture, holding their tentacles well up

[3] The term "elastic" is often applied only to materials that have a linear stress-strain curve. Here we relax this requirement. Elasticity can alternatively be defined in terms of the storage of strain energy, as discussed later in this chapter.

into flow to filter food from the water
(Fig. 2.1). Occasionally an eddy may stir
the water around an expanded anemone,
placing a stress of about 10^4 Pa on the
mesoglea of the body wall (Koehl 1977a,b).
Koehl (1977c) has shown that meosoglea
is relatively stiff ($E = 1.7 \cdot 10^5$ Pa) when
loaded for periods of only a second or two,
as would be the case with a passing eddy.
The strain resulting from this force is
small, on the order of

$$\epsilon_t = \frac{\text{stress}}{\text{modulus}} = 0.06.$$

Consequently, the anemone is not
stretched out of the appropriate feeding
posture by such forces. As we will see in
Chapter 13, this calculation is a gross oversimplification, but the conclusion remains
valid.

If an anemone is sufficiently bothered
(by a series of waves during a storm, or
by an inquisitive biologist), it contracts
drastically. The water in the coelenteron is
expelled as muscles in the body wall contract, and the animal hunkers down to a
shape resembling a wad of chewing gum
stuck to the rock. This shape reduces the
hydrodynamic force on the creature, and
in this posture *M. senile* manages to
weather severe wave action. After the disturbance has passed, however, the animal
is faced with the problem of reexpanding
itself. Because of the arrangement of its
variable-volume hydrostatic skeleton, the
anemone cannot use the muscles of the
body wall for reexpansion. The only force
available is that due to the pressure
created by the cilia of the syphonoglyph as
they pump water into the coelenteron.
This pressure is quite small (10–1000 Pa;
Koehl 1976), and the stress placed on the
body wall mesoglea is correspondingly
small (10^2–10^4 Pa). If the mesoglea were
an elastic solid with the stiffness required
to resist wave forces, the stress from the
reexpansion pressure would be insufficient
to inflate the animal. However, the mesoglea is a somewhat viscous material. Its

stiffness in resisting forces applied for a
long period of time is twenty to thirty
times less than that resisting a rapidly applied force—sufficiently low to allow for
a slow reexpansion.

This example serves to illustrate the important fact that in viscoelastic biological
materials, mechanical properties depend
on the rate at which the material is deformed. Just as we must make certain,
when specifying a drag coefficient, that it
applies to the particular Reynolds number
found in a certain habitat, we must also be
sure that the mechanical properties measured for a material are measured at a
strain rate resembling the one found in the
natural environment.

The viscous nature of viscoelastic materials affects the strain energy characteristics of a material. The faster a viscoelastic
material is deformed, the more force is required to reach a certain deformation, and
the larger the strain energy. Unlike the energy used to extend an elastic solid, the
energy used to deform the fluid component of the material is not stored. Instead,
it ultimately ends up as heat. The reason
for this can be shown as follows. Consider
Figure 12.8a, a cyclic force-deformation
curve for an ideal elastic solid. Force is
required to deform the sample, with each
force uniquely corresponding to a particular deformation. As the force is removed,
the material follows the same force-
deformation path back to its undeformed
dimensions. Consequently, the area under
the unloading curve is precisely equal to
the area under the loading curve—all the
strain energy put into the material is
stored and can be regained. This ability to
store strain energy can be used as an alternative definition of elasticity.

Figure 12.8b shows the force-
deformation behavior of a fluid. Here, a
given force does not uniquely determine a
particular deformation. Instead, as long as
the force is applied, the material deforms
at a constant rate. Because energy is the
product of force and deformation, strain

energy constantly increases. When the force is removed from a viscous fluid, the deformation rate goes to zero—in other words, the material does not return to its original shape. The strain energy that goes into deforming the fluid is not stored and cannot be recovered. Figure 12.8c shows the cyclic force-deformation behavior of a viscoelastic material. Although the material may return to its original dimensions, strain energy has been lost to viscosity. The energy lost, expressed as a percentage of the total energy expended (energy lost +

energy recovered), is the material's *hysteresis*. Conversely, the energy recovered, expressed as a percentage of the total energy, is the *resilience*. The energy dissipation capacity of viscoelastic materials is important in the present context primarily because it contributes to the damping of oscillating systems constructed from these materials, helping to determine the number and amplitude of oscillations a structure undergoes after being disturbed by a rapidly applied force. However, the consideration of oscillating systems takes us far afield from the assumption of static forces on which this discussion is based, and this subject is deferred to Chapter 14.

Figure 12.8. The contrast between elastic solids and viscous liquids. (a) The force required to deform a solid is proportional to the amount of deformation. Strain energy is stored. (b) The force required to deform a viscous liquid is proportional to the rate of deformation. No strain energy can be recovered. (c) A viscoelastic material exhibits a combination of the properties of solids and liquids.

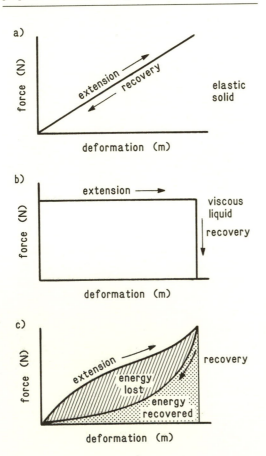

The Spectrum of Biological Materials

Having thoroughly defined our terms, we now turn to their use as applied to biological materials. From what materials are wave-swept organisms constructed? What are their mechanical properties? The answers to these simple questions are in some cases known in great detail, and in other cases not at all. What follows is a general overview of the mechanical properties of biological materials, with some of the details left out (where such details are known but irrelevant), and some of the gaps filled in by educated guesswork. The material properties that are most often of interest in the study of wave-swept organisms are stiffness, breaking strength, and breaking strain energy.

Stiffness

Table 12.2 lists values of modulus for a variety of biological materials. The most noteworthy fact regarding this list is its great range, more than eleven orders of magnitude. It is useful as a mnemonic to break this range down into four representative classes: slimes, rubbers, fibers, and crystalline composites.

Table 12.2
Typical Mechanical Properties of Biological and Man-made Materials

	Tensile Strength Pa · 10^5	Breaking Strain	Breaking-Strain Energy Density J/m^3 · 10^4	E Pa · 10^6	Density Kg/m^3 · 10^3
Slimes					
Water	—	—	—	0	1.0
Pedal mucus[a]	1–3	—	—	—	1.0–1.1
Barnacle cement[a]	1–3	—	—	—	1.2
Rubbers					
Resilin	40–60	2–3	400–900	2	1.2
Abductin	80–120	2–3	800–1,800	4	1.2
Elastin	40–60	2–3	400–900	2	1.2
Octopus arterial elastomer[b]	40–60	2–3	400–900	2	1.2
Fibers					
Silks	5,000–10,000	0.2–0.35	5,000–18,000	5,000–10,000	1.2
Collagen	500–1,000	0.08–0.10	200–500	2,000	1.2
Cellulose	5,000–10,000	0.02–0.10	500–5,000	20,000–80,000	1.2
Chitin	1,000	0.01–0.02	50–100	40,000	1.2
Keratin	1,000–2,000	0.3–0.8	1,500–3,000	4,000	1.2
Crystalline Composites					
Coral skeleton[c]	400	.0003	0.6	60,000	2
Mussel shell[d]	560	.0018	5	31,000	2.7
Bone	1,900	.01	950	18,000	2
Man-made Materials					
Steel	30,000	0.015	2,000	200,000	7.9
Glass	1,000	<0.001	<5	100,000	2.5
Cement	40	<0.001	<2	4,000	2.8
Fiberglass	3,000–10,000	0.01	150–500	30,000–100,000	1.5–2

	Shear Strength Pa · 10^5	Breaking Strain (Shear)	Breaking-Strain Energy Density (Shear) J/m^3 · 10^4	G Pa · 10^6	Density Kg/m^3 · 10^3
Pedal mucus[e]	0.001–0.003	1–5	0.1–0.5	0.0001–0.0005	1.0–1.1

[a] Denny et al. 1985.
[b] Shadwick & Gosline 1983.
[c] Vosburgh 1982.

[d] Currey 1980.
[e] Denny & Gosline 1980.
All other data from Wainwright et al. 1976.

Included in the slimes are all materials having a shear modulus less than 10^5. The vast majority of these are mucins of one form or another: the slippery mucus that coats macroalgae such as *Macrocystis* and *Laminaria*, the pedal mucins secreted by the foot of gastropods by which they adhere to the substratum, the mucin that forms the matrix of materials such as sea-anemone mesoglea. All of these materials are formed primarily of water—some pedal mucins are more than 95% water. It should not be surprising, then, that these materials are viscoelastic. Water, which has been included here because it is used in hydrostatic skeletons, is very close to being an ideal fluid. It has no measurable elastic modulus, and a low enough viscosity so that it substantially resists deformation only if deformed very rapidly. When used as part of a hydrostatic skeleton, the most relevant property of water is not its elastic modulus, but rather its bulk modulus, its resistance to any change in volume. For all practical purposes water is incompressible. Mucins make use of this incompressibility and employ water primarily as a means of filling space. The mechanical properties of this bulk are modified through the addition of high molecular weight mucopolysaccharides and glycoproteins. The interaction of these large molecules gives the material some semblance of stiffness. For example, at a strain rate of about 10/s the pedal mucus of the limpet *Patella vulgata* has a modulus of about 100 Pa (Grenon and Walker 1980). Although very low on the scale of biological materials, this limited stiffness contributes to the role of mucins as adhesives (Chapter 15). In addition to being used on their own, mucins find wide utility as matrix materials in soft composite solids. A prime example is the body wall mesoglea of sea anemones such as *M. senile* and the open-coast anemone *Anthopleura xanthogrammica*. In both of these animals mucin fills the spaces in a loose reticulum of collagen fibers. Aside from giving the

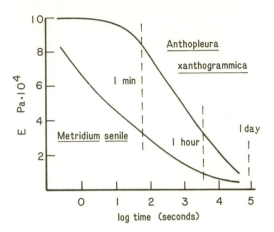

Figure 12.9. The stiffness (= modulus, E) of sea-anemone mesoglea decreases with the time over which a force is applied. (Redrawn from Koehl 1977 by permission of the Company of Biologists, Ltd.)

material some additional stiffness because of their own high modulus, the collagen fibers serve to amplify the shear strain rate in the mucin as the material is deformed. This amplified strain rate reacts with the viscosity of the mucin to produce an increased modulus. Gosline (1971) and Koehl (1977c) showed that the modulus of the *M. senile* body wall is about 10^5 Pa at a strain rate of 10/s, but that this modulus decreases drastically at lower strain rates. The same general pattern is found in the body-wall mesoglea of *A. xanthogrammica*, but this material is somewhat stiffer (Fig. 12.9). As with all modulus values cited for viscoelastic materials, those for slimes cited in Table 12.2 apply only at the strain rates cited.

The rubbery materials found in wave-swept organisms are all proteins, with the polypeptide chains of the material cross-linked to form a network. The reason such cross-linked polymer chains behave in an elastic manner is interesting in itself, and the reader can consult Wainwright et al. (1976) or Gosline (1980) for general reviews, or Flory (1953, 1969) for the full-blown theory. Rubbery materials generally have elastic moduli in the range of 10^5 to

10^7 Pa, and the protein rubbers are no
exception. Shadwick and Gosline (1983)
have shown that the modulus of octopus
arterial elastomer is virtually independent
of strain rate at strain rates between 0.01/s
and 100/s. Thus the material behaves very
nearly as an elastic solid. The modulus of
abductin has been measured only in a
narrow range of strain rates (for a review,
see Shadwick and Gosline 1983); but because of the resilience exhibited by the
material, it should be safe to assume that
it, too, behaves as an elastic solid, with
a stiffness independent of strain rate.
Rubbery materials, although stiffer than
slimes, are still not stiff enough generally
to serve as the primary structural elements
in a skeleton. Their main use is as energy-
storage devices.

With the exception of silks, all the major
fibrous materials found in nature are found
in the wave-swept environment. These fibers are crystalline polymers—some portion of their polymer chains (proteins for
collagen and keratin, polysaccharides for
cellulose and chitin) are tightly bound into
ordered, crystalline regions. This crystalline nature has the effect of making these
materials very stiff. Their elastic moduli
are generally on the order of 10^9 Pa, cellulose being somewhat stiffer with a modulus
of 10^{10}. As one might expect from such
stiff materials, these fibers are used as tensile elements and in the construction of
skeletons. The stiffness of the material is
apparent, however, only when the fiber is
loaded in tension; when bent or loaded
in compression, these fibers simply buckle,
which causes no problem in many cases.
For instance, the byssal threads of mussels
act as a series of guy lines to hold the organism to the substratum. Any force tending to dislodge the animal places a portion
of the byssal threads in tension. Similarly,
collagen fibers used as tendons and liga-
ments are mechanically arranged so that
they act in tension. In order to be useful
in the construction of compression-bearing
skeletal elements, fibers must be packaged

in a way that ensures that they do not
buckle. The usual manner in which this
has been accomplished is to wind the fibers in the shape of the desired structural
element and then embed them in some
sort of matrix. One example, the body-wall
mesoglea of sea anemones, has already
been discussed. In this case the matrix is
some six to seven orders of magnitude less
stiff than the collagen fibers it embeds;
consequently, the properties of the matrix
tend to govern the properties of the material as a whole. Stiffer matrices take more
advantage of the stiffness of fibers. For example, chitin fibers in arthropod cuticles
are embedded in a matrix of cross-linked
protein. The cuticle as a whole has a stiffness ($1.5 \cdot 10^{10}$ Pa) approaching that of
the chitin itself ($8.6 \cdot 10^{10}$ Pa; Ker 1977 as
cited in Alexander 1983). The matrix protein has a modulus of about $1.2 \cdot 10^8$ Pa
(Ker 1977). Cellulose is found in spiral
wrappings of the cell walls of plants, and
the stiffness of the overall material depends greatly on the matrix embedding the
cellulose. In woody plants the cellulose
fibers are tied together by a matrix of
hemicellulose and the protein lignin; the
resulting material has a stiffness about
one-tenth that of the cellulose fibers themselves. The cell walls of most wave-swept
plants (e.g., the macroalgae) are not ligni-
fied and are therefore less stiff. The mechanical properties of macroalgae have not
been well studied, but information provided by Koehl and Wainwright (1977) for
the alga *Nereocystis leutkeana* and Delf
(1932) for *Laminaria digitata* and *Fucus
serratus* suggests that a reasonable value
for the stiffness of algal stipes is 10^7 Pa.

The stiffest materials found in organisms are those formed at least in part from
inorganic crystalline materials. The most
extreme examples are the ossicles found
in echinoderms such as sea stars and sea
urchins. These structural elements are
formed of pure aragonite, a form of calcium carbonate, with an elastic modulus
of about 10^{11} Pa. The skeletons of herma-

typic corals are also formed of nearly pure calcium carbonate. The shells of molluscs, barnacles, and brachiopods use calcium carbonate as a stiffening agent in their shells, but they separate individual crystalline elements with an organic matrix. The resulting materials have a stiffness of about 10^{10} Pa. Some sponges differ from the above organisms by using stiff crystalline ossicles made of silicon dioxide, but these siliceous ossicles have a modulus similar to that of the calcified structures. Crystalline inorganic inclusions are also used to stiffen the chitinous cuticles of arthropods such as crabs.

Strength

Table 12.2 lists the known strengths of relevant biological materials. Note the list's relatively narrow range. While the stiffnesses of biological materials varies over more than eleven orders of magnitude, the stresses at which these materials break covers only about five.

The tensile breaking strengths of slimes are difficult to measure. These materials behave primarily as fluids, and if subjected to continuously applied force they tend to flow instead of break. The value cited here is for a pedal mucin acting as an adhesive in a thin layer between the substratum and the foot of a gastropod. When this adhesive joint is placed in tension, it breaks at some particular stress. It is unclear whether this breakage actually occurs in the mucus itself, or at the interface between the mucus and the substratum; but these materials have an apparent strength in tension of 10^4 to 10^5 Pa. Mucins fail in shear at a stress of 100–500 Pa.

Another material with a similar strength is the basal cement of adult barnacles, a proteinaceous material that may or may not be cross-linked. Unfortunately, no one has ever been able to get a large enough sample of the stuff to perform a full set of mechanical tests, so the only property of the material we know is its breaking strength when used as an adhesive. It is possible that the values cited here do not so much reflect the strength of the adhesive as they do the strength of the substratum to which the barnacle was glued. For instance, of the 347 barnacles dislodged at Shi-Shi Beach on the Pacific coast of Washington, 47% came loose because the rock they were attached to broke before the basal adhesive failed (Denny, unpublished).

Values for the breaking strength of rubbery materials are difficult to find in the literature. The theory of rubber elasticity is so elegant that most experimental work on biological rubbers has been directed toward examining the molecular basis for the materials' solidity, and few people ever bother to tug on a piece of rubber until it snaps (at least few people report their results). Consequently the values cited here are the result of educated guesswork. Most rubbers have a modulus of 10^5 to 10^7 Pa (Table 12.2). It is reasonable to assume that rubbery materials can be stretched to strains of at least 3–5 before they break. Combining these two "facts," we can guess that the breaking strengths of rubbers are at least $3 \cdot 10^5$–$5 \cdot 10^7$ Pa. The actual strengths may be a bit higher.

The strengths of most fibrous materials are quite well known, and the values cited here are presented with some confidence (Table 12.2). The similarity in strength among these widely disparate materials may at first seem surprising, but a brief examination of the macromolecular structure of these fibers helps to explain the situation. As mentioned earlier, a sizable portion of the material of each of these fibers is contained in stiff, crystalline regions. In general, the backbone of the polymer chains forming these crystals are lined up parallel to the axis of the fiber, and tensile forces on the fiber are directed along the polymer backbones. When the fiber is undeformed, the noncrystalline regions contain polymer chains that are arranged more or less at random, but when the fiber is axially deformed these, too, tend to line up with the fiber axis. By the time the fiber is fully extended and ready to break, the tensile force is being applied almost entirely to the backbone of the polymer chains. Thus it is the strength of the bonds holding the backbone together that determines the strength of the material, and these bond strengths are very similar for the peptide bonds of proteins and the glycoside bonds of polysaccharides. Obviously this simplified explanation is not the full story. Strength is also a function of the bonds between polymer chains, the degree of crystallinity, etc., but this basic idea seems valid.

The similarity in strength of the four basic types of fibrous materials means that the composite materials constructed with these fibers also have similar strengths. Materials such as arthropod cuticle and some keratinous structures, where fibers form a high percentage of the volume of the material, have strengths approaching that of the fiber itself. Materials in which fibers form only a small volume fraction may be considerably less strong than the fiber (e.g., sea-anemone mesoglea). A theory allowing the strength of a composite material to be predicted from the volume fraction of fibers and matrix is discussed in detail by Wainwright et al. (1976).

The reader may be surprised at the strengths of crystalline-composite materials (Table 12.2), which are generally no higher, and in some cases lower, than the strengths of fibers. This relative weakness is the price these materials must pay for being very stiff. Inorganic crystals have tensile strengths on the order of only 10^7 Pa, and because materials such as echinoderm ossicle and mollusc shell are almost entirely inorganic, they are relatively weak.

Breaking Energy

The range of breaking energies found in biological materials is, like that of breaking strengths, relatively small. A simplistic explanation notes that breaking energy depends on both the breaking strength and the breaking extension, and those materials with high breaking strengths generally have low breaking extensions. It all tends to even out.

Although the breaking extension of slimes is an ill-defined characteristic because of the fluid nature of these materials, there *is* an example. The pedal mucus of the terrestrial slug *Ariolimax columbianus* has been studied in some detail, and the breaking energy in shear has been determined. When deformed in shear, this material has a linear force-deformation curve up to a strain of 5–6, at which point the material abruptly yields. The stress at yielding is dependent on strain rate, but it is in the general range of 500 Pa. If the abrupt yielding is viewed as a form of breakage, the breaking energy can be calculated to be 3 kJ/m^3. Similar values can be expected for the pedal mucins of marine invertebrates. This value is, on the overall scale of things, quite small. In addition to being flexible and weak, mucins are not "tough."

Values for the breaking energy of biological rubbery materials are difficult to find. Using the same assumptions made when figuring breaking strength and assuming a linear stress-strain curve, the breaking energy of a rubber with a modulus of 10^7 Pa is about 10^7 J/m^3. The actual value will probably be somewhat lower due to nonlinearities in the stress-strain curve. If the rubber is deformed very quickly, its viscosity may increase the breaking energy to 10^8 J/m^3. On the scale of such things, these values are high. Rubbers are very tough.

The fact that the breaking energy of rubbers is high is generally of less biological consequence than the fact that most of the energy that goes into deforming a rubber is stored and can be recovered. This energy storage capacity is used in numerous ways. A classic example of the

Figure 12.10. A pad of abductin opens the shell of a scallop.

utility of energy storage is provided by the scallop (Fig. 12.10). Like most bivalves, scallops encase themselves in a rigid shell. When the animal is bothered, the two valves of the shell are clamped tightly together by the contraction of the adductor muscles (the part of the scallop you eat). When "clammed up" in this fashion, no living portion of the scallop is open to the environment or to predators. This strategy allows the scallop to avoid being eaten, but it presents a mechanical problem when the organism wants to open its shell. The force required to abduct the valves cannot be supplied by muscles because there are no muscles outside of the hinge. Instead, the opening force is provided by a pad of rubber (abductin) sandwiched in the inner angle of the hinge. The energy stored in this rubber when the valves are closed is released to reopen the shell.

The breaking energies of fibrous biomaterials are somewhat lower than those of rubbers, from 10^6 to just over 10^7 J/m^3, primarily due to the small breaking extensions of fibers. For example, cellulose fibers typically break at strains of only 1%–3%. Collagen, the most extensible of the fibers present in the ocean, breaks at a strain of around 10%. In a teleological sense, the price the organism pays for the stiffness and strength of a fiber is a small reduction in the breaking energy compared to that of a rubber.

This trade-off is even clearer in the case of crystalline composites. Although fairly strong, crystalline organic materials break at exceedingly small strains, at 0.005 or less. As a result their breaking energies are

likely to be quite small. Breaking energies for most biological crystalline composite materials have not been measured, and we again resort to the sort of estimation procedure used earlier. For example, the shell of *Modiolus modiolus*, a mussel, has a tensile strength of $5.6 \cdot 10^7$ Pa and a breaking strain of about 0.0018 (Currey 1980). Assuming a linear stress-strain curve, this suggests a breaking energy of about $4.2 \cdot 10^4$ J/m^3, one hundred to one thousand times smaller than that of rubber and fibrous composites.

Consequences of Mechanical Properties

Gastropod Crawling

The biological ramifications of these materials' properties are dealt with in Chapters 16 and 17. However, before we leave this discussion, it will be useful to examine two cases where a material is loaded in a simple manner (pure shear or tension).

Wave-swept gastropods (e.g., snails, limpets, and abalones), are faced with a serious mechnical problem. In order to feed they must move, crawling about by means of their single ventral foot. These movements often involve climbing vertical walls against the force of gravity, and the hydrodynamic forces incurred by waves must be resisted. The animal must remain stuck to the substratum while it locomotes; to be dislodged entails, in all likelihood, becoming a meal for an anemone. Adhesion alone is not a particular problem (see Chapter 15)—the real problem comes in moving and remaining adherent. How does an organism that has only one foot contrive to walk on glue? The answer lies in the peculiar mechanical properties of the pedal mucus.

We have noted that the pedal mucus of *A. columbianus* abruptly yields at a stress of about 500 Pa (depending on strain rate) and a strain of 5–6 (independent of strain rate). When the mucus yields, however, it does not break in the normal sense of the word, but it turns into a viscous fluid, and (after yielding) the stress required to deform the mucus is solely a function of the strain rate. However, this particular mucin has a surprising property. Once the deformation of the material is halted, the material "heals" within 0.1 s or so—that is, it regains the properties of a solid (Fig. 12.11a).

This "yield-heal" capability of the mucus is coupled with the movements of the foot to allow the adhesive locomotion required by gastropods. The mucus is sandwiched in a thin layer between the ventral surface of the foot and the substratum; for *A. columbianus* the layer is 10 to 20 μm thick. The foot moves by a series of muscular waves (Fig. 12.11b). The waves in *A. columbianus* move from the rear of the foot forward, and a single wave stretches most of the way across the foot. Eleven to seventeen waves are present on the foot at any one time. Other wave patterns are possible, and common. For example, the waves of limpets such as *Patella vulgata* and *Collisella pelta* move from the front of the foot rearward, each wave extending only halfway across the foot. In these cases the waves alternate sides, and two or three waves are present at one time. An extensive description of pedal wave patterns is provided by Miller (1974a, b).

As the portion of the foot contained in a wave moves forward, it shears the mucus beneath it. During the passage of a wave the foot may progress a millimeter or more; consequently the strain of the mucus under the moving foot is quite high, on the order of 100. Now consider a wave progressing toward a portion of the foot that is stationary and resting on mucus in its solid form. As the wave reaches the stationary portion of the foot, the foot begins to move. The mucus beneath it is sheared, and after a movement of 50–60 μm ($\gamma =$ 5–6) the mucus reaches its yield point and turns into a fluid. As a result, the mucus under the rest of the wave is in a fluid

state, and the force required to move this portion of the foot is simply that due to viscosity. After the wave has passed, the foot becomes stationary and the mucus beneath it quickly heals.

In applying a force to push the moving parts of the foot forward, an opposite force is created that tends to push the stationary parts of the foot back. For the animal to move, this reaction force must be resisted by the stationary parts of the foot. As long as the animal is not moving too fast, the force due to sliding the foot forward, divided by the area of the stationary parts of the foot, is less than the stress at which the solid mucus yields. Thus the stationary parts stay stuck in place, the moving parts advance, and the animal crawls along.

This explanation of gastropod locomotion is based on the locomotion of *A. columbianus* (Denny 1981, 1984; Denny and Gosline 1980), but it seems reasonable that a similar mechanism operates in limpets and snails. Abalone may use a different mechanism. When viewed in the plane of the substratum, these animals appear to lift the foot during the passage of a wave. The moving portions of the foot are thus moving over water, a low viscosity fluid, and the force due to movement is predicted to be correspondingly low. However, this mechanism may entail a severe reduction in the adhesive capability of the animal when moving (Chapter 15).

Kelp Extensibility

Nereocystis leutkeana, the bull kelp, lives in nearshore waters on both exposed and protected Northwest Pacific shores. The plants can grow to lengths of 40 m, their holdfasts attached to the bottom and their fronds held at the surface by means of a gas-filled float (Fig. 12.12). Water movements due to tidal currents or waves impose drag forces on the fronds, which in turn place the stipe in tension. The breaking strength of the stipe is not high, about $4 \cdot 10^6$ Pa (Koehl and Wainwright 1977).

Figure 12.11. (a) The "yield-heal" cycle of slug pedal mucus. Initially the stress rises in proportion to the strain as the mucus is sheared—the behavior of a solid. At a strain of 5–6, the mucus yields and behaves as a liquid upon further deformation. After a brief unstressed period, the mucus has "healed" and again behaves like a solid. (b) The slug crawls by moving portions of the foot (the waves and rim) forward over mucus in its liquid state. The force required for this movement is resisted by the stationary portions of the foot (the interwaves) held in place by mucus in its solid state. (Redrawn from Denny 1981 by permission of the Company of Biologists, Ltd.)

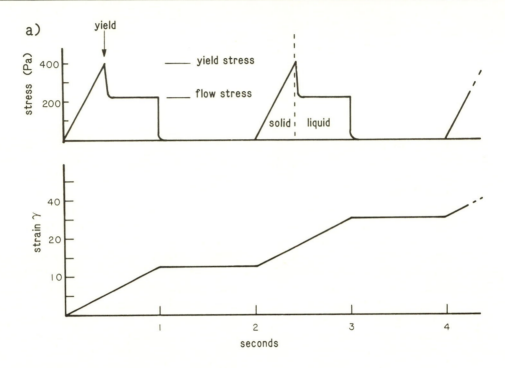

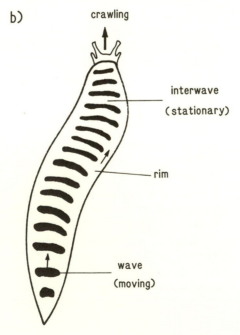

Figure 12.12. A *Nereocystis* plant sways in response to wave-induced water motion. Only when fully extended in one direction is a tensile stress placed on the stipe (see text).

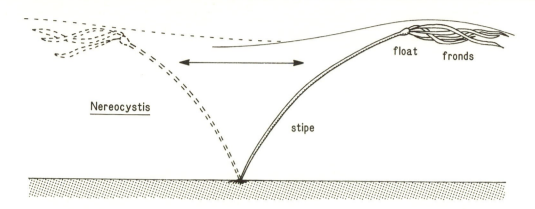

Koehl and Wainwright found that the drag force at a water velocity of only 1 m/s resulted in a tensile stress of $4 \cdot 10^5$ Pa, 10% of the breaking stress. If the drag on the float and fronds is totally due to friction drag rather than form drag (and therefore directly proportional to water velocity), we predict that the stipe will break at a water velocity of around 10 m/s. If some of the drag is due to form drag, breakage would occur at a lower water velocity. Although steady tidal currents of 10 m/s are uncommon, velocities of 10 m/s are not uncommon at the surface in shallow water waves (Chapters 5, 7). Do *N. leutkeana* plants break every time a large wave comes by? The answer of course is no, and the reason lies in the stipe's extensibility. *N. leutkeana* stipes extend to strains of nearly 0.4 before they break. Thus any force attempting to break such a plant must be applied over a very large distance—for a 40 m plant, 16 m. This extension takes time. During the passage of a wave, the plant must first assume a posture in which the stipe is entirely lined up with the prevailing water velocity. Only then can an effective tensile stress be placed on the plant. Assuming that the plant had lined itself up with the backwash of the preceding wave, the float and fronds must travel 60–70 m shoreward before they are appropriately lined up. At a water velocity of 10 m/s, this would take 6–7 s. Once lined up, the plant cannot extend at the same velocity as the water creating the drag force. If it did, the water's velocity relative to the plant would be zero; hence no force tending to extend the stipe would be imposed. Just as a guess, assume that the stipe extends at a velocity one-quarter that of the water around it. It would thus take something over 6 s for the stipe to extend to its breaking strain in addition to the 6–7 s required to align the stipe—a total of 12–13 s. Since water travels shoreward for only half of a wave period, breakage of this plant would require a large-amplitude swell or surf beat with a 24–26 s period. This analysis is only a gross approximation, but the message should be clear: *N. leutkeana* survives not because it is strong but because it is large and extensible.

Despite its flexibility, *N. leutkeana* often breaks, and large masses of its coiled corpses are common sights on Pacific beaches after storms. Koehl and Wainwright (1977) have shown that the majority of these breakages occur because the stipe is weakened by urchin grazing or because several stipes become twisted in such a way that the entire drag force from the bundle was applied to a single stipe.

Appendix 12.1

True Strain and Poisson's Ratio

A cube of volume = 1 is stretched to twice its length while maintaining the same volume (Figure 12.13).

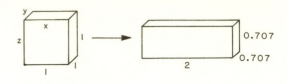

We calculate Poisson's ratio using both ϵ and ϵ_t:

ϵ	ϵ_t
$e_x = \dfrac{\Delta x}{x_0} = \dfrac{1}{1} = 1$	$e_x = \ln \dfrac{x}{x_0} = \ln \dfrac{1}{2} = 0.693$
$e_y = \dfrac{\Delta y}{y_0} = \dfrac{-0.293}{1} = -0.293$	$e_y = \ln \dfrac{y}{y_0} = \ln \dfrac{0.707}{1} = -0.347$
$e_z = \dfrac{\Delta z}{z_0} = \dfrac{-0.293}{1} = -0.293$	$e_z = \ln \dfrac{z}{z_0} = \ln \dfrac{0.707}{1} = -0.347$
$\therefore v_{yx} = v_{zx} = 0.293$	$\therefore v_{yx} = v_{zx} = 0.500$

Static Beam Theory

Examples of simple loading regimes are difficult to find in nature. As a rule, organisms come in rather bizarre and complex shapes, and hydrodynamic forces can cause the application of complex stresses. To tie together our knowledge of applied forces and material properties, we must first account for the role of *shape* in determining the capabilities of structures. This is the province of a large body of engineering endeavor cumulatively known as *beam theory*.

Beam Theory

Pure Bending

Consider a piece of material loaded in tension. If the force is applied uniformly, the stress near the end is the same no matter where in the cross section it is measured. This applied stress is transferred through the material, and if the cross-sectional area varies gradually along the length, the stress remains uniform. Only if there is a rapid change in the cross-sectional area does the stress in the sample vary across a section. Note that no mention has been made of the shape of the cross section; for a sample in pure tension with stress uniformly applied, shape is not important.

This simple situation does not hold for samples loaded in shear. Consider the sample shown in Figure 13.1a. A cube of material is glued to a surface and a force is applied to its free end in a direction parallel to the surface. The applied force and its reaction in the wall form a couple, and the sample is sheared. The average

shear stress on the free end of the sample is simply the applied force divided by the area of the free end. However, this average stress cannot apply evenly across the area. Recall from Chapter 12 that at static equilibrium, shear stresses must be equal

Figure 13.1. Shear stresses in a bent beam. (a) The applied force imposes a shear stress on the material of the beam. (b) The prism shown in (a) is removed to show the orthogonal shear stress required for equilibrium (see text). (c) The shear stress varies parabolically across the thickness of a uniform beam. (Redrawn from Gere and Timoshenko, Mechanics of Materials, 2d ed., 1984, by permission of Wadsworth, Inc.)

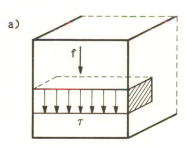

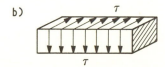

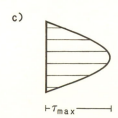

in perpendicular planes. Consequently, if we examine a prismatic portion of the material (as shown in Fig. 13.1b), the applied stress on the free end of the prism must be accompanied by an equal shear stress on the upper surface. If the prism is in the interior of the cube, this horizontal shear stress is provided by the tug of the material above the prism; but if the upper surface is at the top surface of the cube, there is no material above it to provide this horizontal stress. As a result, the horizontal stress must be zero at the upper surface. Consequently, the vertical shear stress must also be zero at the upper edge of the cube. The end result of this reasoning is that the shear stress varies across the surface of what one would think to be a uniformly loaded sample. The stress distribution is shown in Figure 13.1c for an object of rectangular cross section; the shear stress is zero at the edges and rises parabollically to a maximum in the center of the sample. This particular shear stress distribution holds only if the cross section of the sample is constant along its length; objects of different shape have different distributions. (For a more complete examination of shear stress distributions, see Gere and Timoshenko 1984).

Shape also affects stress distributions when shear and direct forces are applied in combination. Because it is very unusual for an organism in a wave-swept environment to be subjected solely to a tensile, compressive, or shearing force, it is important that we examine the general role of shape.

The simplest example of combined forces consists of tensile and compressive forces of equal magnitude acting on the ends of a beam so that one side of the beam is compressed and shortens, while the other side is in tension and lengthens (Fig. 13.2a). As a result, the beam bends. This particular situation, with opposite couples applied to the ends of the beam, is termed "pure" bending because nowhere is the sample placed in shear.

Figure 13.2. Pure bending. (a) The action of applied couples places the beam in pure bending, the top is stretched and the bottom compressed with an undeformed neutral surface in between. (b) Calculation of strain for a beam in pure bending (see text). (c) Calculation of the distribution of moments within a beam—the derivation of the second moment of area, I (see text).

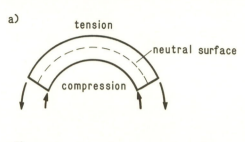

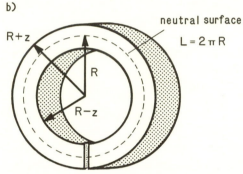

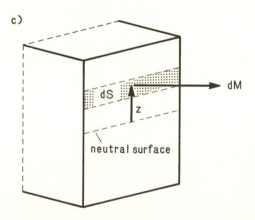

In the following analysis of pure bending, the goal is to relate the properties of the beam material to the amount of bending caused by a set of applied forces. We assume, arbitrarily, that the beam has a square cross section and, at rest, a length of $2\pi R$. Imagine now that the beam is bent until it forms a complete circle (Fig. 13.2b). In doing so, the outside of the beam is stretched and the inside is compressed. If we assume here that the stiffness of the beam is the same in tension as it is in compression, the part of the beam lying half-way between the inner and outer surfaces remains the same length. This unstretched, uncompressed layer is called the *neutral surface*, and by definition has length $2\pi R$. It should be apparent that the circle formed by the beam has a radius of R, suggesting a handy method for determining the strain at any point in the beam. For this derivation we use engineers' strain (change in length per original length).

The portion of the beam a radial distance z away from the neutral surface has a length of $2\pi(R \pm z)$. This strain of the material at this point in the beam is thus

$$\epsilon = (x/x_0) - 1 = \left\{ \frac{2\pi(R \pm z)}{2\pi R} \right\} - 1 = \pm z/R.$$

$$(13.1)$$

The material of the beam has a modulus, $E = \sigma_n/\epsilon$. Thus

$$E = \pm \frac{\sigma_n}{(z/R)};$$

$$\sigma_n = \pm Ez/R. \qquad (13.2)$$

The $\pm$ sign designates tension or compression, respectively. Thus the stress at any point in the beam varies directly with the distance away from the neutral surface and inversely with the radius of curvature.

The smaller the radius, the more sharply the beam is bent, and we can define a value called the curvature, ξ,

$$\xi \equiv 1/R. \qquad (13.3)$$

Stress varies directly with the curvature:

$$\sigma_n = \pm Ez\xi. \qquad (13.4)$$

This analysis allows us to specify the stress distribution resulting from a certain bending, but it does not say anything about the magnitude of the couple required. How can we calculate the required bending

moment? Note that stress is force per area. If for each point in the beam we can specify an area, knowing the stress at that point we can then calculate a force. Because stress varies with distance away from the neutral axis, we must keep the specified area small. In practice, we make it infinitesimally small and assign an area dS to each segment of the beam a distance z from the neutral surface (Fig. 13.2c). The force acting at each point in the beam is

$$f = \sigma \, dS = \pm Ez\xi \, dS. \qquad (13.5)$$

The force due to the stress acting at each infinitesimal area lies some distance away from the neutral surface, and applies a bending moment equal to $f|z|$. Thus the bending moment at each point is

$$dM = f|z| = Ez^2\xi \, dS \qquad (13.6)$$

and the total bending moment is equal to the integral of all these infinitesimal moments:

$$M_{\text{total}} = \int_{-z_{\text{max}}}^{z_{\text{max}}} Ez^2\xi \, dS = E\xi \int_{-z_{max}}^{z_{\text{max}}} z^2 \, dS.$$

$$(13.7)$$

The value of the integral $(\int_{-z_{\text{max}}}^{z_{\text{max}}} z^2 \, dS)$ is given the name *second moment of area*, symbolized by I. Eq. 13.7 can thus be expressed simply as

$$M_{\text{total}} = E\xi I. \qquad (13.8)$$

The stiffer the material (E) or the greater the curvature (ξ), the larger the moment required, as might be expected. The required moment also varies with this mysterious, nonintuitive value I, the second moment of area. This turns out to be the very measure of shape that is the focus of this discussion. By examining eq. 13.7 we can see that I accounts for shape by weighting the contribution to M_{total} of each infinitesimal area according to the square of its distance away from the neutral surface. Consequently, the areas farthest from the neutral surface contribute disproportionately to the moment required to bend the beam. Shapes that have a high *I* (such as hollow tubes) are, per cross-sectional area, much stiffer in bending than shapes with small *I*'s (a flat rectangle, for instance). Table 13.1 lists the values of *I* for various shapes. The second moment of area can be estimated for any biological cross section by the cut-and-paste approach to integration as explained in Appendix 13.1.

It will be useful to backtrack slightly and derive a relationship between stress and applied moment. From eq. 13.8 we know that

$$M/I = E\xi.$$

Expressing modulus as σ_n/ϵ and curvature as $1/R$,

$$M/I = \frac{\sigma_n}{(\epsilon R)}.$$

From eq. 13.1 we recall that $\epsilon R = z$. Thus

$$\sigma_n = Mz/I. \qquad (13.9)$$

Therefore, given the applied moment and the second moment of area, the stress can be found at any point in a beam.

Cantilever Bending

This analysis has been based on pure bending. As with pure tension or pure shear, it is unlikely that we will encounter a case of pure bending anywhere in nature. However, we have reached the point where we can analyze a type of deformation that *is* common in nature—cantilever bending. This form of bending occurs when a beam of any shape is held rigidly at one end and is subjected to a force either at the other end or along its length (Fig. 13.3). This loading regime bears a striking resemblance to any number of wave-swept organisms: sea anemones, coral blades, macroalgae, limpets, barnacles, snails, hydroids, and tube worms. Because of its universality among wave-swept organisms, we will examine this form of bending in some detail.

Table 13.1
Formulas for Calculating the Second Moment of Area, *I*, for Various Cross Sections

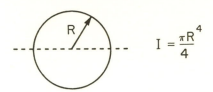

$$I = \frac{\pi R^4}{4}$$

$$I = \frac{\pi}{4}(R^4 - R_i^4)$$

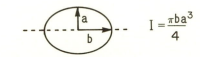

$$I = \frac{\pi b a^3}{4}$$

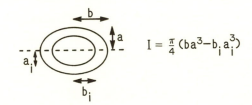

$$I = \frac{\pi}{4}(ba^3 - b_i a_i^3)$$

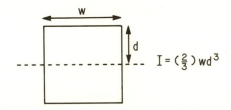

$$I = \left(\frac{2}{3}\right)wd^3$$

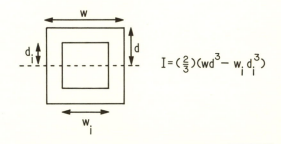

$$I = \left(\frac{2}{3}\right)(wd^3 - w_i d_i^3)$$

Cantilever bending differs from pure bending in that a shearing force is applied to the beam in addition to the application of a bending moment. This is best seen by examining a small section of the beam as if it were floating in space at equilibrium under the applied forces—a so-called *free body* (Fig. 13.4). To obtain a free body, imagine making two cuts through the cantilever perpendicular to the neutral surface, the first a distance x from the free end of the beam, and the second at $x + dx$. If these cuts were actually to be made, the beam segment between cuts (the free body) could not support itself in midair and would fall downward. Because this does not occur in the intact beam, we know that forces opposing these movements must have existed. We now restore these forces, thereby rebuilding equilibrium.

We begin by examining the shear forces acting on the free body. This task would be facilitated if we could legitimately slide the applied force f to the area of the free body. Easily done. As shown in Figure 13.4, an upward force equal to f is added at the end of the beam, and a corresponding downward force is added at x. In this maneuver we have added no net force and are left with a downward force acting on the right-hand surface of the free body. In order for equilibrium to be maintained in a vertical plane, this force must be resisted by an upward force at $x + dx$. The action of these two oppositely directed forces is to shear the free body. In sliding f down the beam, we never had to specify the magnitude of x; the process would be the same anywhere on the beam. We can conclude, then, that the shear force is constant along the beam.

However, this shear force is not the only force acting on the free body. To see this, we first restore f to its original position at the end of the beam. This downward force acts over a moment arm of length x, imposing a moment of magnitude $M = fx$ tending to rotate the free

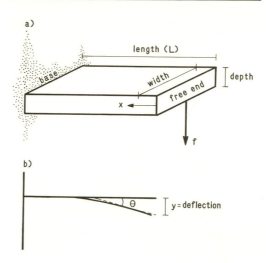

Figure 13.3. Definition of terms applied to cantilevers. (a) Dimensions of the beam. (b) Deflection.

Figure 13.4. Calculation of the stress distribution along the length of a cantilever (see text).

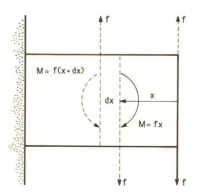

body clockwise. This bending moment is a function of position on the beam; it is zero at the free end, and reaches a maximum (fL) at the base. To counteract the rotation caused by this moment, we install at the left end of the free body a bending moment with magnitude M but acting in the opposite sense. Actually, this second moment must be a bit larger than M because of the additional moment arm due to the thickness of the free body, dx. This combination of a constant shear force and varying bending moment restores the free body to both rotational and translational equilibrium, and we can conclude that these are all the forces that act on the free body when it is incorporated into the beam.

The stress distributions in the free body are a combination of those caused by pure bending and pure shear. The bending moment causes tensile and compressive stresses, and these vary across the thickness of the beam, reaching maxima at the upper and lower surfaces. For a beam of constant section, the shear stress, although invariant along the length of the beam, varies across the thickness, reaching a maximum at the neutral surface. Thus shear stress is maximal where tensile and compressive stresses are zero, and vice versa.

As a consequence of shear stresses occurring in orthogonal pairs, a cantilever tends to shear along the neutral surface—that is, at 90° to the direction of force application. This effect is easily demonstrated by using a paperback book as a cantilever with the spine forming the base. When the book is flat, the ends of the pages line up. In bending the book, the pages slide past each other, in effect shearing. Trying to clamp the free end of the book between thumb and forefinger while bending the book illustrates how important shear stiffness can be in resisting cantilever bending.

Because of the varying bending moment, the curvature of a cantilever increases linearly along its length, being zero at the free end and maximal at the base. To arrive at the final deformed shape of the beam, the deformation due to shear is added to that due to bending. The complete solution to the displacement of the end of a cantilever is given by Gere and Timoshenko (1984) as

$$y = L - L \left\{ \frac{4EI}{FL^2} \right\}^{1/2} \cdot \{ H(k) - H(k, \phi) \},$$

$$(13.10)$$

where y is the deflection in the direction

of the applied force, L is the length of the beam, f is the applied force (here assumed to be applied at the end of the beam), $H(k)$ is the complete elliptic integral of the second kind,

$$H(k) = \int_0^{\pi/2} (1 - k^2 \sin^2 t) \, dt,$$

and $H(k, \phi)$ is the elliptic integral of the second kind,

$$H(k, \phi) = \int_0^{\phi} (1 - k^2 \sin^2 t) \, dt.$$

In solving these integrals,

$$k = \left\{ \frac{1 + \sin \theta}{2} \right\}^{1/2}$$

$$\phi = \arcsin \left\{ \frac{1}{k\sqrt{2}} \right\},$$

and θ, the angle to which the beam is bent (Fig. 13.3), is solved by trial and error from the trancendental equation

$$B(k) - B(k, \phi) = \left\{ \frac{FL^2}{EI} \right\}^{1/2},$$

where $B(k)$ is the complete elliptic integral of the first kind,

$$B(k) = \int_0^{\pi/2} \left\{ \frac{1}{1 - k^2 \sin^2 t} \right\}^{1/2} dt,$$

and $B(k, \phi)$ is the elliptic integral of the first kind,

$$B(k) = \int_0^{\phi} \left\{ \frac{1}{1 - k^2 \sin^2 t} \right\}^{1/2} dt.$$

Values for the above integrals are obtained most simply by looking them up in the appropriate tables (e.g., Abramowitz and Stegun 1965).

The use of this set of equations is straightforward but very laborious. Fortunately, one seldom needs the accuracy of this exact solution, and simple approximations are available. These approximations are based on the assumption that, θ, the angle to which the beam is bent, is small, requiring that the end deflection be small. The general rule of thumb is that if the deflection of the end of the cantilever is less than 10% of the length of the beam, the small deflection approximations are valid.

The simplest of these approximations is for a cantilever bending under the action of a single force acting at its end. The deflection $y(x)$ at distance x from the end of the beam is

$$y(x) = \left\{\frac{-fL^3}{3EI}\right\}\left\{1 - \frac{3x}{2L} + \frac{x^3}{2L^3}\right\}. \quad (13.11)$$

When $x = 0$, that is, at the free end,

$$y(0) = \frac{fL^3}{3EI}. \quad (13.12)$$

As one would expect, the stiffer the material of the beam (E) and the larger its second moment of area (I), the smaller the deflection. These two terms commonly occur together in beam equations, and the product EI is termed the *flexural stiffness*.

The deflection is very sensitive to the length of a beam. Given two beams similar in all respects except that one is twice the length of the other, the end of the longer beam deflects eight times as much under an equal load.

Table 13.2 lists other small-deflection approximations for cantilever deflections depending on the type of load applied. For large deflections, we must use exact solutions similar to the one given above, or at least a higher-order series approximation. For example, in their study on the bending of the flexible macroalga *Eisenia arborea*, Charters et al. (1969) needed to use the appropriate large-deflection equation to obtain reasonable results. Corrections should also be made for cantilevers whose width is large relative to their length. For these sorts of adjustments, see Roark and Young (1975).

The flexural stiffness of a beam has been shown to be proportional to its second moment of area, I, and I varies with the beam's shape and dimensions. Given a fixed volume of material, how can the stiffest beam be constructed? Assume, for example, that the beam is tubular. From

Table 13.1 we can deduce that I for such a shape is $(\pi/4)(R^4 - [R - t]^4)$, where R is the radius of the tube and t is the thickness of its wall. It would seem that by increasing the radius of the tube while appropriately decreasing the wall thickness, we could increase the flexural stiffness without limit. Followed to its logical end, this reasoning would result in a tube with infinite radius, an infinitely thin wall and an infinite flexural stiffness. Everyday experience should lead one to doubt the validity of this conclusion. For example, consider a common, thin-walled tubular structure—a soda straw (Fig. 13.5a). If a straw is loaded as a cantilever, it does indeed bend, but after a small amount of bending it kinks (usually near the base) and abruptly loses its stiffness. The same phenomenon occurs in tubular organisms such as sea anemones (Fig. 13.5b). This sort of kinking behavior is known as *local buckling*, and it sets the upper limit to the stiffness of structures in bending. Local buckling is a catastrophic occurrence, it happens all at once at a critical value of stress in the wall of the structure. For tubular structures, the appropriate equation for calculating critical stress is

$$\sigma_c \simeq 0.5E\frac{t}{D}, \quad (13.13)$$

where t is the wall thickness, D is the diameter, and 0.5 is an empirically determined coefficient (Wainwright et al. 1976). As the ratio of thickness to diameter decreases, so does the critical stress, effectively limiting the maximum stiffness of structures of defined volume.

Applications of Static Beam Theory

To illustrate the usefulness of beam theory in the study of wave-swept organisms, we will examine two examples: the bending of sea anemones, and the shapes of coral blades.

Table 13.2
Formulas for Calculating the End Deflection, z, of Cantilevers under Various Loading Regimes
(formulas assume that z is less than 10% of the beam length, L)

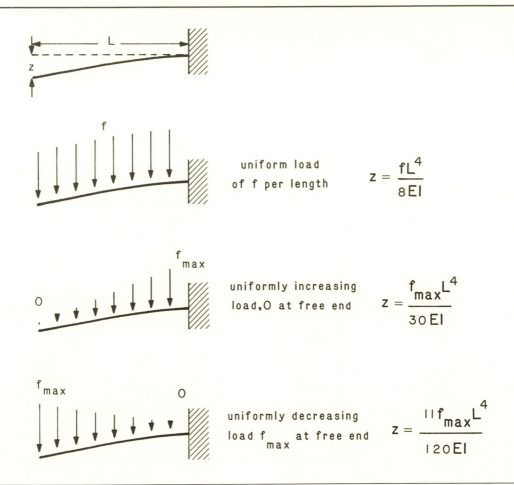

uniform load
of f per length

$$z = \frac{fL^4}{8EI}$$

uniformly increasing
load, 0 at free end

$$z = \frac{f_{max}L^4}{30EI}$$

uniformly decreasing
load f_{max} at free end

$$z = \frac{11f_{max}L^4}{120EI}$$

Sea Anemones

A fully expanded *M. senile* consists of a long, tapered cantilever beam with a crown of tentacles on its free end (Fig. 13.6a). Koehl (1977c) has shown that the modulus of the body-wall mesoglea when subjected to a stress for a long period of time (e.g., 6 hours, about the duration of a tidal current) is about 10^4 Pa. At shorter times (a few seconds) the material has a modulus of about 10^5 Pa. For this analysis we assume that water accelerations are negligible (Koehl 1977b) so that the only force acting on the body column is that due to the drag on the crown, f. Using the values for C_d and S_ρ given in Figure 13.6a,

$$f = \tfrac{1}{2}\rho u^2 S_\rho C_d$$

$$= \tfrac{1}{2}(1025 \text{ kg/m}^3)(u, \text{ m/s})^2(0.02, \text{ m}^2)0.8$$

$$= 8.8 \, u^2. \tag{13.14}$$

The maximum water velocity recorded by Koehl (1977a) near the crown of *M. senile* was 0.2 m/s, causing the application of 0.35 N.

Figure 13.5. Local buckling in (a) a soda straw, and (b) a sea anemone, *Metridium senile* (redrawn from Koehl 1977b, by permission of the Company of Biologists, Ltd.).

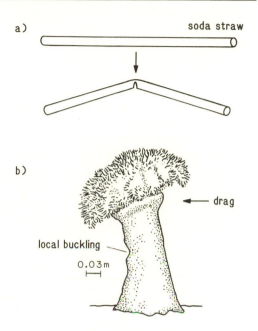

the column buckles under long-term loads ($E = 10^4$ Pa) is

$$\sigma_c = 0.5E \frac{t}{D} = 0.5(10^4)(0.01) = 50 \text{ Pa},$$

the ratio of thickness to diameter being 0.01 both near the crown and at the base. This critical buckling stress is well below the actual wall stress at a velocity of 0.2 m/s. Thus the anemone should, in theory, buckle, and this proves to be the case in nature (Fig. 13.5b). A critical water velocity can be calculated at which the column near the crown should buckle. Inserting the equation for applied force as a function of velocity (eq. 13.14) into eq. 13.15 and setting this equal to the critical buckling stress,

$$\sigma_c = 50 \text{ Pa} = \frac{(8.8 \ u^2)(0.12 \text{ m})(0.025 \text{ m})}{3 \cdot 10^{-8} \text{ m}^4}$$

$$= 8.8 \cdot 10^5 \ u^2$$

$$u = \left\{ \frac{50}{8.8 \cdot 10^5} \right\}^{1/2} = 7.5 \cdot 10^{-3} \text{ m/s}.$$

Seven and a half millimeters per second is quite slow, and these animals are buckled most of the time. Even if the force causing buckling is only applied briefly (and the modulus is thus increased to 10^5 Pa), the organism buckles at a velocity of only 2.4 cm/s. Koehl (1977a) points out that buckling may actually serve a useful function by allowing the crown of tentacles to bend over perpendicular to flow. This posture increases the filtering area.

In contrast, the great green anemone *A. xanthogrammica* is commonly found in areas that are subjected to much higher water velocities and accelerations than those experienced by *M. senile*. The animals are short, fat cylinders (Fig. 13.6b). Again we start by assuming the most of the flow force on the creature is due to the tentacles (Koehl 1977a,b). Using values cited in Fig. 13.6b, we calculate the force acting at the end of this short cantilever:

$$f = \frac{1}{2}\rho u^2 S_p C_d + \rho C_m V \, du/dt$$

$$= 0.978u^2 + 0.155 \, du/dt.$$

What happens to the anemone under the action of this force? To begin, we calculate the wall stress at two points where I is known. We use a version of eq. 13.9, modified to reflect the fact that the wall is one radius distance, R, from the neutral axis:

$$\sigma = \frac{FxR}{I}, \qquad (13.15)$$

where x is the distance from the application of the force to the particular point on the body wall. Near the crown ($x = 0.12$ m, $R = 0.025$ m, $I = 3 \cdot 10^{-8}$ m^4), $\sigma_n = 3.5 \cdot 10^4$ Pa. A similar calculation for the wall near the base ($x = 0.33$ m, $R = 0.06$ m, $I = 6.9 \cdot 10^{-7}$ m^4) yields a value of 10^4 Pa. Even though the column of the anemone is quite long, the taper is such that the maximum stress occurs not at the base, but near the crown.

Does the column primarily buckle, bend, or shear? The critical wall stress at which

Figure 13.6. Force coefficients and dimensions for the calculation of wall stress in two sea anemones. (Data from Koehl 1977b by permission of the Company of Biologists, Ltd.).

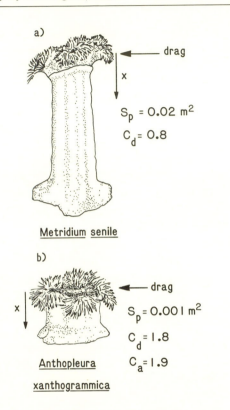

a)

drag

x

$S_p = 0.02$ m^2

$C_d = 0.8$

Metridium senile

b)

x

drag

$S_p = 0.001$ m^2

$C_d = 1.8$

$C_a = 1.9$

Anthopleura

xanthogrammica

Koehl (1977a) cites maximum values for u and du/dt of 0.2 m/s and about 2 m/s^2, resulting in an applied force of 0.34 N. We have assumed here that the maximum velocity and the maximum acceleration occur at the same time—an absurd situation. However, the velocities and accelerations cited by Koehl were measured under relatively benign conditions, and it seems quite likely that the values used here could co-occur during stormy periods (see Chapter 7).

The wall stress for *A. xanthogrammica* is calculated in the same fashion as for *M. senile*. The result indicates a stress of about 10^3 Pa for this more "exposed" organism, somewhat lower than that experienced by the more "protected" anemone.

Does *A. xanthogrammica* locally buckle?

The body-wall material has a modulus of 10^5 Pa for the time scale at which surge-channel forces are imposed. Applying eq. 13.13, we calculate the critical stress to be $1.8 \cdot 10^3$ Pa. This value is somewhat higher than the wall stress calculated above. Under these conditions the anemone should not buckle; rather, it should bend, shear, or both. The maximum tensile and compressive stresses in the wall were calculated above to be 10^3 Pa, resulting in a strain of about 0.01. The average shear stress is simply the applied force divided by the wall area ($4.5 \cdot 10^{-4}$ m^2), a value of 792 Pa. Given that the shear modulus of the mesoglea is likely to be about a third that of the tensile or compressive modulus (eq. 12.8), the column of the anemone will reach a strain about twice as much in shear as in bending. This tendency to shear rather than bend is useful to the organism. By shearing it maintains a posture with the oral disk oriented horizontally, the best posture for catching prey falling to the bottom of the surge channel. In either case, the strain is quite small, amounting to at most a deflection of 0.002 cm at the oral disk for the load calculated here.

These examples should indicate the usefulness of beam theory in supplementing one's subjective impression of biological mechanics. Although it might be expected that a thin-walled column such as a *M. senile* should buckle when subjected to flowing water, it is quite counterintuitive that the open-coast *A. xanthogrammica* should, under normal environmental forces, have a lower wall stress than its more protected cousin.

Corals

The skeletons of plants and animals are metabolically costly and time-consuming structures to build. It seems reasonable to suppose that successful organisms do not waste time and energy building skeletons that are stronger than necessary. For the present argument, we interpret time and energy in terms of volume—a skeleton that functions with the smallest possible volume uses the least amount of expensive skeletal material, and probably takes the least time to build. How would one build a cantileverlike skeleton with the least amount of material required to resist a certain force? In answering this question, we invoke what might be termed the "one hoss shay" theorem: If, when the beam is stressed to the point of breaking every bit of the material breaks at once, that beam is formed of the minimum possible volume of material.[1] The proof of this theorem is simple: If one part of the beam breaks while another part does not, the part that does not break is stronger than required. Consequently, material could have been removed from that overly strong section, and the volume thereby reduced. Only when the beam is whittled down so that every portion is at its breaking stress can this argument no longer be applied; at that point the minimum volume has been reached.

This principle is applied here to the case of hermatypic corals. Many shallow-water, reef-building corals form cantileverlike skeletons. Examples include *Acropora cervicornis*, *Acropora palmata*, and *Millepora complanata*. All three of these species grow within 2–3 m of the surface on Caribbean reefs, and are therefore subjected to the hydrodynamic forces accompanying waves. In addition to the forces from moving water, these animal colonies must resist the impact of wave-borne objects that are commonly tossed about during storms, and they must resist the force applied by their own weight.

In what shape would these colonies be expected to grow if they were to follow the "one hoss shay" principle? The answer depends on the particular type of force applied and on the cross-sectional shape of the beam. Corals form solid rather than tu-

bular skeletons, and accordingly we use beams with solid cross sections. There are two simple ways to model coral cantilevers. First, we could give the beam an elliptical cross section (Fig. 13.7a). If the major and minor axes are the same length, the beam has a circular cross-section. If one axis is much longer than the other, a platelike beam is approximated. In this model, the ratio of beam width to depth is held constant at a value of h; beams that follow this model get wider as they get deeper. Alternatively, one can use a model in which the width of the beam is constant and only the depth is allowed to vary (Fig. 13.7b). Between these two models one can approximate the shapes of most cantileverlike corals.

Point Loading. When a beam is struck by a solid object, a localized ("point") load is applied. For the sake of simplicity, we assume that this load is applied at the free end of the beam and is sufficient to place the top and bottom of the beam under a stress equal to the breaking stress of the material, σ_b. Using eq. 13.15, we write an expression relating the applied force to

Figure 13.7. Definition of the dimensions of shapes used to model corals (see text).

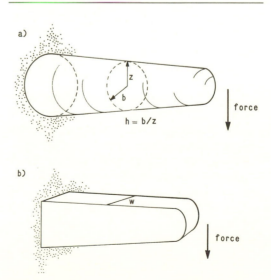

a)

h = b/z

force

b)

w

force

the stress at the top and bottom of the beam, and we define the half-depth of the beam at any point along its length to be some (as yet undefined) function, z, of the distance, x, from the free end:

$$\sigma_b = fxz/I. \qquad (13.16)$$

For this example we arbitrarily use a beam with an elliptical cross section, for which the second moment of area is

$$I = \frac{\pi h z^4}{4}. \qquad (13.17)$$

Here z is the half-depth of the beam, in this case the length of the elliptical axis parallel to the applied force (Fig. 13.7a). Thus

$$\sigma_b = \frac{4fxz}{\pi h z^4} = \left\{\frac{4fx}{\pi h}\right\} z^{-3}$$

$$z^3 = \left\{\frac{4fx}{\sigma_b \pi h}\right\}$$

$$z = \left\{\frac{4fx}{\sigma_b \pi h}\right\}^{1/3}. \qquad (13.18)$$

To keep the stress constant on its upper and lower surfaces, the beam must get deeper in proportion to the cube root of the distance from the free end. Because a constant cross-sectional shape is maintained, the beam gets wider as well. The greater the applied force, f, the deeper the beam must be. Conversely, the stronger the beam material (the greater σ_b), the less deep the beam need be.

Figure 13.8 is a graph of relative depth versus relative length for a number of blades of the hydrocoral, *Millepora complanata*. This coral apparently grows such that its depth and width increase approximately in proportion to the cube root of length as we would predict for a cantilever subjected to point loading. This coral is found in the surf zone of reefs in Panama, and it seems reasonable that it is often hit by flying debris, which would apply a point load.

The same sort of calculation can be carried out for a cantilever of constant width

Figure 13.8. The shape of the fire coral *Millepora complanata*. The open dots were measured on samples of *M. complanata* collected near Colón, Panamá. The solid line is the shape predicted for an optimally shaped beam subjected to a point load; thickness is proportional to the cube root of the distance from the free end (see text).

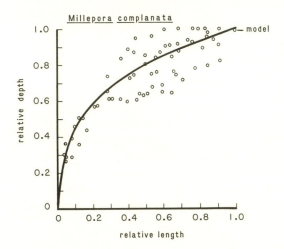

subjected to a point load at its free end. In this case the depth of the beam must increase in proportion to the square root of the distance from the free end.

Loading from Drag. The velocity of moving water places a force on coral blades that is proportional to the projected area of the blade and to the square of the velocity (Chapter 11). What shape must a blade be so that all parts of the beam are equally likely to break under this form of load? An exact solution is possible (Appendix 13.2) but requires some basic equations from beam theory that we have not derived here. This question, however, provides an excellent opportunity to introduce an alternative approach to problems in beam theory—one which, since the advent of the digital computer, has become a standard research tool.

The idea behind this approach is illustrated in Figure 13.9. We assume the cantilever to be formed from a series of blocks. We start with a cantilever that is only one block long and then cause it to "grow" by

Figure 13.9. (a) An incremental model for the growth of a coral (see text). (b) The shape of the coral *Acropora palmata*. The data points were taken from one specimen collected near Colon, Panama. The solid line is the shape predicted by a computer model for the optimal shape of a beam loaded by drag; the depth of the beam increases in linear proportion to the distance from the free end.

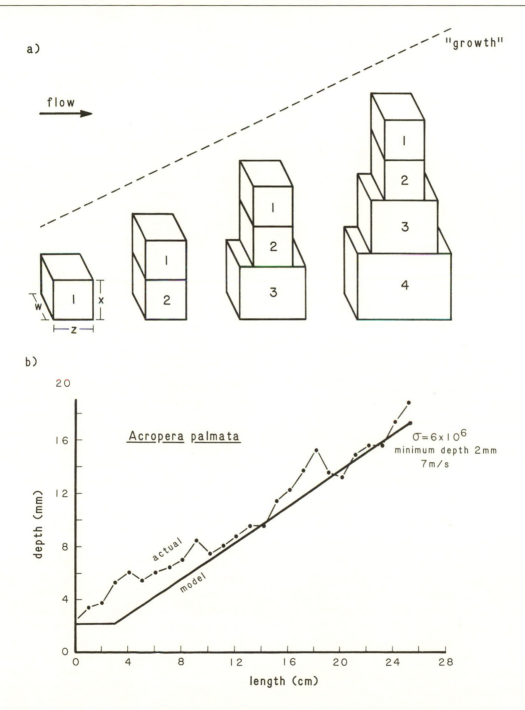

adding blocks at the free end and allowing the older blocks to increase in depth (Fig. 13.9a). As each block is added, we compute the bending moment acting on each of the older blocks, and calculate what thickness each of these blocks must be to keep the bending stress less than the breaking stress. We can grow the blade to any size we want by adding a sufficient number of blocks, and then examine the shape of the beam to deduce the "growth law" to which it adheres.

This approach is not unlike what must occur in the growth of an actual coral blade. A line of newly formed polyps at the free end of the coral cantilever amounts to the addition of a block. Polyps nearer the base of the blade increase in depth, although it is not known whether this is in response to fluid-dynamic forces.

For the sake of simplicity, we model a coral such as *A. palmata* in which blades maintain a nearly constant width as they increase in length. This allows us to model a blade using blocks with a rectangular cross section. The width of each block is W, its length along the beam is x, and the depth, z (measured from the neutral surface), is the quantity we vary. At any one time, blocks are numbered from the free end, and the newly added block is number 1. The drag force acting on each block is calculated assuming that the direction of the water velocity is perpendicular to the width of the beam:

$$f = \tfrac{1}{2}\rho u^2 S_p C_d,$$

where $S_p = Wx$, and C_d is 1.2. The moment acting on the endward surface of block i is the sum of the moments applied by blocks distal to it. Using the subscript j to number these distal blocks, the moment is

$$M_i = \sum_{j=1}^{i} \tfrac{1}{2}\rho u^2 (Wx) C_d (x[i-j]). \quad (13.19)$$

The depth of block i required to keep the bending stress at the top and bottom surfaces equal to the breaking stress is determined using eq. 13.15:

$$\sigma_b = M_i z_i / I,$$

$$z_i = \sigma_b I / M_i.$$

The second moment of area, I, for a rectangular cross section is $I = 2Wz^3/3$. Thus,

$$z_i = \left\{ \frac{3M_i}{2W\sigma_b} \right\}^{1/2}. \quad (13.20)$$

The results of performing this calculation are shown in Figure 13.9b for a blade formed from thirty blocks each 1 cm long and 10 cm wide. A breaking strength of $6 \cdot 10^6$ Pa (typical for *A. palmata*) and a water velocity of 7 m/s were used. The blade grows without increasing in depth until sufficient blocks have been added so that their moment applies to the basal block a stress equal to the breaking stress. As the blade grows beyond this point, the depth must increase in direct linear proportion to the distance from the free end. This is the same result obtained from the more rigorous approach used in Appendix 13.2. Data taken from an actual *A. palmata* blade are plotted in Figure 13.9b and fall reasonably near the line predicted by the model. The velocity used in this calculation (7 m/s) is probably higher than that usually encountered by *A. palmata*, suggesting that these corals incorporate a "safety factor" into their growth pattern.

A similar line of reasoning may be followed to calculate the shape of a beam with an elliptical cross section. If the beam maintains a constant cross sectional shape along its length, it must increase in width in proportion to its increase in depth. One might think that this would lead to a different dependence of depth on length. However, in addition to increasing the drag force, the increase in width also increases the second moment of area. The two factors cancel, and the depth must still increase in direct proportion to the length of the beam. The blades of *Acropora cervicornis* have a circular cross section and increase in half-depth (=radius) in direct linear proportion to the distance from the free end. It is interesting to speculate that the shapes of corals such as *A. palmata* and *A. cervicornis* have evolved in response to the water velocities they encounter.

Loading from Self Weight. Coral skeletons are made from calcium carbonate and have a density of 1,700–2,000 kg/m³. Consequently, these structures are negatively buoyant in seawater, and their own weight places a force on the skeletal cantilever. This force is proportional to the volume of material in the skeleton and is distributed along the cantilever, the exact distribution depending on the shape of the beam.

What shape would these cantilevers take to minimize their volume while resisting their own weight? The answer for a beam of circular cross section is arrived at in the same fashion as the calculations shown in Appendix 13.2, with the result that

$$z = \left\{\frac{2(\rho_c - \rho_w)g}{15\sigma_b}\right\} x^2, \quad (13.21)$$

where ρ_c is the density of the coral skeleton, ρ_w is the density of seawater, and g is the acceleration of gravity. The depth of the beam increases as the *square* of the distance from the free end. No cantilever-like coral follows this growth pattern, and the reason becomes apparent when values for ρ_c and σ_b are inserted into eq. 13.21. A meter-long beam with a circular cross section and a breaking strength of $2 \cdot 10^6$ Pa need only be 0.7 mm in diameter at its base to resist the forces from its own weight. This is much smaller than the diameter needed to resist drag forces. Thus any evolved response to loading by self weight is very likely to have been swamped by the much greater stress due to hydrodynamic forces.

A growth form similar to that predicted for loading by self weight would be appropriate for a coral blade that grows in response to the force placed on it by an accelerating fluid. In this case the appropriate equation is

$$z = \left\{\frac{2\rho_w C_m \, du/dt}{15\sigma_b}\right\} x^2, \quad (13.22)$$

where du/dt is the acceleration of the water past the blade. Unless the acceleration is very high and the water velocity quite low (a difficult situation to imagine for the environment on a reef), the forces due to the water's acceleration are small relative to drag, and it seems unlikely that the growth form of corals has responded to this force.

A final note must be added to this discussion of beam shapes. All of these calculations have been made with the sole purpose of keeping the tensile and compressive stress constant along the beam, and we have tacitly assumed that these stresses are the only ones of importance. What about the shear stresses? A variation in beam depth can have some strange effects on the shear stress (Gere and Timoshenko 1984). Depending on the precise shape of the beam, the maximum shear stress can occur somewhere other than the neutral surface, and the distribution of shear stresses can vary along the length of the beam. For short squat beams where shear stresses are likely to be of the same order as tensile and compressive stresses, we may need to take these shear effects into account when specifying the ''correct'' shape of the beam. We have also assumed that a beam is subjected to only one type of load at any one time. In the real world a cantilever is likely to be simultaneously acted upon by several types of load, and any full treatment of the ''correct'' shape of the beam should take into account the interactions among these loads. This effect will be discussed in Chapter 17.

Appendix 13.1

Calculating I *for Complex Shapes*

An estimate of the second moment of area can be found for any beam cross section using the cut-and-paste approach to integration. Before we can apply this method we need two pieces of information:

1. In what direction is the beam to be bent? The direction of bending is determined by the direction of the applied force and therefore depends on the circumstances. The direction of bending is usually easily determined, but it is also very important—the calculated I often strongly depends on the direction of bending.

2. Where is the neutral surface? The neutral surface is located using physical subterfuge. First, we can easily show that the neutral surface for most beams passes through the centroid of the cross section. This is strictly true only if the beam has the same stiffness in compression as it does in bending, but small differences in modulus do not unduly bias the calculation. To find the centroid of the cross section, we draw an accurate outline of the cross section on a piece of cardboard and cut out the outline. We then note that for a two-dimensional piece of cardboard, the centroid is the same as the center of mass. We easily find the center of mass by hanging the cardboard from a pin stuck near its periphery; the center of mass must lie directly below the pin (see Fig. 13.10A). Sticking the pin in three or four different spots gives us a good estimate of the location of the centroid.

Knowing the centroid and the direction of bending, we can draw a line on the cardboard corresponding to the neutral surface (Fig. 13.10B). We draw a series of lines parallel to the neutral surface, dividing the cross section into a number ($= N$) of areas. For each area we note the distance, z_i, between its midpoint and the neutral surface (Fig. 13.10B). We then cut along each of our lines, ending up with N pieces of cardboard. We carefully weigh and record the weight of each piece. We then cut a standard piece of cardboard (say 10 cm by 10 cm) out of the same kind of

Figure 13.10. Estimating I for a complex shape. (a) The centroid of a cross section is the same as its center of gravity and will lie directly below the point from which the section is suspended. (b) The cross section is cut into parallel strips (see text).

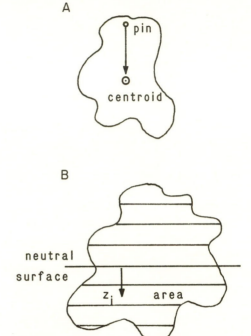

cardboard and weigh it. Knowing the weight of this known area, we can then calculate the area of each of our N pieces of cardboard. Thus, for each piece, we know its area and its average distance from the neutral surface. An estimate of the second moment of area is then

$$I \simeq \sum_{i=1}^{N} \text{area}_i (z_i)^2.$$

The more areas into which the cross section is divided, the more accurate this estimate. If our beam is too small or too big for the cross section to be treated directly in this fashion, we can perform the method on a scale drawing of the cross section as long as we use the scaled distances and areas in the calculations. An analogy of this method is easily adapted for use on a digital computer equipped with a digitizer pad.

Figure 13.11. The forces and moments acting on a segment of a cantilever (see text).

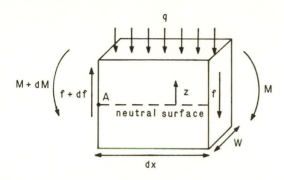

Appendix 13.2

An Exact Solution for the Optimal Drag-Loaded Cantilever (Figure 13.11)

Drag imposes a force per length, q,

$$q = \tfrac{1}{2}\rho u^2 W C_d. \tag{1}$$

For vertical equilibrium,

$$f + df = f + q\,dx$$

$$\therefore \frac{df}{dx} = q. \tag{2}$$

To calculate the moments necessary to maintain rotational equilibrium, we take moments about point A:

$$M + dM = M + f\,dx + q\,dx\,\frac{dx}{2}.$$

Ignoring products of differentials, we calculate that

$$\frac{dM}{dx} = f. \tag{3}$$

Combining results (2) and (3), we get

$$\frac{d^2M}{dx^2} = q. \tag{4}$$

We now calculate M. From eq. 13.9 we know that

$$\sigma = \frac{Mz}{I}$$

For a rectangular cross section, $I = \dfrac{2Wz^3}{3}$ (Table 13.1). Thus

$$M = \frac{2Wz^2\sigma}{3}. \tag{5}$$

Setting $\sigma = \sigma_b$, the breaking stress, and combining results (1), (4), and (5), we get

$$\frac{d^2}{dx^2}\left\{\frac{2Wz^2\sigma_b}{3}\right\} = \tfrac{1}{2}\rho u^2 W C_d.$$

To solve this equation we try setting $z = \alpha x$, a linear increase in depth with length:

$$\frac{d^2}{dx^2}\left\{\frac{2W\alpha^2 x^2\sigma_b}{3}\right\} = \tfrac{1}{2}\rho u^2 W C_d$$

$$\frac{4\alpha^2\sigma_b}{3} = \tfrac{1}{2}\rho u^2 C_d$$

$$\alpha = \tfrac{1}{2}\sqrt{\frac{3\rho u^2 C_d}{2\sigma_b}}.$$

Therefore αx is a correct solution to the shape of the cantilever beam, which maintains a constant stress (equal to the breaking stress) along its length.

CHAPTER 14

~ ~ ~ ~ ~ ~ ~ ~ ~ ~ ~

Dynamic Beam Bending

$\mathbf{W}$e have assumed that the forces we dealt with in the preceding chapter act at static equilibrium. But when a force is applied to a cantilever, the beam has to bend to provide the stresses required for equilibrium. For any real beam, a finite velocity is acquired during the course of this deflection, giving the beam some inertia, and we have yet to take this factor into account. To do so we need a method for describing the dynamic responses of cantilevers.

Harmonic Motion

All descriptions of the dynamics of oscillating systems are based on the idea of *harmonic motion*, and it will be useful to examine briefly the properties of harmonic motion before relating these properties to particular examples in the wave-swept environment.

We begin by considering a mass suspended from a spring (Fig. 14.1a). The system is set in motion by pulling the weight down and then releasing it. The mass bounces up and down. Imagine that there is a pen attached to the moving weight, writing on a piece of paper that moves to the right with a constant speed (Fig. 14.1a). Clearly, the mass moves up and down with a sinusoidal motion.

We can think of the displacement of the mass as the projection on the displacement axis (here taken to be the y-axis) of a line of length equal to the maximum displacement, y_0 (Fig. 14.1b). As this line rotates at constant angular velocity about an axis located at $y = 0$, its projection

travels back and forth between $\pm y_0$. y at time t is thus

$$\text{displacement, } y = y_0 \sin \omega t, \quad (14.1)$$

where ω is the angular velocity (in radians/s) at which the line rotates. ω is known as the *circular frequency*.[1] Given this interpretation of the time course of the deflection, we can determine the velocity at any time by taking the derivative of y with respect to time:

$$\text{velocity, } v = \frac{d(y_0 \sin \omega t)}{dt} = \omega y_0 \cos \omega t.$$

$$(14.2)$$

Recall that $\cos \omega t = \sin(\omega t + \pi/2)$. Thus,

$$\text{velocity, } v = \omega y_0 \sin(\omega t + \pi/2), \quad (14.3)$$

and the velocity is seen to be out of phase with the displacement by $\pi/2$ radians (90°; Fig. 14.1b).

The acceleration of the weight is calculated by taking the second derivative of the displacement with respect to time:

$$\text{acceleration, } a = d^2 y/dt^2 = \frac{d^2(y_0 \sin \omega t)}{dt^2}$$

$$= -\omega^2 y_0 \sin \omega t. \quad (14.4)$$

Recalling that $-\sin \omega t = \sin(\omega t + \pi)$, we arrive at the conclusion that acceleration leads the displacement by π radians (180°; Fig. 14.1b). The logic used here, and the results obtained, are directly analogous to the computations we made in Chapter 5 regarding surface waves with a sinusoidal form.

[1] Frequency (in cycles per second [Hz]) is obtained from circular frequency by dividing ω by 2π.

Figure 14.1. The harmonic motion of a mass on a spring. (a) A pen on the mass traces a sinusoid on a moving chart. y_0 is the amplitude of the oscillation. (b) The phase relationship among displacement, velocity, and acceleration in harmonic motion (see text).

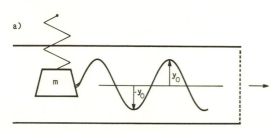

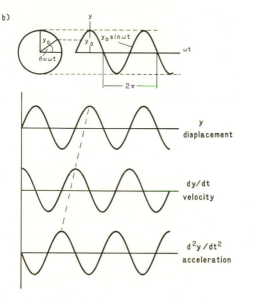

The Oscillating Cantilever

An oscillating weight on a spring is a handy tool for introducing harmonic motion, but bears little resemblance to most wave-swept plants and animals. As we have seen, many wave-swept organisms can be described as cantilevers, and we concentrate on the oscillatory properties of these beams. While much of the theory discussed here applies to oscillating systems in general, the specific equations are designed for cantilevers, and we should be careful in using them for other systems.

Consider a slender vertical cantilever with a mass m attached to its free end (Fig. 14.2a). A horizontal force has been applied to the end of the cantilever and the beam is bent—the greater the deflection, the greater the resistive force due to bending. For small deflections, the resistive force increases linearly and may be described by the equation

$$f = ky, \qquad (14.5)$$

where f is the resistive force, y is the displacement of the end of the cantilever, and k is a constant of proportionality. From reference to eq. 13.12, we can see that in this case

Figure 14.2. (a) The oscillation of a mass on a massless cantilever—definition of terms (see text). (b) Terms used in calculating the effective mass of an oscillating cantilever (see text).

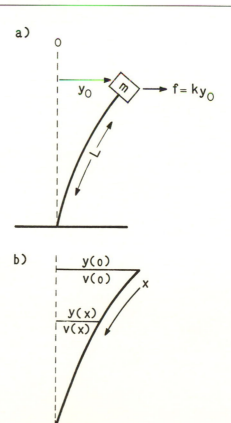

$$k = \frac{3EI}{L^3}, \qquad (14.6)$$

where L is the length of the cantilever. For the moment, we assume that the external force has been applied at some time in the past and the system is now at static equilibrium, the beam being deflected a distance y_0. The internal force of bending exerts a force

$$f = ky_0 = \frac{-3EIy_0}{L^3}, \qquad (14.7)$$

the minus sign indicating that the force is directed to the left. The energy expended in bringing the end of the beam to this deflection is the integral of force times deflection, and it is the potential energy stored in the equilibrium system,

$$\text{potential energy} = \int_0^{y_0} ky \, dy = \tfrac{1}{2} ky_0^2. \qquad (14.8)$$

We now suddenly remove the externally applied force. At this instant the mass feels a net force and accelerates to the left. Recalling that force is equal to mass times acceleration, we see that

$$m \frac{d^2y}{dt^2} = ma = -ky_0. \qquad (14.9)$$

As the end of the cantilever moves to the left, its velocity increases, but because the deflection (and therefore the bending force) is decreasing, the rate of acceleration decreases. In general, the acceleration at any deflection can be expressed as

$$a = \frac{-ky}{m}. \qquad (14.10)$$

When the end of the beam reaches its undeflected position ($y = 0$), the bending force is zero and the acceleration is therefore also zero (eq. 14.10). Because the deflection is zero, the potential energy stored in the system is also zero (eq. 14.8). However, at this point the mass is traveling at its maximum velocity, and therefore has its maximum kinetic energy,

$$\text{kinetic energy} = \tfrac{1}{2} mv^2, \qquad (14.11)$$

where v is the velocity of the mass at the end of the beam. The inertia of the moving mass causes it to continue on past the point of zero deflection. As it does so, a bending force is again exerted on the mass; in this case the force is directed toward the right. The acceleration caused by this force gradually decreases the velocity of the mass, until at a distance $-y_0$ the velocity is once again zero. At this deflection the potential energy is again at a maximum, and the whole process starts all over. The mass oscillates back and forth, trading deflection for velocity, and potential energy for kinetic energy. If the beam material is perfectly elastic (i.e., it has no hysteresis) and there are no viscous fluid-dynamic forces opposing motion, the mass continues to oscillate forever.

Having explained why cantilever beams oscillate, we must now explain why this oscillation should be harmonic. We begin by making what at first appears to be an arbitrary definition. We define the *natural circular frequency, ω_n*:

$$\omega_n \equiv \sqrt{\frac{k}{m}}. \qquad (14.12)$$

Thus

$$k = m\omega_n^2. \qquad (14.13)$$

Substituting this expression for k into eq. 14.10 and rearranging, we arrive at a differential equation that describes the motion of the beam:

$$\frac{d^2y}{dt^2} + \omega_n^2 y = 0. \qquad (14.14)$$

One solution to this equation is

$$y = A \sin \omega_n t + B \cos \omega_n t, \qquad (14.15)$$

where A and B are the two constants necessary for a unique solution. We can show that this is indeed a proper solution by inserting eq. 14.15 into equation 14.14:

$$\frac{d^2y}{dt^2} = -\omega_n^2 A \sin \omega_n t - \omega_n^2 B \cos \omega_n t,$$

$$\omega_n^2 y = \omega_n^2 A \sin \omega_n t + \omega_n^2 B \cos \omega_n t,$$

$$\therefore \frac{d^2y}{dt^2} + \omega_n^2 y = 0.$$

A and B can be expressed in terms of the velocity at time zero and the deflection at time zero:

$$v(0) = A\omega_n \cos \omega_n 0 - B\omega_n \sin \omega_n 0$$

$$\therefore A = v(0)/\omega_n$$

$$y(0) = A \sin \omega_n 0 + B \cos \omega_n 0$$

$$\therefore B = y(0).$$

Thus 14.15 becomes

$$y = \left\{ \frac{v(0)}{\omega_n} \right\} \sin \omega_n t + y(0) \cos \omega_n t, \quad (14.16)$$

where y is the deflection at time t.

We can see from eq. 14.16 that every time $\omega_n t = 2\pi$ or a multiple of 2π, y returns to the same value. Thus the natural period of the oscillating system, T_n (the time between identical states of the system) is

$$T_n = \frac{2\pi}{\omega_n}. \quad (14.17)$$

Referring to the definition of ω_n (eq. 14.12), we see that

$$T_n = 2\pi \sqrt{\frac{m}{k}}. \quad (14.18)$$

Inserting into eq. 14.18 the value of k for a cantilever, the natural period of the cantilever shown in Figure 14.2a can be calculated:

$$T_n = 2\pi \sqrt{\frac{mL^3}{3EI}}. \quad (14.19)$$

The greater the mass at the beam's end and the greater the beam's length, the greater the natural period. We can easily test this prediction by taping a mass to the end of a ruler, clamping the ruler to the edge of a table and plucking the mass.

The longer the ruler or the greater the mass, the longer the period. Conversely, the greater the stiffness E of the material from which the ruler is constructed or the greater the second moment of area I, the shorter the period of oscillation.

This example has been presented primarily as an illustration. We have tacitly assumed that the only mass that oscillates is the mass at the end of the beam, a reasonable assumption only if the end mass is much larger than the mass of the beam itself. This is seldom the case with natural cantilevers, where the mass of the beam is substantial and the end mass is usually small. We have also assumed that the beam material is perfectly elastic, an assumption that clearly does not apply to biological beams. Fortunately, methods have been devised to handle these complications.

The Effective Mass of a Beam

The effective mass of a cantilever can be deduced by examining the contribution of the distributed mass of the beam to the overall kinetic energy of the system. Consider the following example. The free end of a cantilever with constant cross section is deflected a distance $y(0)$ (Fig. 14.2b), where the 0 denotes distance from the end rather than time. For any deflection of the free end we can, by reference to eq. 13.11, specify a close approximation of the deflection of any point on the beam, $y(x)$, and these deflections can be expressed as a proportion of the end deflection, $y(x)/y(0)$. If the end of the beam at deflection $y(0)$ travels with a velocity $v(0)$, the velocity at any point along the beam has a velocity $v(x)$ which is the same proportion of $v(0)$ as $y(x)$ is of $y(0)$. Using this equality of ratios, the kinetic-energy contribution of any small length, dx, of the beam can be written as

$$d(\text{kinetic energy}) = \tfrac{1}{2}m'v^2(0)\left\{\frac{y(x)}{y(0)}\right\}^2 \, dx,$$

(14.20)

where m' is the mass/length of the beam. By integrating this expression for kinetic energy from the free end of the beam $(x = 0)$ to the base $(x = L)$, the overall kinetic energy contribution of the beam's mass can be determined:

kinetic energy

$$= \tfrac{1}{2}\int_0^L m'v^2(0)\left\{\frac{y(x)}{y(0)}\right\}^2 \, dx. \quad (14.21)$$

From eq. 13.11, we see that the deflection at a point a distance x from the free end of a cantilever, expressed as a proportion of the deflection at the end, is

$$\frac{y(x)}{y(0)} = 1 - \left\{\frac{3x}{2L}\right\} + \left\{\frac{x^3}{2L^3}\right\}, \quad (14.22)$$

and the kinetic energy of the beam is

kinetic energy

$$= \tfrac{1}{2}\int_0^L m'v^2(0)\cdot\left\{1 - \frac{3x}{2L} + \frac{x^3}{2L^3}\right\}^2 \, dx.$$

(14.23)

By expanding the squared expression within the integral, we can carry out this integration simply (if somewhat laboriously), with the result that the kinetic energy of the beam is

kinetic energy $= (33/140)\tfrac{1}{2}m'Lv^2(0).$ (14.24)

Now, $m'L$ is the total mass of the beam, m_b. So

kinetic energy $= (33/140)\tfrac{1}{2}m_b v^2(0).$ (14.25)

But $[\tfrac{1}{2}]m_b v^2(0)$ is what the kinetic energy would be if the entire mass of the beam were traveling with velocity $v(0)$. Thus the mass of the beam contributes an effective mass equal to 33/140 of its total mass. The effective overall mass of the system (m as used in eqs. 14.9, 14.13, 14.18, and 14.19) is the sum of m_b and the concentrated mass at the end of the beam, m_e.

$$m = (33/140)m_b + m_e. \quad (14.26)$$

Incorporating this result into eqs. 14.12 and 14.19, we have

$$\omega_n = \sqrt{\frac{3EI}{L^3[(33/140)m_b + m_e]}} \quad (14.27)$$

$$T_n = 2\pi\sqrt{\frac{L^3[(33/140)m_b + m_e]}{3EI}}. \quad (14.28)$$

This derivation assumes that m' and I are constant along the beam. For beams with cross sections that vary, the equations must be suitably adjusted.

Damped Oscillations

Eqs. 14.27 and 14.28 provide a method for accurately estimating the natural circular frequency and natural period of a cantilever, but still assume that the beam material is perfectly elastic. To account for viscous losses in the beam material, we return to the equation of forces with which we began this discussion (eq. 14.14). Imagine that a beam has just passed through the undeflected point of its oscillation. Its velocity is at a maximum but is decreasing as the bending force (proportional to the amount of deflection) builds up. Kinetic energy is being traded for potential energy. If the beam material has a viscous component, the movement of the beam is also resisted by a force proportional to the rate of deflection. This *viscous damping force* decreases the kinetic energy of the system, but because energy is not stored in viscous interactions the potential energy is not correspondingly increased. As a result, the kinetic energy of the system reaches zero before the beam has deflected a distance equal to the initial deflection. The system gradually winds down, the amplitude of oscillation decreasing with time. This behavior may be described mathematically by incorporating the viscous damping force into equation 14.14:

$$ma + cv + ky = 0, \quad (14.29)$$

where cv is the viscous damping force and

c is the *damping coefficient*. We assume a solution of the form

$$y = e^{St}. \tag{14.30}$$

Substituting this into eq. 14.29 gives us

$$(mS^2 + cS + k)e^{St} = 0, \tag{14.31}$$

whichs turns out to be true for all values of t if

$$S^2 + (c/m)S + k/m = 0. \tag{14.32}$$

This result can be verified by substitution.

Eq. 14.32 is a quadratic equation. Recalling the formula for finding the roots of such an equation,[2] we conclude that

$$S = -\left\{\frac{c}{2m}\right\} \pm \sqrt{\left\{\frac{c}{2m}\right\}^2 - \frac{k}{m}}, \tag{14.33}$$

the two roots being labeled S_1 and S_2. The general solution we seek can then be written as

$$y = Ae^{S_1 t} + Be^{S_2 t}, \tag{14.34}$$

where A and B, as before, are constants determined by the initial velocity and displacement. Substituting eq. 14.34 into 14.30, we arrive at the solution that

$$y = \exp - \left\{\frac{c}{2m}\right\} t \left\{ A \exp \sqrt{\left\{\frac{c}{2m}\right\}^2 - \frac{k}{m}} t \right.$$
$$\left. + B \exp \sqrt{\left\{\frac{c}{2m}\right\}^2 - \frac{k}{m}} t \right\}. \tag{14.35}$$

This solution requires a bit of explanation. The initial term $[\exp(-c/2m)t]$ is an expression for exponential decay: as t increases this term gets smaller. This accounts for the effect we noted earlier—that the amplitude decreases through time for a damped oscillation. The larger the damping coefficient, the faster the amplitude decays.

The nature of the term in brackets in eq. 14.35 depends on whether $[c/(2m)]^2$ is greater than, equal to, or less than k/m.

When $[c/[2m)]^2 - (k/m) = 0$, the system is said to be *critically damped*. Solving

[2] $(x = [-b \pm (b^2 - 4ac)^{1/2}]/2a)$

this equality for c we define a *critical damping coefficient*, c_c:

$$\left\{\frac{c_c}{2m}\right\}^2 = \frac{k}{m}, \qquad c_c^2 = 4km, \qquad c_c = 2\sqrt{km}. \tag{14.36}$$

Recall from eq. 14.13 that $k = m\omega_n^2$, so

$$c_c = 2\sqrt{km} = 2\sqrt{m^2\omega_n^2} = 2m\omega_n. \tag{14.37}$$

As we will see, it is useful to express any particular damping coefficient as a fraction of the critical damping coefficient. Thus

$$\zeta \equiv \frac{c}{c_c} \tag{14.38}$$

$$\therefore \frac{c}{2m} = \zeta \frac{2m\omega_n}{2m} = \zeta\omega_n. \tag{14.39}$$

The expressions allow us to write equations such as 14.35 in a simpler form. Substituting $\zeta\omega_n$ for $c/2m$ in eq. 14.33 and noting that $k/m = \omega_n^2$ (eq. 14.12), we can restate eq. 14.35 as

$$y = \exp(-\zeta\omega_n t)$$
$$\cdot \{A \exp \sqrt{\zeta^2\omega_n^2 - \omega_n^2}\, t + B \exp \sqrt{\zeta^2\omega_n^2 - \omega_n^2}\, t\}$$
$$= \exp(-\zeta\omega_n t)$$
$$\cdot \{A \exp \sqrt{\zeta^2 - 1}\,\omega_n t + B \exp \sqrt{\zeta^2 - 1}\,\omega_n t\}. \tag{14.40}$$

Expressions using ζ can be converted to the form containing $c/2m$ and k/m by substituting the following equalities:

$$\exp(-\zeta\omega_n t) = \exp\left\{\frac{-c}{2m}\right\} t \tag{14.41}$$

$$\sqrt{\zeta^2 - 1}\,\omega_n = \sqrt{\left\{\frac{c}{2m}\right\}^2 - \frac{k}{m}} \tag{14.42}$$

$$\sqrt{1 - \zeta^2}\,\omega_n = \sqrt{\frac{k}{m} - \left\{\frac{c}{2m}\right\}^2}. \tag{14.43}$$

When ζ is greater than 1, eq. 14.40, solved in terms of the initial deflection $y(0)$ and initial velocity $v(0)$, becomes

$$y = A \exp \{[-\zeta + \sqrt{\zeta^2 - 1}]\omega_n t\}$$
$$+ B \exp \{[-\zeta - \sqrt{\zeta^2 - 1}]\omega_n t\},$$

where

$$A = \frac{v(0) + (\zeta\omega_n y_0) + \sqrt{\zeta^2 - 1}\,\omega_n y_0}{2\sqrt{\zeta^2 - 1}\,\omega_n}$$

$$B = \frac{-v(0) - (\zeta\omega_n y_0) - \sqrt{\zeta^2 - 1}\,\omega_n y_0}{2\sqrt{\zeta^2 - 1}\,\omega_n}.$$

$$(14.44)$$

This expression is graphed in Figure 14.3a. The deflection gradually returns to zero but never overshoots. In other words, when $[c/2m]^2$ is greater than k/m, the beam is nonoscillatory; the damping is so great that harmonic oscillations are prohibited. In this situation the system is said to be *overdamped*.

When $[c/2m]^2$ is less than k/m ($\zeta < 1$), each radical in eq. 14.33 can be rewritten in the form

$$\pm\sqrt{\zeta^2 - 1}\,\omega_n = \pm i\sqrt{1 - \zeta^2}\,\omega_n. \quad (14.45)$$

Substituting into eq. 14.40, we find that $A \exp \sqrt{\zeta^2 - 1}\,\omega_n t$ and $B \exp \sqrt{\zeta^2 - 1}\,\omega_n t$ can be expressed as

$$A \exp(S_1 t) = A \exp i\sqrt{1 - \zeta^2}\,\omega_n t$$

$$B \exp(S_2 t) = B \exp -i\sqrt{1 - \zeta^2}\,\omega_n t. \quad (14.46)$$

These results may appear to do nothing but complicate the situation, but with a bit of mathematical legerdermain all becomes simple again. We note the fact (due to Euler) that

$$N \exp(i\omega t) = N \cos \omega t + Ni \sin \omega t \quad (14.47)$$

Thus

$$A \exp i\sqrt{1 - \zeta^2}\,\omega_n t = A \cos \sqrt{1 - \zeta^2}\,\omega_n t$$
$$+ Ai \sin \sqrt{1 - \zeta^2}\,\omega_n t.$$

$$(14.48)$$

This equation has exactly the same form as eq. 14.15, which was shown to represent harmonic motion. Here the place of B in eq. 14.15 is taken by Ai and the position occupied by ω is taken by $\sqrt{1 - \zeta^2}\,\omega_n$. The term $B \exp -i\sqrt{1 - \zeta^2}\,\omega_n t$ can be expanded in the same manner and is likewise representative of an oscillatory motion.

Both the oscillations just described have the same circular frequency, $\omega_d =$

$\sqrt{1 - \zeta^2}\,\omega_n$, and when added together the result is an oscillation of the form shown in Figure 14.3b. The conclusion, then, is that when ζ is less than 1—i.e., $k/m < (c/2m)^2$—the motion of the system is oscillatory but the oscillations are damped, and the maximum deflection decreases exponentially with time. The larger ζ (i.e., the larger c), the faster the deflection decays. Solved in terms of the initial conditions, eq. 14.40 becomes

$$y = \exp(-\zeta\omega_n t)\left\{\frac{v(0) + \zeta\omega_n y_0}{\sqrt{1 - \zeta^2}\,\omega_n} \sin \sqrt{1 - \zeta^2}\,\omega_n t \right.$$
$$\left. + y_0 \cos \sqrt{1 - \zeta^2}\,\omega_n t\right\}. \quad (14.49)$$

It should be noted that the circular frequency of oscillation of the damped system is less than that of the undamped system. In the undamped case (eq. 14.34), $\omega_n = \sqrt{k/m}$. In the damped case, the circular frequency $\omega_d = \sqrt{1 - \zeta^2}\,\omega_n$. Eq. 14.49 can thus be rewritten as

$$y = \exp(-\zeta\omega_n t)\{[(v(0) + \zeta\omega_n y_0)/\omega_d] \sin \omega_d t$$
$$+ y_0 \cos \omega_d t\}. \quad (14.50)$$

Written in a slightly different fashion, eq. 14.50 becomes

$$y = \exp(-\zeta\omega_n t)[(v(0) + \zeta\omega_n y_0)/\omega_d] \sin(\omega_d t + \phi),$$
$$(14.51)$$

where ϕ is a phase angle that appropriately adjusts the timing of the wave form. This alternate version of eq. 14.50 will be useful later on.

When $\zeta = 1$, the beam is critically damped: S_1 and S_2 both equal $\zeta\omega_n$, and eq. 14.34 reduces to the form

$$(A + B)e^{-\zeta\omega_n t} = Ce^{-\zeta\omega_n t}. \quad (14.52)$$

The single constant is insufficient to define the motion uniquely, but by letting ζ approach (but not equal) 1—i.e., the difference between $[c/(2m)]^2$ and k/m approaches, but does not equal, zero—the solution can be found in the limit and expressed in terms of the initial velocity and deflection (Thomson 1981):

Figure 14.3. Damped oscillation. (a) The "oscillation" of an overdamped beam. The two exponentially decreasing terms in eq. 14.44 are added to yield the time course of the beam's deflection. In this example the beam has an initial velocity in addition to an initial deflection. The deflection gradually decreases and never crosses zero. (b) Damped oscillatory motion. The amplitude of the oscillation gradually decreases. (c) Critically damped oscillation. The motion of the beam depends on the initial velocity. The deflection may cross 0, but can do so only once.

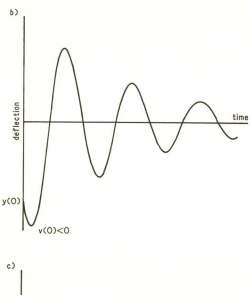

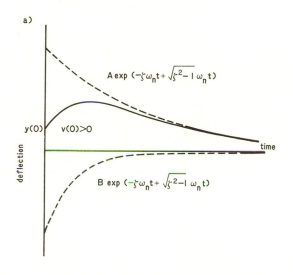

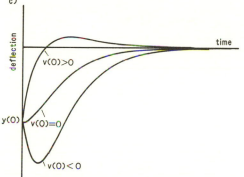

$$y = \exp(-\omega_n t)\{[v(0) + \omega_n y_0]t + y(0)\}. \tag{14.53}$$

This type of response is shown in Figure 14.3c. This behavior is different from that of the nonoscillatory system in that, given the proper initial conditions ($v(0) > 0$ while $y(0) < 0$, or vice versa), the beam will recover from an initial deflection on one side of its resting position by overshooting to a deflection on the other side. It is different from the oscillatory system in that this overshoot occurs once at most.

To this point the equations for damped harmonic motion have been presented with the tacit assumption that c, the damping coefficient, is a known quantity. In practice, one is seldom handed a cantilever for which the damping coefficient is known, and it is necessary somehow to determine c. The simplest method is to measure the rate at which the amplitude of oscillation decreases (Fig. 14.4). Amplitude y_1 is measured at one amplitude peak, and amplitude y_2 is measured one cycle later at time $t + T_d$, where T_d is the damped period defined analogously to eq. 14.17. From eq. 14.51 we can express the ratio of these two amplitudes as

$$\frac{y_1}{y_2} = \exp(-\omega_n t)[(v(0) + \zeta\omega_n y_0)/\omega_d]\sin(\omega_d t + \phi)$$
$$\div \exp(-\omega_n[t + T_d])[(v(0) + \zeta\omega_n y_0)/\omega_d]$$
$$\cdot \sin(\omega_d[t + T_d] + \phi). \tag{14.54}$$

Noting that the values of the two sines must be equal ($\sin t + T_d = \sin t$, by definition), we reduce eq. 14.54 to

Figure 14.4. Terms used in defining the logarithmic decrement (see text).

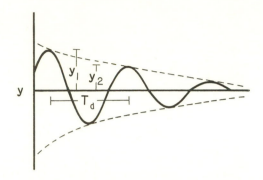

$$\frac{y_1}{y_2} = \frac{\exp(-\zeta\omega_n t)}{\exp(-\zeta\omega_n[t + T_d])} = \exp(\zeta\omega_n T_d).$$

$$(14.55)$$

We now define a value, δ, the *logarithmic decrement*:

$$\delta = \ln\frac{y_1}{y_2} = \ln\exp(\zeta\omega_n T_d) = \zeta\omega_n T_d = \frac{c}{2m}T_d$$

$$(14.56)$$

$$\therefore c = \frac{2m\delta}{T_d}.$$

This is the relationship we are after. By plucking a cantilever and measuring successive amplitudes, we can calculate the logarithmic decrement. From the logarithmic decrement we obtain the damping coefficient.

Forced Harmonic Motion

All this is valuable information, but in an examination of the wave-swept environment we are less concerned with what happens to a structure after an applied force is released (as the above description requires) than in how the system behaves when a force is applied periodically or impulsively.

We first examine a case in which the cantilever is subjected to a sinusoidally varying periodic force, a situation that could occur for an organism submerged beneath surface waves. We begin our analysis by returning to the differential equation relating the forces acting on the beam,

$$ma + cv + ky = 0. \qquad (14.29)$$

This equation is true only if no external force is applied, and we are now dealing with a case in which an external force is assumed. Thus eq. 14.29 must be rewritten as

$$ma + cv + ky = f_0 \sin\omega t, \qquad (14.57)$$

where f_0 is the amplitude of an applied sinusoidally varying force and ω is its circular frequency. This circular frequency can take any value and is totally independent of the natural circular frequency of the beam. We assume that the solution to this equation is an oscillation with the same frequency as the driving force and that the solution has the form

$$y = y_{max}\sin(\omega t - \phi), \qquad (14.58)$$

where y_{max} is the maximum deflection and ϕ is the phase difference between the displacement and the driving force. Substituting eq. 14.58 into eq. 14.57, we arrive at the conclusion that

$$\overbrace{-m\omega^2 y_{max}\sin(\omega t - \phi)}^{\text{accelerational force}} + \overbrace{c\omega y_{max}\cos(\omega t - \phi)}^{\text{damping force}}$$

$$+ \overbrace{ky_{max}\sin(\omega t - \phi)}^{\text{bending force}} = f_0\sin(\omega t).$$

$$(14.59)$$

Simply stated, eq. 14.59 tells us that the sum of the forces due to acceleration, damping, and bending (corrected for phase) must equal the applied driving force. We also know from the properties of harmonic motion described earlier (Fig. 14.1b) that the damping force, $c\omega y_{max}\sin(\omega t - \phi)$, leads the bending force by $\pi/2$ radians (i.e., 90°) and that the acceleration force, $-m\omega^2 y_{max}\sin(\omega t - \phi)$, leads the bending force by π radians (180°). Thus, if we draw vectors with lengths and directions appropriate to all of these forces, we arrive at the diagram shown in Figure 14.5.

The vector representing the maximum beam deflection (y_{max}) rotates with angular velocity ωt. It is the projection of this vector on the displacement axis that gives us the deflection at any time t. The bending force vector (ky_{max}) is in phase with the displacement. From Eqs. 14.58 and 14.59 we know that the applied force differs from the displacement by a phase angle ϕ. This is shown by a vector of length f_0 at angle ϕ to vector y_{max}. The projection of this vector on the displacement axis represents the variation in applied force through time. The vectors representing the maximum amplitude of the damping ($c\omega y_{max}$) and acceleration ($m\omega^2 y_{max}$) forces are added to the bending force, maintaining the proper phase relationships so as to sum to f_0. Invoking the Pythagorean theorem,

$$(ky_{max} - m\omega^2 y_{max})^2 + (c\omega y_{max})^2 = f_0^2$$

$$y_{max}^2 = \frac{f_0^2}{(k - m\omega^2)^2 + (c\omega)^2}$$

$$y_{max} = \frac{f_0}{\sqrt{(k - m\omega^2)^2 + (c\omega)^2}}.$$

$$(14.60)$$

Similarly, by purely trigonometric arguments we can show the phase ϕ to be

$$\phi = \arctan\left\{\frac{c\omega}{k - m\omega^2}\right\}. \quad (14.61)$$

These expressions can be rewritten in terms applicable to cantilever beams by substituting for k the value $3EI/L^3$ (eq. 14.6):

$$y_{max} = \frac{f_0}{\sqrt{\left\{\frac{3EI}{L^3} - m\omega^2\right\}^2 + (c\omega)^2}} \quad (14.62)$$

$$\phi = \arctan\left\{\frac{c\omega}{\frac{3EI}{L^3} - m\omega^2}\right\}. \quad (14.63)$$

Recall that m, the effective overall mass, is the sum of the concentrated end mass and the effective mass of the beam (eq. 14.26).

Multiplying both sides of eq. 14.60 by k/f_0 results in an expression whose form is easily understood (Fig. 14.6a):

$$\frac{ky_{max}}{f_0} = \frac{k}{\sqrt{(k - m\omega^2)^2 + (c\omega)^2}}$$

$$\frac{ky_{max}}{f_0} = \frac{3EI}{L^3 \sqrt{\left\{\frac{3EI}{L^3} - m\omega^2\right\}^2 + (c\omega)^2}}.$$

$$(14.64)$$

The ordinate of Figure 14.6a is the ratio of the maximum force the beam experiences due to its bending (ky_{max}) to the maximum externally applied force (f_0). At static equilibrium this value is 1, so this axis is in essence the ratio of the dynamic force experienced by the beam to the force it would experience at static equilibrium. For simplicity, the abscissa is drawn as the ratio of the circular frequency of the applied driving force ω to the beam's natural frequency ω_n.

Several points should be noted from Figure 14.6a. At driving frequencies well above the natural frequency of the beam, large applied forces result in negligible deflection of the beam, and hence in neglig-

Figure 14.5. A graphical analysis of the vector relationship among the applied force, the elastic restoring force, the acceleration, and the force due to damping (see text). (Redrawn from T. Thomson, Theory of Vibration with Applications, 1981, pg. 49, by permission of Prentice-Hall, Englewood Cliffs, N.J.)

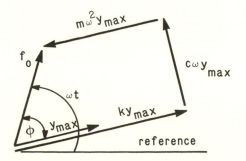

Figure 14.6. Characteristics of forced oscillation. (a) The relative amplitude (ky_{max}/f_0) as a function of damping and frequency. Lightly damped beams undergo large oscillations when the driving frequency ω equals the beam's natural frequency ω_n. (b) The phase shift between the driving force and the resulting beam deflection. For lightly damped beams, when $\omega < \omega_n$ the deflection is nearly in phase with the driving force. When $\omega > \omega_n$, the deflection is nearly out of phase.

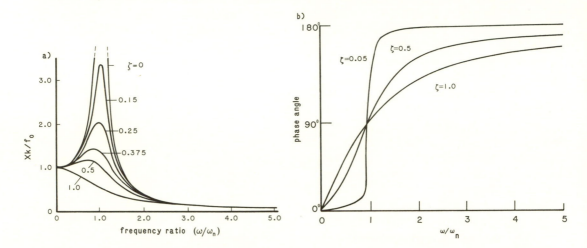

ible force on the beam material. At the natural frequency of the beam (the resonant frequency), the deflection of the beam can be very large unless the damping coefficient is substantial. The effects of damping are much less obvious at frequencies away from the beam's natural frequency.

The biologically relevant conclusion to be drawn from this derivation is that damping is necessary in a structure lest its deflection (and hence the beam stress) become very large when the beam is forced to vibrate at its resonant period. It seems somewhat unlikely, however, that the period of nonturbulent oscillatory motions of waves will correspond to the resonant period of most wave-swept organisms. For example, even a large cantileverlike organism such as the sea palm *Postelsia palmaeformis* has a resonant period of 1s or less, while the period of all but the shortest wave-length water waves are 5 s or greater. It is still possible, however, that fluctuations in force caused by the shedding of vortices could bend a biologi-

cal cantilever at its resonant frequency, thereby causing large deflections. For example, consider a cylindrical algal stipe with a diameter of 1 cm and a natural circular frequency of 63 radians per second (= 10 Hz). If we assume a Strouhal number of 0.2 (see Chapter 11), a water velocity of 0.5 m/s will cause vortices to be shed at the stipe's natural frequency, and violent oscillations might ensue. Alternatively, in fast turbulent flow the fluctuation in the mainstream velocity might have a substantial component at the stipe's natural frequency, which could similarly lead to large-amplitude oscillations.

The phase difference between the driving force and the displacement (eq. 14.63) is graphed in Figure 14.6b. For small damping coefficients the displacement is very nearly in phase with the driving force when $\omega < \omega_n$. If $\omega > \omega_n$, the displacement is very nearly 180° out of phase with the driving force. This finding substantiates the conclusion drawn in Chapter 8 regarding forced tidal oscillations.

Figure 14.7. Impulse. (a) For a constant force, impulse is simply the product of force and time. (b) For a variable force, impulse must be calculated as the integral of force · time.

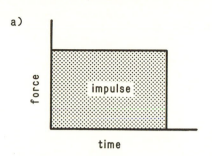

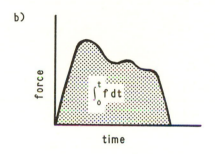

Impulsive Forces

There are other types of loading that can cause problems for biological cantilevers. Wave-swept organisms are continually subjected to short-duration forces of large amplitude—for instance, when the advancing bore of a broken wave first impacts an organism. Such short-duration forces are termed *impulsive*. What is the response of a cantilever to such impulsive forces?

Before we can answer this question, we need to digress briefly to define a new term. Impulse, $\hat{f}$, is the product of force and time (Fig. 14.7a). If the force varies through time (Fig. 14.7b) this value is best expressed as the time integral of force:

$$\hat{f} = \int_0^t f(t)\,dt. \qquad (14.65)$$

The definition of impulse leads us to a bit of mathematical trickery known as the delta function. The impulse can be kept constant as the time of force application is decreased by increasing the force applied. Taken to its logical extreme, a near infinite force applied for an infinitesimal period of time results in a finite impulse. If this finite impulse is equal to unity, one has a unit impulse of infinitesimal duration; this is the delta function. For reasons that will become clear, this function is given the symbol $\delta(t - \epsilon)$ and has the property that it equals 1 at time equal to ϵ and is zero at all other times. Thus $\int_0^\infty \delta(t - \epsilon)\,dt = 1$. The utility of the delta function becomes apparent when we note that the integral of the product of force $f(t)$ and the delta function at time $t = \epsilon$ is equal to $\hat{f}(t)$ at ϵ:

$$\hat{f}(\epsilon) = \int_0^\infty f(t)\,\delta(t - \epsilon)\,dt. \qquad (14.66)$$

Thus the concept of the delta function allows us to think of force applied at any time in terms of the impulse it imparts. What does an impulse do when applied to a cantilever? Recall once again that force equals mass times acceleration. Thus

$$f = m\,dv/dt$$

where v is dy/dt, the velocity of the system. When we rearrange this equation,

$$\frac{f\,dt}{m} = dv, \qquad (14.67)$$

it is clear that an impulse ($f\,dt$) causes a change in velocity (dv) inversely proportional to mass. If the impulse is applied over a very short period of time, it causes what amounts to a step change in the velocity of the system without having time to change the deflection appreciably.

This idea can be applied to a cantilever initially at rest. First we return to eq. 14.16, the expression that gives the deflection $y(t)$ as a function of the initial deflection and velocity for an undamped system:

$$y = \left\{\frac{v(0)}{\omega_n}\right\} \sin \omega_n t + y(0) \cos \omega_n t. \qquad (14.16)$$

By specifying that the system is undeflected $[y(0) = 0]$ and at rest $[v(0) = 0]$ until the impulse is applied (whereupon a velocity equal to $\hat{f}/m$ is imparted), we arrive at the conclusion that

$$y = \left\{\frac{\hat{f}}{m\omega_n}\right\} \sin \omega_n t. \qquad (14.68)$$

For an undamped cantilever, $\omega_n = \sqrt{3EI/[L^3 m]}$. Thus

$$y = \frac{\hat{f}}{m\sqrt{\frac{3EI}{L^3 m}}} \sin \sqrt{\frac{3EI}{L^3 m}}\, t. \qquad (14.69)$$

For the case of a damped cantilever, we begin with eq. 14.50. If we use as the initial conditions a deflection of zero and the velocity imparted by the impulse, we have

$$y = \exp(-\zeta\omega_n t)\left\{\frac{\hat{f}}{m\omega_d}\right\} \sin \omega_d t. \qquad (14.70)$$

Both of these equations (14.68 and 14.70) have a similar form where the deflection y equals $\hat{f}$ multiplied by some function (a delta function) that specifies the response of the cantilever to a unit impulse. We call this response function $h(t)$. Thus, for an undamped beam,

$$h(t) = \frac{1}{m\omega_n} \sin \omega t, \qquad (14.71)$$

and for a damped beam

$$h(t) = \exp(-\zeta\omega_n t)\frac{1}{m\omega_d} \sin \omega_d t. \qquad (14.72)$$

These equations describe the response of a cantilever to one single impulse of infinitesimal duration. Although the force caused by a breaking wave may last only a fraction of a second, it is not infinitesimal. Rather, even a short-duration natural force must be viewed as the sum of many infinitesimal impulses. For example, take a time function of force such as that shown in Figure 14.8. We want to know what the displacement of the cantilever is at time t some time after the imposition of this complex force. At time ϵ_i a force $f(\epsilon_i)$ is applied for time $\Delta\epsilon$. We know that the ap-

Figure 14.8. The response of a beam to a varying force. (a) At time ϵ, force $f(\epsilon)$ acts on the beam. If only this force had been applied to the beam, the time course of deflection would be as shown in (b). At time t ($t - \epsilon$ after the imposed impulse), the deflection is $y = f(\epsilon)h(t - \epsilon)$. The actual deflection of the beam is the sum of all the individual deflections (eq. 14.76).

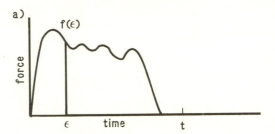

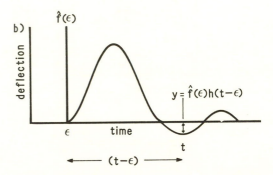

plied impulse $\hat{f}_i$ is equal to $f(\epsilon_i)\Delta\epsilon$. What is the deflection at time t ($t - \epsilon$ s after the impulse is applied) as a result of this individual impulse? This response is calculated in the same manner as above:

$$y(t) = \hat{f}(\epsilon_i)h(t - \epsilon_i)$$
$$= f(\epsilon_i)\Delta\epsilon\, h(t - \epsilon_i). \qquad (14.73)$$

This procedure can be repeated for any ϵ_i during the applied-force sequence, each $\hat{f}(\epsilon_i)$ resulting in a certain deflection at time t. The *overall* deflection at time t is simply the sum of the individual deflections resulting from all $\hat{f}(\epsilon_i)$.

$$y(t) = \sum_{i=0}^{t} f(\epsilon_i)\Delta\epsilon\, h(t - \epsilon_i), \qquad (14.74)$$

which in the integral form becomes

$$y(t) = \int_0^t f(\epsilon)h(t - \epsilon)\, d\epsilon. \qquad (14.75)$$

Figure 14.9. The response of a cantilever to a constant force suddenly applied at time $t = 0$. An undamped beam deflects twice as far as it would in static response to the force.

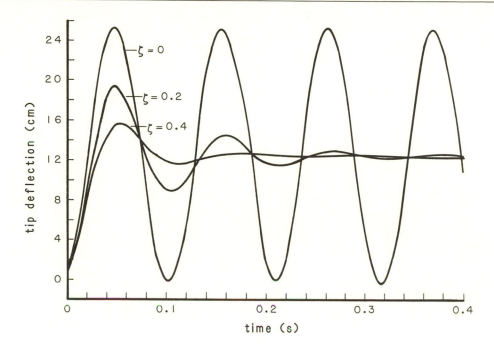

Inserting the expression $h(t - \epsilon)$ for a damped cantilever beam, this expression becomes

$$y(t) = \int_0^t \exp[-\zeta\omega_n(t - \epsilon)] \frac{f(\epsilon)}{m\omega_d} \sin \omega_d(t - \epsilon) \, d\epsilon$$

$$y(t) = \frac{1}{m\omega_d} \int_0^t \exp[-\zeta\omega_n(t - \epsilon)] f(\epsilon)$$

$$\cdot \sin \omega_d(t - \epsilon) \, d\epsilon. \qquad (14.76)$$

This is the answer we seek. For any time after the imposition of a force we can specify the deflection of the cantilever, and from the deflection we can calculate the stress at any point in the beam.

Before we discuss an example from the wave-swept environment, it will be useful to examine a simplified case. We assume that a cantilever, initially at rest, experiences a force, f, which is instantly applied and then maintained constant. Thus for all times greater than zero $f(t) = f$, and eq. 14.76 becomes

$$y(t) = \frac{f}{m\omega_d} \int_0^t \exp[-\zeta\omega_n(t - \epsilon)] \sin \omega_d(t - \epsilon) \, d\epsilon.$$

$$(14.77)$$

This result is graphed in Figure 14.9. For damped beams the beam reaches an equilibrium deflection equal to f/k at times well after the imposition of force. The maximum deflection undergone in the process of establishing this equilibrium is determined by the damping of the beam. An undamped beam experiences a deflection *twice* the equilibrium deflection, and would be subjected to bending stresses double those experienced due to the imposition of f as a static force. As the damping coefficient is increased, the maximum excursion decreases. For an overdamped, nonoscillatory beam the maximum deflection is equal to the equilibrium deflection.

The response of a cantilever to an actual environmental force is shown in Figure 14.10. The water velocity fluctuates such

Figure 14.10. Deflection of a cantilever in response to a varying force. (a) The hydrodynamic force applied by a turbulent bore. (b) The response of the cantilever for various values of the damping coefficient. The nearer the beam is to being critically damped, the better it tracks the applied force.

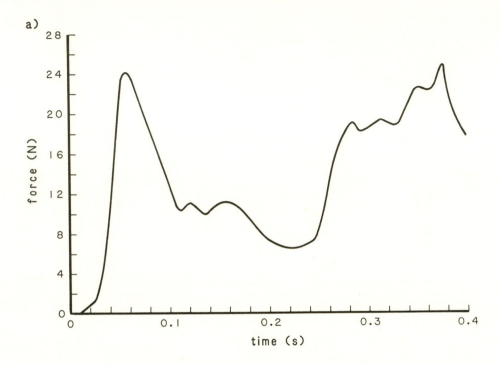

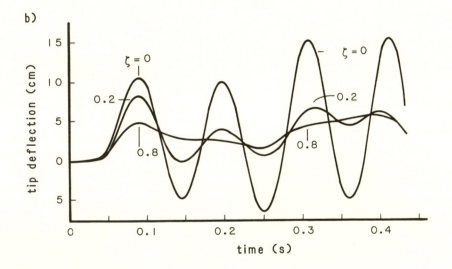

that the force record of Figure 14.10a is imposed on the beam. The response of the beam (at several levels of damping) is shown in Figure 14.10b. The undamped beam experiences large-amplitude fluctuations, and its bending record bears little resemblance to the force record. In other words, the undamped beam follows the dictates of its own inertia, and the resulting oscillations are large, placing large forces on the beam, material. As the damping coefficient is increased, the amplitude of oscillation, and thereby the force on the beam material, decreases. Again, damping would seem to be advantageous for waveswept organisms.

The information presented here is far from complete. For example, we have assumed that the shape a beam takes when dynamically bent is the same as when it is statically bent. This particular shape is actually only one of many modes in which the beam can bend, and each mode has a different natural frequency. If in practice more than one mode is present at one time, the beam's motions are complicated, and the analysis presented here may not be a valid representation. We have also tacitly assumed that except for the viscosity required for damping, the beam material behaves elastically and linearly. It is on this basis that we equate the beam deflection with stress in the beam material. If the beam material is substantially viscous, the stress placed on the beam can be determined by the rate of bending as well as the amount of bending. For a thorough introduction to the dynamics of beams, see Thomson (1981).

Adhesion

The term "adhesion" is used here in a very general sense: adhesion is the force that tends to hold two objects together. This term could be applied to all the component parts of an organism (cell-cell adhesion, etc.), but the two objects we are most interested in are an organism as a whole and the substratum to which it adheres.

Mechanical Adhesion

Perhaps the simplest form of adhesion is mechanical adhesion. For instance, if a crab wants to adhere to a rock, it wraps its legs around some protuberance. To dislodge the animal one must forcibly unbend its legs. This type of adhesion is a simple application of beam theory and can be treated by the concepts outlined in Chapters 13 and 14. We will not deal with mechanical adhesion further here, and the interested reader should consult Nachtigall (1974).

Close Physical Contact

The forces that hold together nonionic, noncovalently bonded materials are the *van der Waals forces*. These are the forces that hold a solid piece of steel or glass together, and the strength of these materials is quite high. Why then don't van der Waals forces serve as the universal, all-purpose adhesive? The problems involved become evident when we try to assemble an object from its constituent parts. For instance, a shattered water glass cannot be reconstructed just by putting the pieces

back together. The reason lies in the fact that van der Waals forces operate effectively only over very short distances. Their strength decreases as the seventh power of distance. In practice this limits the useful range of these forces to a distance of 1–10 nm, and if two separate pieces of material can be made to fit together within these tolerances, they adhere quite well. For example, steel can be polished to sufficient flatness so that when two pieces are placed face to face and squeezed together, they adhere over a substantial portion of their surface and become "cold welded." But when a water glass is broken, the fracture surfaces are quite complex on a microscopic scale. It is reasonable to assume that the process of fracture has imparted some permanent deformation to each face such that, on the scale of 1–10 nm, two pieces that appear to fit together quite nicely are actually mismatched. The landscape of one piece is not the precise mold-image of the other, and the two pieces touch only where a high spot on one piece hits a high spot on the other. The van der Waals forces between these restricted areas are not sufficient to cause the overall adhesion of the two pieces.

It has been speculated that this microscopic mismatch is the physical basis for friction. We have long known that the friction force between two solids depends on the force pushing them together. This fact allows the definition of a coefficient of static friction, k_f

$$k_f = \frac{f_s}{f_n}, \qquad (15.1)$$

where f_n is the normal force pushing two

pieces together and f_s is the shear force required to initiate the sliding of one piece over the other. The greater the normal force, the more the high spots on each piece are deformed, and consequently the greater the area of actual close contact between pieces. As a result, adhesion is enhanced. In order for one piece to slide, the high spots (which are in effect welded together) must be sheared off. The force required for this shearing is the friction force. The validity of this concept has been experimentally demonstrated for the friction between various metals (see Wake 1982 for a review).

Of course this explanation for friction need not be confined to a microscopic scale. Any macroscopic roughness of two contacting surfaces leads to friction between the surfaces as one slides past the other, and in this case one need not invoke van der Waals forces at all.

Unless the relative humidity is very low, a broken piece of material is quickly coated with a monomolecular layer of water, and this layer is sufficient to abolish effectively adhesion via the van der Waals forces. This fact, in addition to the permanent deformation caused by fracture, ensures that adhesion through close physical contact is not a practical way of reconstructing a broken object.

However, in special cases close physical contact does result in adhesion. When two soft, flowable materials (e.g., pieces of wax or modeling clay) are squeezed together they can adhere quite well. The reason is clear in light of the discussion in Chapter 12. If two mismatched pieces of a viscoelastic material are placed together and a stress is applied, the effective contact between pieces is confined to the touching of "high spots." However, because the entire force pushing the two pieces together is applied to the small area of these high spots, the local stress is quite high and the material flows. The effect is to lower the original high spots until more area is brought into contact between the

two pieces, and the process continues (at least in theory) until the two pieces match up exactly. At that point, van der Waals forces cause effective adhesion. Again, there are limits to this process. Any coating of water prevents close contact, and air trapped when the materials are first brought into contact can similarly prevent the materials from adhering.

The astute reader may be wondering at this point why a coating of water should inhibit adhesion. Certainly, the water might prevent the two coated surfaces from coming into close contact, but why can't the water itself adhere, in essence acting as a glue? This is a very real question, one with wide-ranging implications, and it will be treated later in this chapter in the discussion of surface energy.

Pressure Difference Adhesives

Consider the apparatus shown in Figure 15.1. A cylinder containing a piston is sealed to a substratum. The cross-sectional area of the piston is S and the ambient pressure in the fluid (either air or water) outside of the cylinder is p_0. If the fluid beneath the piston is at ambient pressure, there is no force other than gravity that will tend to hold the piston to the substratum.

Now consider the consequences of a force being applied between the cylinder and the piston, tending to move the piston upward (Fig. 15.1b). If the fluid beneath the piston is a gas, the piston moves upward, increasing the volume of the gas and decreasing its pressure, p_i. The decrease in pressure results in a pressure difference, Δp,

$$\Delta p = p_o - p_i \qquad (15.2)$$

across the piston. This pressure difference causes a force, ΔpS, that pushes the cylinder and piston onto the substratum. This is an adhesive force in the sense that an oppositely directed force of at least this

Figure 15.1. Adhesion by a difference in pressure. (a) When no force acts on the piston, the enclosed fluid is at the same pressure as the surrounding air or water; there is no adhesion. (b) A force exerted on the piston lowers the pressure in the enclosed fluid and adhesion is affected. (c) A limpet. The force provided by contracting pedal muscles lowers the pressure in the pedal mucus and the animal adheres firmly to the substratum

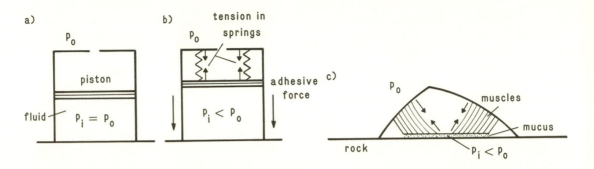

magnitude must be applied if the cylinder is to be dislodged from the substratum. This is the principle behind suction cups. The "piston force" of a suction cup (the type found on darts) is provided by the elasticity of the rubber. When the suction cup is forced against a surface, the air beneath it is squeezed out and the lip of the cup forms a seal. The rubber then attempts to return to its original shape, in the process pulling up on the gas space beneath the cup and decreasing the pressure. The pressure difference holds the cup in place.

The adhesive capabilities of a pressure-difference mechanism are limited if a gas is used as the fluid beneath the piston. Imagine that before a force is applied, the piston fits flush with the substratum. If the piston moves upward, it creates a perfect vacuum, $p_i = 0$, and the pressure difference is p_o. At sea level p_o is approximately 10^5 Pa, and this represents the maximum adhesive force for such a mechanism under these conditions. Underwater the adhesive force increases with depth, doubling with each 10 m below the surface.

This adhesive mechanism is limited in another fashion as well. The force due to pressure difference acts in a direction normal to the surface of the substratum and does not provide any inherent resistance

to forces attempting to slide the apparatus along the substratum. The only resistance to forces directed in shear is due to the friction between the base of the cylinder and the surface on which it sits. If the friction coefficient were zero, the adhesive capability of this mechanism in shear would likewise be zero.

The first of these problems can be circumvented by using a liquid—water—as the fluid beneath the piston. Under the application of a piston force, the piston attempts to move upward and the pressure in the liquid is reduced. Due to the high bulk modulus of water, the actual movement accompanying this decrease in pressure is very small, but this does not alter the effect of the piston force—a pressure difference is created and an adhesive force is applied. When confined to a closed space such as this, water behaves as if it has a tensile strength, which may be as high as 600 atmospheres ($6 \cdot 10^7$ Pa). The piston force (and consequently the adhesive strength) can thus be quite large.

A pressure-difference mechanism using mucus (about 90% water) as the enclosed fluid is apparently used by limpets. The foot of the limpet is admirably arranged to act as a piston. Muscles run from the foot's ventral surface upward to the shell (Fig. 15.1c), and a thin layer of mucus is sandwiched between the foot and the sub-stratum, serving both as the enclosed fluid and as a seal at the edge of the foot.

When a limpet is sitting undisturbed on the rock substratum, the foot muscles do not apply any substantial upward force on the foot. In this condition the animal is easily dislodged. Any indication of trouble, however, causes the limpet to "hunker down." Presumably the foot muscles con-tract, pulling upward on the foot and creating a negative pressure in the pedal mucus. Thus the limpet quickly increases its adhesive capability, and once clamped down in this fashion it can often not be dislodged without breaking the shell or ripping the muscles loose from their ten-dons. Measurements of the force required to dislodge limpets show that the stress placed on the foot can be as high as $5 \cdot 10^5$ Pa, demonstrating that the adhe-sive mechanism is probably not "suction" in the sense of having a gas as the en-closed fluid. Some of this tenacity may be due to the mucus acting as a Stefan adhe-sive, a mechanism explored later in this chapter.

The use of water as the enclosed fluid in a pressure-difference adhesive system does not alleviate this mechanism's inher-ent lack of shear resistance. If the coeffi-cient of friction is low, the force required to slide the apparatus along the substra-tum is not substantially increased merely by the fact of having water beneath the piston. For example, consider a layer of water 10 μm thick beneath a piston with a 1 cm radius. It would take an applied shear force of only $3 \cdot 10^{-4}$ N to slide the apparatus along the substratum at a veloc-ity of 1 cm/s. The fact that limpets use mucus instead of water can account for a finite but modest resistance to statically applied shear forces. Using the yield stress of *A. columbianus* mucus (Chapter 12) as an example, a stress of about 0.1 N would be required to initiate sliding for a limpet with a circular foot 1 cm in radius.

Despite this problem, the actual shear resistance of limpets on natural substrata is quite large. For instance, the force re-quired to dislodge *Collisella digitalis* in shear is roughly equal to that required for dislodgement by a normal force (Denny, unpublished). This can be attributed to a high friction coefficient between the edge of the shell and the rock. The edge of the shell is often butted directly against grains in the rock, and the animal is dislodged only when the lip of the shell breaks.

Pressure differences leading to adhesion can also be caused by the phenomenon known as *capillarity*. Before describing capillarity, we must digress briefly to dis-cuss the concept of surface energy.

Consider a drop of water. A water mole-cule in the interior of the drop is attracted to all the molecules around it. Because this attraction acts (on average) equally in all directions, the net force on the mole-cule is zero. A molecule at the surface, however, is more strongly attracted by the water in the drop than by the air outside, resulting in a net force pulling the surface molecule in. For the molecule to have got-ten to the surface to begin with, it must have been moved there against this net force, requiring that work be done. This work is the surface energy. Water has a surface energy of about 70 mJ/m^2 when in contact with air. Note that 1 J/m^2 = 1 Nm/m^2 = 1 N/m, a tension. Thus sur-face energy is the same as surface tension. It is because of this energy that a drop of water suspended in air tends to form a sphere, a sphere being the shape with the minimum surface area (and therefore, sur-face energy) per volume.

Solids also have surface energies, but because the molecules in a solid are more or less fixed in space, we cannot imagine them being moved to the surface layer. Instead we should visualize the surface energy of a solid as the minimum energy that must be expended in creating a new surface when we break the solid into two pieces. As we will see later in this chapter, this minimal energy may be considerably less than the actual energy required to break an object. The surface energies of solids vary considerably. Solids such as wax have surface energies that are lower than that of water, and solids such as glass have surface energies much higher than water (Table 15.1).

When a water drop comes in contact with a solid, the result depends on the surface energy of the solid. Consider a case where the surface energy of the solid is less than water. If the higher-surface-energy water were to spread over the lower-surface-energy solid, it would only increase the total surface energy of the system. As a result, the water "beads up," minimizing its contact with the solid, thereby also minimizing the total surface energy. In contrast, if the solid has a higher surface energy than water, the total surface energy is lowered if the water spreads over the solid. In this case the water is said to "wet" the solid surface.

Figure 15.2. (a) Two disks of radius R, separated by a distance z, adhere due to the pressure difference created by the capillarity of the fluid between the disks. (b) Stefan adhesion. As two disks of radius R are separated, fluid must flow into the space between them. Thus, separation is opposed by the fluid's viscosity (see text).

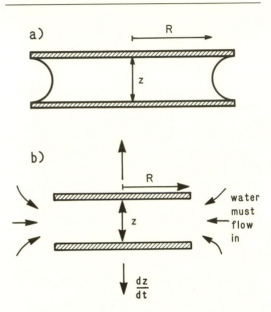

Now consider a system in which water is sandwiched between two disks made of a wettable solid, the sandwich sitting in air (Fig. 15.2a). Because the disks have a higher surface energy than water, the water spreads over the disks. If the distance between the disks is fixed, the tendency of the water to spread by capillary action results in a tension in the water (equal to the surface tension of water against air) and creates a negative pressure in the water. The pressure difference created is

$$p_o - p_i = \gamma_{lv}\left\{\frac{2}{z} - \frac{1}{R}\right\}, \quad (15.3)$$

where z is the spacing between the plates, R is the radius of the circular layer of waters and γ_{lv} is the surface energy of water against air. Although the surface energy of water is a fixed value, the dimensions

Table 15.1
Surface Energies of Some Solids and Liquids (from Waite 1982)

Solids	Critical Surface Energy (mJ/m²)
Polytetrafluoroethylene	18.5
Polyethylene	31
Polyglcine	45–51
Wool	42.5
Starch	39
Tooth enamel	38–40
Cellulose	45
Glass	170
NaCl	300

Liquids	Surface energy (mJ/m²)
Water	72.8
Ethanol	22.8
Benzene	28.9
Phenol	40.9
1% Gelatin	8.3

of the system can be changed to affect the magnitude of the pressure difference. The pressure difference is large if the seperation between the disks is small or if the radius of the disks is large. For example, disks 1 cm in radius separated by a layer of water 10 μm thick experience a pressure difference of $1.4 \cdot 10^4$ Pa, an adhesive capability only one-seventh of that which could be attained by a well-designed suction cup using air as the enclosed fluid.

There are two practical problems in using capillary adhesion. Like other pressure-difference adhesives, it has no inherent resistance to shear force. Capillarity also only works when the fluid between the disks is in contact with another fluid with a different surface energy—for example, water and air. Thus an organism in the wave-swept environment could adhere by capillarity when the tide is out, but unless a layer of air is maintained around the adhesive apparatus, when the tide comes in and the organism is immersed, the surface energy difference would go to zero—and the pressure difference as well. All things considered, capillary adhesion does not appear to be a practical adhesive mechanism for the wave-swept environment.

Stefan Adhesion

Although the fluid sandwiched between immersed disks exerts no capillary adhesive force, these disks may still be firmly adherent. Consider two circular disks immersed in water and separated by a small distance (Fig. 15.2b). In trying to separate these disks, water must be drawn into the gap between them. As water flows into the gap, it is sheared—the faster the rate of separation, the higher the shear rate. Because water is a viscous fluid, it resists being sheared, and a force is required to separate the disks. The analysis of this system was first carried out by J. Stefan in 1874, and the equation relating the applied force to the rate of separation is

$$\text{force} = 1.5\pi R^4 \mu \frac{dz}{dt} \frac{1}{z^3} \qquad (15.4)$$

for disks immersed in the fluid that separates them (Wake 1982). For disks separated by a liquid but immersed in air, the appropriate equation is

$$\text{force} = 0.75\pi R^4 \mu \frac{dz}{dt} \frac{1}{z^3}, \qquad (15.5)$$

half that for disks immersed in a liquid (Wake 1982). In both cases, the force increases in direct proportion to the rate of separation, and like capillarity it is profoundly affected by the dimensions of the system. The larger the radius and the smaller the initial separation, the greater the adhesive force. For example, two disks with 1 cm radius, separated by a 10 μm layer of water and immersed in water, require a stress of $7.5 \cdot 10^5$ Pa to be separated at a velocity of 1 cm/s. This adhesive capability is as high as any discussed so far, and is considerably higher than most. Furthermore, the requirements for this sort of adhesion (water sandwiched between two surfaces) are so simple that it is difficult to see how organisms in the wave-swept environment *avoid* using this adhesive mechanism.

Although Stefan adhesion can resist large loads applied for short periods of time, we should not forget that this adhesive force is due solely to the viscosity of the fluid. A small force applied for a considerable time will slowly separate the plates. As the separation between plates grows, the rate of separation increases, and small forces may eventually cause the system to fail. Thus the usefulness of this mechanism is confined to organisms that are not subjected to long term forces (and given the continuous nature of gravitational acceleration, these are difficult to find) or to organisms that use a fluid with an exceedingly high viscosity.

Because Stefan adhesion is not due to a pressure difference, there is no inherent tendency for some part of the organisms to be forced down onto the substratum. This lack of a normal force can result in a lowered frictional resistance to shear. In the absence of friction (such as the shell of a limpet on the rock) the resistance of a Stefan adhesive to shear is simply that due to the viscosity of the liquid, as explained above. Again, the usefulness of this type of adhesive would seem to be confined to organisms using a very viscous fluid. Crisp (1960) suggests that acorn barnacles may use a Stefan adhesive. Individual *Balanus balanoides* were found to be capable of slowly sliding across the substratum when subjected to the pressure created by the growth of a neighboring barnacle. This observation suggests that the basal adhesive of these barnacles is not a cross-linked solid but rather a viscoelastic material that can flow. Crisp did not measure the force that caused his barnacles to slide, so we do not have any way of guessing at the viscosity of the basal adhesive.

Solid Glues

The lack of shear resistance inherent in pressure difference and Stefan adhesives can be remedied by causing the fluid separating two pieces of material to solidify. If the resulting solid has a high shear strength, the adhesive joint has the capability of resisting large, statically applied forces. Moreover, the solidification of the fluid does not appreciably affect its intrinsic capability of adhering to the solids it separates—if the fluid wets the solids, the solidified glue forms a firm bond. All of our everyday man-made glues use this principal. The more venerable adhesives such as library paste, model airplane glue, and Elmer's glue are liquids that solidify when a solvent evaporates. Glues such as the cyanoacrylates and epoxies change from liquids to solids as a result of a catalyzed cross-linking of the liquid. Perhaps the oldest example of such a glue is ice. Anyone who has inadvertently left a wet

pair of mittens outside on a freezing day can attest to the adhesive capabilities of ice.

However, the very fact that glues are solids leads to problems. Imagine a situation such as that shown in Figure 15.3, a small bubble trapped in a liquid that separates two surfaces. If a normal force is applied, the stress from one surface is transferred through the liquid to the other surface. The air in the bubble does not have an appreciable tensile strength, so stress in the vicinity of the bubble must be routed around it. Consequently, the stress in the fluid immediately around the bubble is

Figure 15.3. Stress concentrations. (a) Because a bubble cannot resist a tensile shear stress, the stress is concentrated in the material beside the bubble. (b) The multiplication of stress concentration in a sharp-ended crack. (c) The dimensions of an elliptical imperfection determines its stress-concentrating effects (see text).

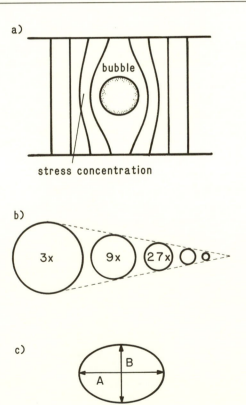

somewhat higher than elsewhere. This more highly stressed fluid flows faster than the other fluid, and the bubble deforms to a shape that minimizes the effects of its presence—a long, thin ellipsoid aligned with the stress axis.

When a fluid solidifies, this ability to flow disappears. Stress still cannot be transferred through the bubble, but the solid around the bubble cannot flow to allow the bubble to change shape. As a result, the stress in the glue around the bubble is increased and remains high—the bubble acts as a stress concentrator. The factor by which the stress is concentrated depends on the shape of the bubble; for a spherical bubble the factor is about 3. The consequences of such stress concentrations in a glue can be drastic. If the glue is loaded to only one-third its breaking stress, the presence of a spherical stress concentration would cause the material to break anyway. Matters can be considerably worse if nonspherical imperfections are considered. For example, a sharp-ended crack (Fig. 15.3b) can be approximated by a series of bubbles of decreasing radius. The largest bubble increases the stress near it by a factor of three. The next smaller bubble is placed in this stress concentration and increases the stress near it by another factor of three, an overall factor of nine. Another smaller bubble can be added, increasing the local stress to twenty-seven times the average stress, and so on. As the radius at the tip of the crack decreases and the length of the crack increases, the stress concentration can become gigantic. Wainwright et al. (1976) cite a formula for calculating the stress concentration of a sharply ended, elliptical imperfection:

$$\sigma = \bar{\sigma}\left\{1 + \frac{2A}{B}\right\}, \qquad (15.6)$$

where A and B are the major and minor axes, respectively, of the ellipse (Fig. 15.3c). For this calculation it is assumed that the minor axis of the ellipse lies parallel to the stress axis. For example, a crack 1 mm

long but only 1 μm high would lead to a stress concentration of about 2000. The equation shown here applies strictly only to holes in a thin plate, and values calculated for other situations should be taken only as rough estimates of the actual value of stress concentration. However, the point should be clear: small imperfections can lead to large stress concentrations in solid glues.

Why then don't all structures collapse? Any real material is bound to be full of small imperfections, and if each one of these leads to a stress concentration, surely the application of any load at all will cause the material somewhere to exceed its breaking strength. The theory of stress concentrations was proposed in 1913 by C. E. Inglis, and Gordon (1978) amusingly describes the nightmares that faced engineers in the years thereafter. Imagine walking up to the Brooklyn Bridge with a diamond stylus, scoring a tiny scratch in one of the beams, and having the entire structure collapse as a consequence.

The fact is, however, that stress concentrations do not lead to global structural collapse, and the reason was discovered by A. A. Griffith in 1921 while working on the properties of glass. Consider a piece of material loaded in tension as shown in Figure 15.4. As the material is stretched, work is done and strain energy is evenly distributed throughout the volume of the sample. Imagine that a surface crack is then introduced, extending into the material a distance L. The stress trajectories that previously ran through the crack are diverted around it, and the material in the vicinity of the crack becomes unstressed. As a consequence, the strain energy of this volume of unstressed material is decreased. As the crack propagates into the material, new areas of surface are formed, and because the material has a surface energy, energy must be provided from somewhere to create this new surface. Griffith proposed that this energy is provided by the release of strain energy as

Figure 15.4. Crack propagation and the critical crack length. (a) As a crack extends, strain energy is released from an area around the crack. (b) When the energy released by extending the crack exceeds the energy required to create new crack surface, the crack may propagate catastrophically. (Redrawn from Gordon 1978 by permission of the author.)

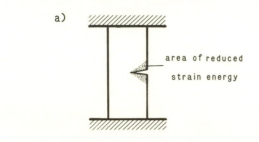

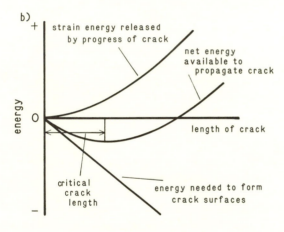

the crack extends. If the sample shown in Figure 15.4 has a constant depth, the volume of material in which stress is reduced is proportional to the area shown, and therefore to the square of the length of the crack. Similarly, the new surface created as the crack extends is proportional to the length of the crack. The relationship between these two energies is shown in Figure 15.4b. For very short cracks, the energy required to extend the crack is greater than that released when the crack extends. Under these conditions the crack is not self-propagating—external energy must be supplied to lengthen the crack. At a certain, critical crack length the energy

released when the crack extends is sufficient to create new crack surface area, and the crack becomes self-propagating. As the crack lengthens beyond this point, the ratio of energy provided to energy required increases, and the rate of crack propagation speeds up. Eventually, the crack propagates at nearly the speed of sound in the material. It is the existence of a critical crack length at which a crack becomes self-propagating that supplies the solution to the engineer's nightmares: stress concentrations can cause the collapse of a structure only if they are associated with cracks of sufficient size.

The dimensions of a critical crack are governed by the properties of the material through which the crack must propagate. The relationship calculated by Griffiths is that the critical crack length, L_c, is

$$L_c = \frac{2WE}{\pi\sigma^2}, \qquad (15.7)$$

where E is the Young's modulus of the material in the vicinity of the crack, σ is the average stress applied to the material near the crack's tip, and W is the *work* (or *energy*) of *fracture*, the energy required to create new crack surface. E and σ are independently measurable, as has been shown in Chapter 12, but W is not often determined independently. Instead, a crack of known dimensions is introduced into a material of known modulus and strength, and the stress required to cause crack propagation is determined. The measured values of L, E, and σ are then inserted into eq. 15.7 to calculate W. Values of W for several materials determined in this fashion are shown in Table 15.2.

The energy required to create new crack surface is not the same as the surface energy of the new surface itself. This equality could exist only if all the energy of crack propagation went directly to creating new surface area. In reality the surface energy of the new area is only a small fraction of the overall energy required to extend the crack; the rest goes to plasti-

Table 15.2
Very Approximate Work of Fracture for
Various Materials (from Gordon 1978)

Material	Work of Fracture (J/m²)
Glass, pottery	1–10
Cement, brick, stone	3–40
Polyester and epoxy resins	100
Nylon, polyethylene	1,000
Bones, teeth	1,000
Wood	10,000
Mild steel	100,000–1,000,000
High-tensile-strength steel	10,000

cally deform the material or to do work against viscosity.

Using a value for breaking strength appropriate for a cross-linked protein (10^7 to 10^8 Pa) and a modulus and work of fracture appropriate for a cross-linked polymer such as an epoxy resin (10^8 Pa, 100 J/m²), respectively (Gordon 1978), the critical crack length for a biological glue can be estimated to be on the order of 1–100 μm. It is difficult to imagine an organism attaching itself to a wave-swept substratum without having an imperfection of these dimensions somewhere in the adhesive joint. A crack length of only a milli-meter would limit the effective adhesive strength of this hypothetical glue to under 10^4 Pa, a value that is much nearer the adhesive strengths observed in nature. Thus, although glues are an excellent idea in theory, their adhesive strength may be limited via the action of stress concentrations to values that are no higher than those attained by other adhesive mechanisms. The theory of crack propagation has important implication not only for glues, but for most other biological structures as well (for a discussion, see Gordon 1978 or Wainwright et al. 1976).

Solid glues have another problem in addition to cracks. Just as a high-surface-energy liquid will not spread over a low-surface-energy solid, a high-surface-energy glue will not bind well to a low-surface-energy substratum. If all surfaces were clean, this might not pose a problem, since most naturally occurring substrata (rocks) have surface energies higher than most glues. However, in the wave-swept environment all surfaces are effectively coated by water and are likely to have a coating of diatoms and bacteria as well. These coatings reduce the surface energy of the substratum, and most glues do not stick well. This problem can be avoided to a certain degree if the glue used is soluble in water. In this case the glue can diffuse into the water that coats the surface.

CHAPTER 16

~ ~ ~ ~ ~ ~ ~ ~ ~ ~ ~

Structural Wave Exposure

In the preceding fifteen chapters we have examined the mechanics of the wave-swept environment in considerable detail, so we are now in a position to apply our knowledge in a coordinated exploration of the biological consequences of the physical environment. These consequences are traditionally lumped together as *wave exposure*, a term that is a bit too all-encompassing for present purposes. In this chapter we confine ourselves to what is more appropriately called *structural wave exposure*, the environmental determinants of a plant's or animal's structural integrity. For example, from our knowledge of waves we can (at least in theory) specify the flow regime at a particular point on the substratum, and we can measure or deduce the forces exerted by the moving water on an individual living at this spot. Similarly, we can measure the strength and tenacity of the plant or animal. Our task here is to tie together these separate bits of information to provide an estimate of whether the physical environment will allow the individual to survive, and if so, for how long.

Assume for the moment that we can measure an organism's strength or tenacity with precision and can specify the maximum force it experiences. From this maximum force we can calculate the stress exerted on the organism's skeleton and adhesive. Given these assumptions, we can define an index of structural exposure that compares these two values:

$$\chi = \frac{\text{maximum stress experienced}}{\text{strength or tenacity}}. \quad (16.1)$$

Equivalently,

$$\chi = \frac{\text{maximum force experienced}}{\text{tenacity} \cdot \text{area}}, \quad (16.1b)$$

where area is measured at the point the organism is likely to break (e.g., the basal area of an acorn barnacle). If χ is 1 or greater, the organism breaks or is dislodged.

This sort of exposure index is a useful tool. If it could be specified for each individual at a given site, a biologist would be well along the way to quantifying the impact that the wave-force environment has on the wave-swept community. One could, for instance, pinpoint those organisms that will be dislodged before they have time to reproduce and those that will live to a ripe old age.

In practice, however, the parameters that go into defining the structural exposure index are not as straightforward as they might first appear. From Chapters 13 and 14 we know that the stress exerted on an organism depends on the organism's shape and the time-course of the applied force. For any individual organism the shape of structures that are likely to be broken can be measured easily and accurately. In contrast, it may be very difficult to measure accurately the maximum force applied to the organism because the factors that determine the imposed force are, to a certain extent, stochastic. This indeterminant nature of the physical environment makes the maximum force a time-dependent variable. For instance, wave heights and periods vary through time, so that the longer an individual is present on the shore, the greater the probability of encountering large velocities and rapid accelerations. Similarly, the more waves an individual encounters, the greater the likelihood of being at the chance spot on the substratum where the velocity and acceleration are maximal.

There is a similar problem associated with the indeterminacy of tenacity. Each individual has a well-defined strength or tenacity when subjected to a specific force, but one must first dislodge the animal or plant to measure its tenacity. The only way around this problem is to measure the tenacity of a large number of conspecifics and then specify the distribution of tenacities shown by a particular species at a particular site. This distribution allows us to quantify the *probability* that a plant or animal chosen at random has a given strength. In other words, we can point at an intact barnacle and determine what strength it is *likely* to have, but we cannot know its *precise* strength.

A Simplified Index

Before we deal with these problems, it will be useful to examine a simple index of structural wave exposure (eq. 16.1). By judiciously defining our terms we can come up with some useful rules of thumb.

We start by relating the imposed stress to the wave height. In this instance we circumvent the problem of randomly varying wave heights, and assume that we can specify the largest wave that an organism will encounter. This is not necessarily as gross an assumption as it might seem. While in most cases we will be unable to specify whether a site is subjected to a maximum of a 5- or a 6-meter-high wave, we can usually specify the maximum wave height to within an order of magnitude. This sort of ballpark approach will be extensively used throughout the rest of this book.

Given that we can specify a range in which the maximum wave height lies, how then do we specify the stresses that this wave causes? We begin by specifying the maximum horizontal velocities and accel-

erations.[1] This can be done quite accurately for waves in shallow water up to the point at which the wave breaks. For simplicity we use linear wave theory (eq. 5.25 and Table 5.2):

$$u_{max} = \frac{\pi H}{T} \frac{1}{\sinh kd}, \qquad \text{at } s = 0$$

$$a_{max} = \frac{2\pi^2 H}{T^2} \frac{1}{\sinh kd}, \qquad \text{at } s = 0, \quad (16.2)$$

where d is the water's depth, H the wave height, T the wave period, and k the wave number ($2\pi/L$), where L is the wave length. These equations refer to the velocity and acceleration measured at the bottom. Recall that the maximum velocity occurs when the acceleration is zero, and vice versa.

Once the wave breaks, a different relationship must be used to specify the maximum velocity. For now we will consider the case of a steep shore where the maximum water velocity associated with a breaking wave is roughly equal to the velocity of the water at the breaking wave crest.

$$u_{max} \simeq \sqrt{g2H_b}. \qquad (16.3)$$

Here we have assumed that waves break at a water depth equal to the wave height. Gently sloping shores will be considered later in this chapter.

At present we have no reliable method for calculating the acceleration associated with the turbulent flow of a broken wave. This is a factor we must contend with as we go along—it turns out not to be a problem.

Having specified the maximum water velocity, we now calculate what force this flow exerts on a particular organism. Recalling eq. 11.8, we note that the total force acting on an object is the vector sum of the drag, lift, and acceleration reaction:

$$\text{force} = \sqrt{\{\tfrac{1}{2}\rho u^2 S_p C_d + \rho C_m Va\}^2 + \{\tfrac{1}{2}\rho u^2 S_{plan} C_l\}^2}.$$

$$(16.4)$$

[1] In this simple case we ignore vertical velocities and accelerations. Near the substratum, w and (except in impact situations) dw/dt are small compared to u and du/dt.

In the discussion presented at the end of Chapter 11, we concluded that in a typical case the acceleration reaction is small compared to the forces of drag and lift. Thus we can delete the acceleration dependent term in eq. 16.4 without unduly changing the result, and thereby avoid the problem of not being able to specify the acceleration associated with flows in broken waves. This should cause no problem as long as we remember that the force values we calculate will be slightly smaller than those that are actually occurring. When the expression for the acceleration reaction is dropped, the equation for the overall force becomes

$$\text{force} = \sqrt{[\tfrac{1}{2}\rho u^2 C_d S_p]^2 + [\tfrac{1}{2}\rho u^2 C_l S_{\text{plan}}]^2}$$
$$= \tfrac{1}{2}\rho u^2 \sqrt{[C_d S_p]^2 + [C_l S_{\text{plan}}]^2}. \quad (16.5)$$

This equation can be related to the wave height by inserting the appropriate expression for the velocity, $u_{\max}$:

For subtidal organisms (i.e., those subjected to unbroken waves):

$$\text{force}$$
$$= \tfrac{1}{2}\rho \left\{\frac{\pi H}{T}\right\}^2 \frac{1}{\sinh^2 kd} \sqrt{[C_d S_p]^2 + [C_l S_{\text{plan}}]^2}.$$
$$(16.6)$$

For intertidal organisms subjected to newly broken waves:

$$\text{force} = \rho g H \sqrt{[C_d S_p]^2 + [C_l S_{\text{plan}}]^2}. \quad (16.7)$$

The stress that this force imposes on an organism depends on the organism's shape and flexibility. Because shapes and flexibilities vary widely, it is not possible to examine every case here, and the calculation of an exposure index is illustrated using three typical organisms: a limpet, a coral, and a kelp.

A Limpet

The drag and lift coefficients for *Lottia gigantea*, an intertidal limpet, have been discussed in Chapter 11. C_d is about 0.45 for the Reynolds number associated with wave flows, and C_l is around 0.2. A typical

L. gigantea has a projected area S_p of around $3 \cdot 10^{-4}$ m^2 and a planform area S_{plan} of about $11 \cdot 10^{-4}$ m^2. Inserting these values into eq. 16.7, we arrive at the conclusion that the total force exerted on a limpet is

$$\text{force} = \rho g H$$
$$\cdot \sqrt{[0.45 \cdot 3 \cdot 10^{-4}]^2 + [0.2 \cdot 11 \cdot 10^{-4}]^2}$$
$$= \rho g H \cdot 2.6 \cdot 10^{-4}$$
$$\simeq 2.6 H.$$

This force is exerted at an angle of 58° ($\simeq 1$ radian) to the substratum (eq. 11.9).

To calculate an exposure index for *L. gigantea*, we need only specify the limpet's tenacity. We simplify matters by ignoring interindividual variation in tenacities, and deal with a hypothetical "average limpet" having a tenacity equal to the mean tenacity observed for the species. The average tenacity of stationary *L. gigantea* is $1.38 \cdot 10^5$ Pa (Porter, unpublished). Thus the structural wave exposure index for *L. gigantea* is

$$\chi = \frac{2.6 H}{1.38 \cdot 10^5 \cdot S_{\text{plan}}} = 0.063 H.$$

This index is plotted in Figure 16.1 as a function of wave height. Even when the wave height is very large (10 m), χ is less than 1, indicating that a stationary limpet of this size is well capable of coping with typical wave forces. Remember that the stresses we have used in this calculation are underestimates of the real values. If we had accounted for the stress due to the acceleration reaction, the imposed force could be 15%–20% higher than that used here ($\chi \simeq 0.075 H$). Even when this additional force is taken into account the exposure index is still considerably less than 1. Thus, even though we cannot exactly specify the maximum wave height with which this limpet must contend, we are able to say that in all probability the average stationary limpet will not be bothered by wave forces.

This conclusion may need to be modified, however, when the limpet moves around. Miller (1974a) has shown that the tenacity

Figure 16.1. The structural exposure index χ as a function of wave height for the limpet *Lottia gigantea*. The crawling limpet adheres less tenaciously, and therefore has a higher risk of dislodgement.

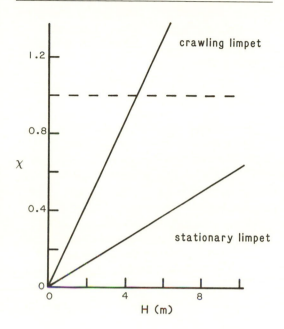

may be able to feed only during moonless nighttime low tides and during high tides. If in addition they cannot move about to feed when the waves are large, this could represent a serious constraint on their overall feeding time.

A Coral

We now turn our attention to an example of a subtidal organism. For the sake of simplicity we use *Acropora cervicornis*, a shallow-water coral with a spikelike blade (Fig. 2.2). The process of calculating the exposure index for this organism follows the same general path as that for the limpet but involves a different set of simplifying assumptions. We begin by examining the total force placed on the coral by the water's velocity and acceleration:

force

$$= \sqrt{\{\tfrac{1}{2}\rho u^2 S_p C_d + \rho C_m V a\}^2 + \{\tfrac{1}{2}\rho u^2 S_{\text{plan}} C_l\}^2}.$$

$$(16.8)$$

We can again ignore the acceleration reaction, in this case not only because the size of the organism in the direction of flow is quite small (the diameter of the blade is 1–2 cm), but also because the accelerations beneath unbroken waves are small. For example, at a depth 3 m (typical of *A. cervicornis*) the acceleration caused by even a very large wave (3 m height, 10 s period) amounts to only 5 m/s² (eq. 16.2). Furthermore, this maximum acceleration occurs when the water velocity is zero, and the acceleration is zero when the water velocity (and therefore the drag) is at a maximum. The cylindrical shape of the coral blade is such that its steady lift coefficient is negligible.[2] This leaves only the drag force to be contended

of many gastropods decreases substantially when they crawl, and this is true of *L. gigantea*. The tenacity of *L. gigantea* when moving averages $3.9 \cdot 10^4$ Pa, only about 35% of its stationary tenacity. Using this figure for the maximum resistible stress, we calculate a higher exposure index ($\chi = 0.22H$; Fig. 16.1), which exceeds 1 at a wave height of about 4.5 m. We can conclude that wave forces can be a disturbing factor for the average crawling *L. gigantea*, and Wright (1978) and Porter (unpublished) have indeed shown that *L. gigantea* move a shorter distance at higher waves.

Limpets must move in order to forage for the diatoms and macroalgae that form their diet. However, the possibility of desiccation and predation by birds can inhibit feeding during daytime low tides, and birds have been observed to prey on limpets during nighttime low tides when there is sufficient moonlight (David Lindberg, Tom Hahn, pers. comm.). As a consequence, limpets

---

[2] A cylinder can have a substantial steady lift coefficient if its axis is parallel to the substratum and the cylinder is less than a diameter or so away from the sea bed (see Sarpkaya and Isaacson 1981). However, this coral grows roughly perpendicular to the substratum, and aside from the momentary lateral lift as vortices are shed (likely to be small at these Re's [see Chapter 11]), should experience negligible lift.

with. Thus, to a first approximation the total force on the coral blade is

$$\text{force} = \tfrac{1}{2}\rho u^2 S_p C_d. \qquad (16.9)$$

What stress does this force place on the coral's skeleton? Recall that for a long slender cantilever of this sort the stresses due to bending are far in excess of the stresses due to shear (Chapter 13). We thus neglect the shear stresses and concentrate on the tensile and compressive stresses imposed by the bending moment imposed by drag. We are helped in this task by the discussion presented earlier (Chapter 13) regarding the growth form of this (and other) corals. Recall that the diameter of *A. cervicornis* increases in direct proportion to its length, and because of this the tensile and compressive stresses imposed by the water's velocity are constant along the length of the blade. This is handy in the present context because it means that the stress imposed by a given water velocity is the same for a small blade as it is for a large one. By analogy to Appendix 13.3, we know that the radius of each coral blade is a function of length:

$$R = \sqrt{\frac{2\rho u^2 C_d}{3\pi\sigma_b}}\, x,$$

where R is the radius at a distance x from the tip. Rearranging to solve for stress, we see that

$$\sigma = \frac{2}{3\pi}\rho u^2 C_d \left\{\frac{x}{R}\right\}^2.$$

Because the blade grows with a linearly increasing radius, the value of x/R is a constant. Substituting eq. 16.2 for the velocity term in this equation and noting that in the shallow water where *A. cervicornis* grows $1/\sinh(kd) \simeq 1/kd = T/2\pi\sqrt{g/d}$ (Table 5.3), we arrive at a relationship among wave height and period, water depth, and imposed stress:

$$\sigma = \frac{2}{3\pi}\rho \left\{\frac{gH^2}{4d}\right\} C_d \left\{\frac{x}{R}\right\}^2. \qquad (16.10)$$

Now, a typical value of x/R for *A. cervicornis* is 10, and we can assume that

$C_d = 1.2$. The average strength of the coral skeleton is $\bar{\sigma}_b = 1.23 \cdot 10^7$ Pa with a standard deviation about this mean of $4.3 \cdot 10^6$ Pa (Denny, unpublished). We can again ignore the variation in strengths and deal with a hypothetical "average coral blade." Thus the estimated structural exposure index is

$$\chi = \frac{2}{3\pi}\rho \left\{\frac{gH^2}{4d}\right\} C_d \left(\frac{x}{R}\right)^2 \frac{1}{\bar{\sigma}_b}. \qquad (16.11)$$

This relationship is shown in Figure 16.2. The exposure index decreases rapidly with depth for waves of a given height. Note that wind waves break when their height is approximately equal to the water depth. If we assume that water velocity does not increase after the wave breaks, breaking places an upper limit on the exposure index for a given H, shown as the dashed line in Figure 16.2. Due to their extremely long wave length, surf beat do not break even in very shallow water. As a consequence their

Figure 16.2. The exposure of a coral. The structural exposure index χ for the coral *Acropora cervicornis* as a function of water depth and wave height. The shallower the water and the higher the wave, the greater the risk of breakage. Seas and swell typically break when the water depth is equal to the wave height (indicated by the dashed line; see Chapter 7), so the higher exposure indices may not be realized. These high exposures would be possible for high-amplitude surf-beat or edge waves, which do not break.

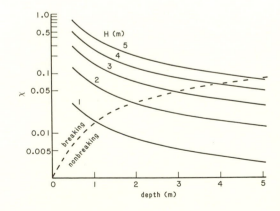

effects may be more important than those of wind waves in very shallow water, although we should keep in mind that the height of surf beat is typically small, often only 10% that of breaking wind waves.

On the basis of these calculations we can tentatively conclude that *A. cervicornis* are not likely to be broken by hydrodynamic forces alone. Tunnicliffe (1981) found that storm waves cause extensive breakage of *A. cervicornis*, but she speculates that much of the damage on the windward side of a reef was due to impact by previously broken pieces of coral. Corals can also be substantially weakened by boring sponges, leading to a higher exposure index.

This approach to calculating the structural wave exposure index is not limited to the two cases presented here. In Figure 16.3 curves are given for the exposure index of various organisms as a function of wave height. Note that there is a very large range in the exposure indices calculated for organisms that inhabit the same shores: some organisms, such as barnacles, seem well adapted to coping with wave forces, while others, such as mussels and predatory snails, are not.

When a predicted exposure index is near to or greater than 1, it behooves us to find some reasonable mechanism that would allow these organisms to survive where we would predict they cannot. The exposure index calculated for the mussel assumes that the organism is solitary and is therefore exposed to the full brunt of the flow regime. In actuality, solitary mussels are very seldom found on wave-swept shores. Rather, mussels clump together in tightly packed extensive beds, each individual shielding its neighbors from the flow. In this situation the velocities calculated on the basis of wave height cannot be expected to apply to each individual mussel, and the exposure index calculated here is a gross overestimate. Predatory snails such as *Thais emarginata* are mobile and probably hide in crevices or in the spaces between mussels when waves cause large water velocities. And finally, we must keep in mind that these calculations are intended only to provide ballpark estimates.

The calculation for a human being assumes that the person stands on the shore and is frontally impacted by the wave. For a wave 5 m high at breaking, the resulting 10 m/s flow would impose a force of about $5 \cdot 10^4$ N, the equivalent of 1.2 tons force. In calculating the exposure index, we assume that even an adept biologist would be knocked off his or her feet by a force of 1,110 N (equivalent to 150 pounds force). The enormous value for the exposure index will perhaps give some tangible appreciation of how well designed most wave-swept organisms are.

Figure 16.3. Exposure indices calculated for shear forces for a variety of organisms.

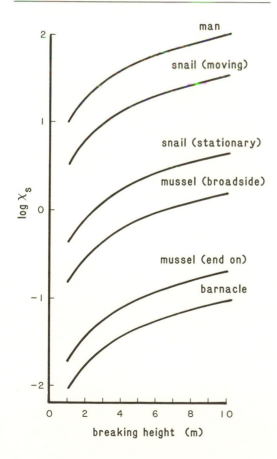

Figure 16.4. (a) A simple model of a flexible alga used to calculate the exposure index for flexible structures (see text. (b) The predicted maximum water velocity imposed on a flexible alga as a function of water depth and stipe length. The longer the stipe and the greater the depth, the less the maximum velocity, and hence, the drag.

a)

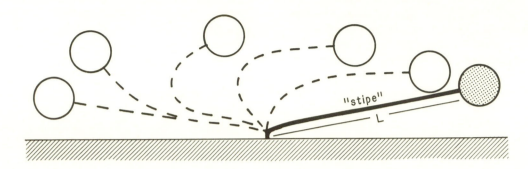

b)

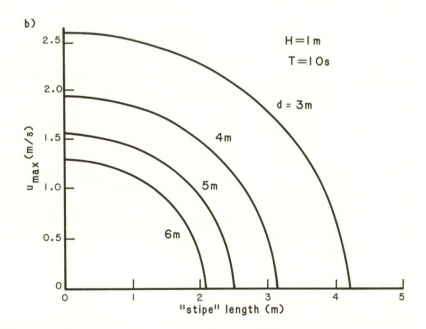

A Kelp

All of these calculations have been made for a rigid organism. If the organism is flexible, corrections may be necessary. To see why, consider the simple, hypothetical kelp depicted in Figure 16.4a. The alga has a flexible, ropelike stipe of length L with a neutrally buoyant sphere attached to its end. If the stipe is thin (and therefore I is small), no appreciable stress is imposed by the bending of the stipe, and we can assume that the stipe can be broken only when pulled upon along its length. This happens only when the sphere has traveled far enough away from the holdfast so that the stipe is extended in the direction of flow. At this point the sphere is held stationary relative to the water, and a drag is exerted.

If the stipe is long enough, the sphere never reaches the end of its tether before the water velocity changes direction. In this case the stipe feels virtually no force. The critical length beyond which the stipe is immune to drag is equal to the maximum horizontal displacement of the water. From Table 5.4 we see that in shallow water

$$\text{displacement} = \left\{\frac{H}{2}\right\} \sin(\omega t) \left\{\frac{1}{kd}\right\},$$

$$\text{maximum displacement} = \pm\left\{\frac{H}{2}\right\}\frac{1}{kd}.$$

$$(16.12)$$

If the stipe length is less than the maximum water displacement the sphere reaches the end of its tether before the flow changes direction, and a drag is imposed. The shorter the stipe, the more rapid the relative velocity it experiences.

This relationship is easily calculated. First we note that the water velocity is greatest when the displacement is zero—that is, when the sphere is at the holdfast. As the displacement increases, the velocity decreases until $u = 0$ at a displacement of $(H/2)(1/kd)$. For a given stipe length, L, we can calculate the time required to extend the stipe by solving eq. 16.12 for t:

$$t_{\text{stretch}} = \frac{1}{\omega} \arcsin\left\{\frac{L2kd}{H}\right\}. \quad (16.13)$$

Because speed steadily decreases until the flow changes direction, the drag force is largest when the sphere is first brought to a halt. By inserting our calculated value for t_{stretch} into eq. 5.36 for the horizontal water velocity in shallow water, we calculate that the velocity when the sphere is first brought to a halt is

$$u = \left\{\frac{\pi H}{T}\right\}\left\{\frac{1}{kd}\right\} \cos(\omega t_{\text{stretch}}). \quad (16.14)$$

This velocity is shown in Figure 16.4b as a function of stipe length and water depth. Spheres attached to short stipes experience flows very nearly equal to those experienced by inflexible organisms. As the stipe length approaches the amplitude of water movement, the velocity experienced (and thereby the drag) rapidly decreases. Insofar as real organisms approximate this model, we can conclude that the exposure index of flexible organisms decreases as size increases.

There are several problems that preclude direct use of this model in the real world. First, we have neglected the inertia of the sphere. As the sphere reaches the end of its tether, a force must be applied by the stipe to bring it to a halt. The magnitude of this force depends on the rate at which the sphere is slowed, which in turn depends on the stiffness of the stipe. A very stiff stipe will abruptly stop the sphere and experience a large inertial force. However, unless the moving mass of the organism is large and its tether very stiff, the inertial force will be small compared to the drag force. For example, a sphere 1 cm in radius experiences a drag of 0.3 N when $u = 2\ m/s$. The mass of a neutrally bouyant sphere of this size is $4.3 \cdot 10^{-3}$ kg, and would have to be decelerated at 70 m/s^2 to equal the drag force. The calculations made here use linear wave theory and therefore cannot be expected to give accurate answers for high waves in very shallow water. These caveats aside, the general message of this discussion holds—the more flexible the organism, the lower the exposure index.

The behavior of a flexible organism may be substantially affected by surf beat. These long waves can cause onshore-offshore water displacements with periods of 20—300 s. On a gently sloping shore these currents may be a substantial fraction of the orbital velocity of the swell and can periodically bias the displacement of the kelp. When the orbital velocity of the swell is in the same direction as that of the surf beat, the kelp reaches the end of its tether sooner than it would otherwise, with a consequent increase in maximum drag.

Accounting for Probability

Although calculating the structural exposure index can be informative, we must now face up to its shortcomings. How can we account for the facts that all waves do not have the same height and all members of a species do not have an average strength?

In this exploration we again use the coral *A. cervicornis* as an example, but the same principles apply to all organisms. We start by measuring the breaking stress of a large number of coral blades, and assume that the distribution of breaking stresses can be reasonably described by the Gaussian, or normal, curve. In this case, if we know the mean breaking stress (σ_b) and the standard deviation ($\mathscr{S}$) of the breaking-stress distribution, we can specify the probability density function:

$$P(\sigma) = \left\{ \frac{1}{\mathscr{S}\sqrt{2\pi}} \right\} \exp - \left\{ \frac{[\sigma - \bar{\sigma}_b]^2}{2\mathscr{S}^2} \right\}, \quad (16.15)$$

where $P(\sigma)$ is the probability that an individual has strength σ.

This function is graphed in Figure 16.5a; it is the typical bell-shaped curve familiar from introductory statistics. Given the probability density function, we can point at random to a coral blade and specify the probability that that blade has a certain breaking stress. Note that the area under the whole probability density curve is equal to 1—that is, it is absolutely certain that corals have a breaking strength between 0 and infinity Pa.

Now imagine that we could apply a known stress, σ', to coral blades in the field. A blade will be broken if its strength is less than this stress. The probability that an individual has a tenacity less than σ' (and will therefore break) is the area under the probability density curve from 0 to σ' (Fig. 16.5b) —the cumulative probability at σ'. This area is the integral

Figure 16.5. Calculating the probability of dislodgement. (a) The normal probability density distribution, a model of the breaking stress distribution. (b) The probability of being dislodged by a known stress is equal to the area under the probability density curve at stresses less than that imposed. (c) When the magnitude of the imposed stress is not precisely known, the probability of dislodgement is the area under the curve that is the product of the probability density of breakage at each stress and the probability that the applied stress exceeds that breaking stress. The stress exceedance curve shown here is a Rayleigh exceedance curve replotted in terms of stress rather than wave height.

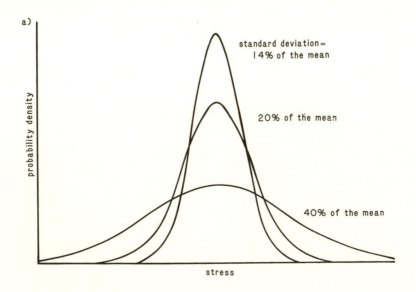

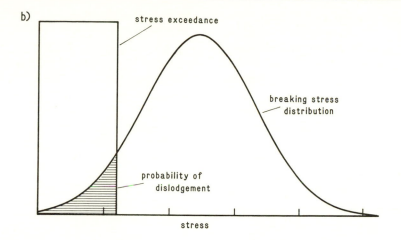

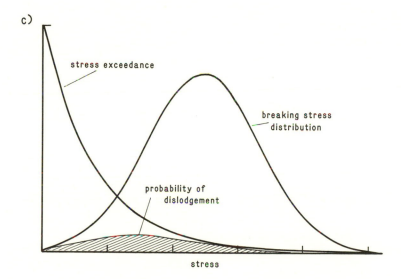

$$\Pi_d = \int_0^{\sigma'} \left\{ \frac{1}{\mathscr{S}\sqrt{2\pi}} \right\} \exp - \left\{ \frac{[\sigma - \bar{\sigma}_b]^2}{2\mathscr{S}^2} \right\} d\sigma$$

$$= \int_0^{\sigma'} P(\sigma)\, d\sigma, \qquad (16.16)$$

where Π_d is an exact measure of the probability of dislodgement. Our ability to use this integral depends on knowing with absolute certainty that a stress σ' was applied; only then can we specify the upper limit to integration. In other words, we must know that the probability that σ exceeds all values less than σ' is 1, and the probability that σ exceeds σ' is zero. This probabilistic inter-

pretation of the applied stress is shown in Figure 16.5b as the step-exceedance curve superimposed on the breaking stress distribution. We can use this exceedance curve to rewrite eq. 16.16 in a more general form:

$$P_d = \int_0^{\infty} P_{(\sigma > \sigma')} \left\{ \frac{1}{\mathscr{S}\sqrt{2\pi}} \right\} \exp - \left\{ \frac{[\sigma' - \bar{\sigma}_b]^2}{2\mathscr{S}^2} \right\} d\sigma'$$

$$= \int_0^{\infty} P_{(\sigma > \sigma')} P(\sigma')\, d\sigma', \qquad (16.17)$$

where P_d is an estimate of the probability of dislodgement, $P_{(\sigma > \sigma')}$ is the probability that $\sigma > \sigma'$, and $P(\sigma')$ is calculated per eq. 16.15. By expressing the integral in this fashion, we avoid the need to specify the upper limit to integration. In the case shown in Figure 16.5b, integrating eq. 16.17 from zero to infinity is the same as integrating eq. 16.16 from zero to σ' because $P_{(\sigma > \sigma')}$ is zero for values above σ'.

The same procedure works when the stress-exceedance curve is not a step function (Fig. 16.5c). For every stress, we multiply the probability that an organism has that particular breaking strength by the probability that the imposed stress exceeds that strength. Integrating all these values we end up with the probability that the organism will be broken.

We now can examine an example of this approach using the coral *A. cervicornis*. We calculated above (eq. 16.10) that the stress imposed on the coral's skeleton varies as a function of wave height. From Chapter 6 we recall that wave heights generally follow a Rayleigh distribution. We can combine these two ideas to calculate the probability that a wave chosen at random will exert a stress exceeding a certain value. From eq. 6.4 we calculate that

$$P_{(\sigma > \sigma')} = \exp - \left\{ \frac{H_{\sigma'}}{H_{\text{rms}}} \right\}^2, \quad (16.18)$$

where $H_{\sigma'}$ is the wave height that causes a stress, σ'. $H_{\sigma'}$ is calculated by solving eq. 16.10 for H and inserting σ' for σ:

$$H_{\sigma'} = \sqrt{\frac{\sigma' 2\pi d}{\rho C_d g}} \left\{ \frac{R}{X} \right\}. \quad (16.19)$$

Thus the probability of dislodgement is

$$P_d = \int_0^\infty \exp - \left\{ \frac{H_{\sigma'}}{H_{\text{rms}}} \right\}^2 P(\sigma')\, d\sigma' = \left\{ \frac{1}{\mathscr{S}\sqrt{2\pi}} \right\}$$

$$\cdot \int_0^\infty \exp - \left[\left\{ \frac{H_{\sigma'}}{H_{\text{rms}}} \right\}^2 + \left\{ \frac{[\sigma' - \bar\sigma_b]^2}{2\mathscr{S}^2} \right\} \right] d\sigma'$$

$$(16.20)$$

This calculation provides a method by which we can express the probability that a coral at a given depth will be dislodged by a randomly chosen wave. Eq. 16.20 is plot-

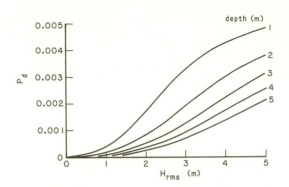

Figure 16.6. The probability of dislodgement, P_d, for the coral *A. cervicornis* as a function of rms wave height and depth.

ted in Figure 16.6 as a function of H_{rms}.

Recalling that H_{rms} is $0.71 H_s$ (eq. 6.5), we can alternatively express this function in terms of the significant wave height. This is useful because the significant wave height is a parameter that a skilled wave observer can readily measure:

$$P_d = \left\{ \frac{1}{\mathscr{S}\sqrt{2\pi}} \right\}$$

$$\cdot \int_0^\infty \exp - \left[\left\{ \frac{H_{\sigma'}}{0.71 H_s} \right\}^2 + \left\{ \frac{[\sigma' - \bar\sigma_b]^2}{2\mathscr{S}^2} \right\} \right] d\sigma'$$

$$(16.21)$$

The probabilities calculated in this fashion concern the interaction between one randomly chosen wave and one randomly chosen individual. But with a wave arriving every 10 s or so, a coral blade of any appreciable age will have experienced thousands upon thousands of waves. How can we calculate the probability that an individual will survive the repeated imposition of wave-induced forces?

We begin by shifting our calculations from the probability of dislodgement to the probability of survival. If there is a probability P_d that the coral will be broken by a wave chosen at random, there must be a probability,

$$P_s = 1 - P_d, \quad (16.22)$$

that it will survive. Let P_{s_1} be the probability

of surviving the first wave encountered and P_{s_2} the probability of surviving the second wave, calculated in exactly the same fashion as for the first. The probability of surviving both these waves is simply

$$P_{s,\text{cumulative}} = P_{s_1} \cdot P_{s_2}. \quad (16.23)$$

As long as the rms wave height and the breaking-stress distribution remain constant, both probabilities are the same, and the probability of surviving these two waves is simply P_s^2. This logic can be extended to any number of waves, again provided that H_{rms} and the breaking-stress distribution are constant. Thus the probability of surviving N waves is P_s^N. This sort of survivorship curve is shown in Figure 16.7a. On average, each wave dislodges a fixed fraction (rather than a fixed number) of the remaining organisms, and the number of survivors asymptotically dwindles.

If we make the assumption that the incident waves have a constant average period, the number of waves a coral experiences is simply the wave frequency (waves/s, $1/T$) multiplied by the time the coral is in the water. Thus the overall probability of surviving is

$$P_{s,\text{cumulative}} = P_s^{(t/T)}, \quad (16.24)$$

where time t is measured in seconds. Recalling that $P_s = 1 - P_d$, we can restate this conclusion in terms of the breakage parameters we discussed earlier:

$$P_{d,\text{cumulative}} = 1 - [1 - P_d]^{(t/T)}. \quad (16.25)$$

Inserting the value for P_d calculated in eq. 16.20, we arrive at a mechanistic expression for the time course of survival:

$$P_{s,\text{cumulative}} = \left[1 - \left\{ \frac{1}{\mathscr{S}\sqrt{2\pi}} \right\} \right.$$

$$\cdot \int_0^\infty \exp - \left\{ \left\{ \frac{[\sigma' - \bar{\sigma}_b]}{2\mathscr{S}^2} \right\} \right.$$

$$\left. + \left\{ \frac{H_{\sigma'}}{H_{\text{rms}}} \right\}^2 \right\} d\sigma' \Bigg]^{t/T}. \quad (16.26)$$

As before, $H_{\sigma'}$ is calculated from eq. 16.19.

This function allows us to calculate the

Figure 16.7. Survival of the coral *A. cervicornis*. (a) Survivorship as a function of time; survivorship approaches 0 asymptotically. (b) The logarithm of survivorship as a function of time. The slope of this curve is a measure of exposure, Ω.

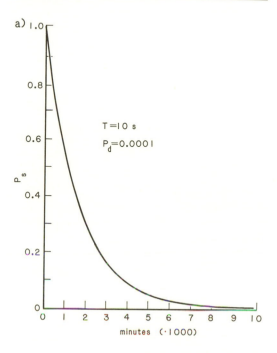

a)

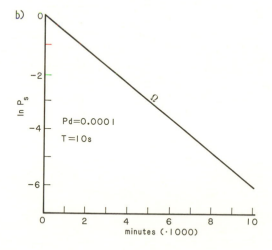

b)

survivorship of a coral as a function of the rms wave height and the observed strength distribution of the coral (Fig. 16.7a). The larger the rms wave height or the lower the average breaking stress, the faster the corals are broken.

The rate at which individuals are broken or dislodged is one measure of the severity of the wave environment as it relates to a particular species. Consequently, we can use the slope of the survivorship curve as an alternative to χ as an indicator of structural wave exposure. Unfortunately, the slope of our predicted curve varies continuously with time (Fig. 16.7a). To arrive at a single number that characterizes the survivorship predicted by this model, we take the time derivative of $ln\, P_s$, and define an alternative index of structural exposure (Fig. 16.7b):

$$\Omega = \frac{d\ln(1 - P_d)^{(t/T)}}{dt} = \frac{d(t/T)\ln(1 - P_d)}{dt}$$

$$= \frac{\ln(1 - P_d)}{T}. \qquad (16.27)$$

There are limitations to the usefulness of this approach. The calculations performed here are only valid as long as H_{rms} remains constant. As we have seen, this condition is not likely to hold for more than a few hours or at most a day. Moreover, the calculation of $H_{\sigma'}$ requires a knowledge of the depth at which the organism lives (eq. 16.19). This depth changes with the tides, an effect we have not taken into account. Finally, these calculations assume that wave heights follow a Rayleigh distribution, and if the water depth is such that waves are near their breaking point, this may not be an accurate representation (Chapter 6). Thus the structural wave-exposure index calculated here should not be taken as an accurate estimate of the actual survivorship of any individual.

We could approach a more accurate estimate by revising the value of P_d whenever the rms wave height or SWL changes.

$$P_{s,\text{cumulative}} = P_{s_1}^{(t_1/T_1)} \cdot P_{s_2}^{(t_2/T_2)} \cdots, \qquad (16.28)$$

where $P_{s_1}^{(t_1/T_1)}$ is the probability of surviving waves with the rms height acting during time interval t_1. During this period, waves are assumed to have period T_1. All of these parameters can be changed for the next time interval, t_2, and so on. This process provides a better estimate of overall survivorship, but would be extremely laborious to calculate for any extended period of time (a year, say) if the rms wave height changed frequently.

This procedure for calculating the probabilistic structural wave-exposure index also has limitations when applied to mobile organisms such as limpets. As we have noted before, the tenacity of a limpet decreases substantially when the limpet crawls. This decreases in mean tenacity shifts the breaking-strength distribution to a lower mean and increases the probability of the limpet being dislodged. It would be an easy matter to calculate a survivorship curve for a crawling limpet if the limpet were crawling all the time, but limpets are not so accommodating. As Wright (1978) and Porter (unpublished) have shown for *L. gigantea*, limpets crawl for only part of the time, and the time spent crawling is dependent on the wave-force environment. This sort of behavioral response makes it difficult to calculate a predictive survivorship curve.

Another problem with the calculations made here concerns the estimation of the breaking-strength distribution. We have used the distribution observed at a site (assumed to be Gaussian) as the best way of estimating the probability that an individual has a certain strength. However, this distribution may change through time as a result of wave forces. For instance, if a large storm causes H_{rms} to rise drastically at a normally protected site, many of the weaker plants and animals will be broken. Unless there is recruitment or immigration of low-tenacity individuals, the mean of the surviving organisms is shifted toward greater tenacity and the distribution is no longer Gaussian. Thus, if no new individuals are introduced into the population, a storm can provide information about those organisms that have survived, telling us something about their strength. If we could incorporate this information into our probability estimation, we would have a better idea of an individual's true survivorship.

Recall from eq. 16.16 that the cumulative probability curve for tenacity is calculated from a probability density function and predicts the likelihood that an organism, chosen at random, has a strength less than a certain value, σ' (Fig. 16.8a). This cumulative breaking-strength curve is calculated on the basis of measurements made at one point in time. How can we modify this curve to account for the historical information provided by waves?

In this example, we will follow the survivorship of a cohort of individuals—for instance, the coral blades that are initiated in a given month. In doing so, we assure ourselves that no new individuals enter the population by recruitment, and we assume that none enter by immigration. We further assume that the tenacity of each individual is constant. Next we note the obvious: an organism that survives the force of a given wave must have a strength capable of coping with the stress imposed by that wave—if its strength were less it would be broken. Thus, if we knew the precise stress imposed by a certain wave, we could say that the surviving animals in the cohort have at least that strength. Unfortunately, we do not know exactly what stress a wave chosen at random will cause, but we can, from the reasoning presented earlier, specify the probability that each wave exerts a stress less than a certain value (Fig. 16.8a):

$$P_{(\sigma < \sigma')} = 1 - \exp - \left\{ \frac{H_{\sigma'}}{H_{\mathrm{rms}}} \right\}^2 . \quad (16.29)$$

Now, the probability that an individual surviving this one wave has a strength less than a certain value σ' is the probability of dislodgement (evaluated at σ'; eq. 16.16)

times the probability that the imposed stress is less than σ' (eq. 16.29). Say that based on our initial measurement of the breaking-strength distribution for the cohort, there is a 10% chance that a coral has a strength less than σ' ($\Pi_d = 0.1$ for σ'). If we subject this coral to a wave known to impose a stress σ' and the coral survives, we can say for sure that its strength exceeds σ', and the probability of dislodgement for a second imposition of the same stress would be 0. However, if this coral is subjected to a wave that has a 50% chance of imposing a stress less than σ' and it survives, we can be only half certain that its strength exceeds σ'. In other words, we can assign it a probability of dislodgement equal to the initial Π_d (0.1) times the probability that the wave exerts a force less than σ' (0.5), for a new estimated value of 0.05. This procedure can be carried out for each value of σ' to produce a new cumulative probability curve for the breaking stress (Fig. 16.8a), and the derivative of this curve with respect to stress gives us a new probability density curve (Fig. 16.8b). This new tenacity distribution has been shifted to a higher mean, has less overlap with the probability distribution of imposed stresses, and results in a lower probability of dislodgement for the next stress encountered.

In this manner, we can correct Π_d and the probability density curve of breaking strength for the effects of one wave chosen at random. To account for the effect of many waves, we follow the same procedure used earlier. The probability that two waves both impose a stress less than σ' is the probability of one wave imposing a stress less than σ' multiplied by the probability of the second wave imposing a stress less than σ':

$$P_{(\sigma < \sigma'),\text{cumulative}} = P_{(\sigma < \sigma')_1} \cdot P_{(\sigma < \sigma')_2}. \quad (16.30)$$

As long as the rms wave height is constant, these two probabilities are the same, and the overall probability of having a stress less than σ' is $P_{(\sigma < \sigma')}^2$. In general, the over-

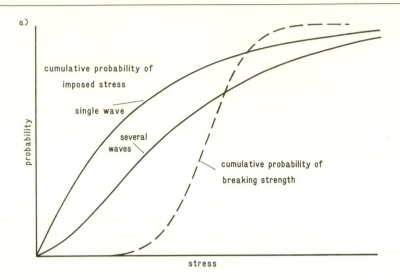

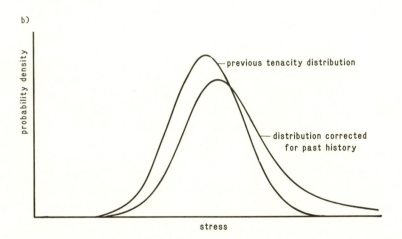

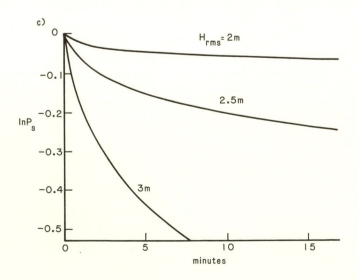

all probability that a stress less than σ' is imposed by N wave is

$$P_{(\sigma < \sigma'),\text{cumulative}} = \left\{ 1 - \exp - \left\{ \frac{H_{\sigma'}}{H_{\text{rms}}} \right\}^2 \right\}^N.$$

(16.31)

Substituting t/T for N as we did previously, we arrive at a final expression for the cumulative probability curve of imposed stress:

$$P_{(\sigma < \sigma'),\text{cumulative}} = \left\{ 1 - \exp - \left\{ \frac{H_{\sigma'}}{H_{\text{rms}}} \right\}^2 \right\}^{t/T}.$$

(16.32)

This value can be evaluated for each σ' and used to calculate the probability density curve for breaking strength at any time t after the initial distribution was measured:

$$P(\sigma') = \frac{d}{d\sigma'} \left\{ \left\{ 1 - \exp - \left\{ \frac{H_{\sigma'}}{H_{\text{rms}}} \right\}^2 \right\}^{t/T} \left\{ \frac{1}{\mathscr{S}\sqrt{2\pi}} \right\} \right. $$
$$\left. \cdot \int_0^{\sigma'} \exp - \left\{ \frac{[\sigma - \bar{\sigma}_b]^2}{2\mathscr{S}^2} \right\} d\sigma \right\}.$$

(16.33)

From this breaking-stress distribution for time t the probability of an individual being dislodged by the next wave can be calculated by analogy to eq. 16.20:

$$P_d = \int_0^\infty \exp - \left\{ \frac{H_{\sigma'}}{H_{\text{rms}}} \right\}^2 \cdot P(\sigma') \, d\sigma'. \quad (16.34)$$

The cumulative values of P_d allow us to construct a survivorship curve for the cohort that accounts for changes in the tenacity distribution through time (Fig. 16.8c). The effect of the gradual shift of tenacities to a higher mean is to decrease the fraction of individuals broken or dis-

lodged by successive waves. As long as the strength of an individual is constant (if unknown), the more waves it survives, the more certain we can be that it is strong enough to withstand the forces imposed by the environment. In other words, the longer an individual survives, the higher its probability of living even longer.

Confined by the assumption of no recruitment or immigration, this analysis paints a picture of an ever-dwindling cohort size and a breaking-strength distribution that gradually shifts to a higher mean. This model is reasonable for a cohort of individuals but is a less-than-compelling description of an entire population. Certainly many populations are not dwindling, and it seems that in nature the tenacity distribution of populations is more or less the same from year to year, with perhaps some seasonal fluctuation. Now, the upward shift in mean breaking strength predicted for a discrete cohort can be offset in an entire population through the introduction of low-tenacity individuals into a population by recruitment. We can also relax our assumptions and allow immigration and the reduction of tenacity in residents. By these mechanisms we can reach an equilibrium between the rate at which susceptible individuals appear in the population and at which organisms are removed by waves. Once this equilibrium is reached, the tenacity distribution is stable. In this sense, the maintenance of a stable breaking-strength distribution is analogous to the maintenance of a stable age distribution in a population where the proportion of individuals in a certain age class depends on the rate at which they are born

Figure 16.8. (facing) Historical correction of the breaking stress distribution. (a) The probability that an organism has less than a certain breaking strength after N waves is the product of its initially having a breaking strength less than that certain value (one point on the measured cumulative breaking strength probability curve, the dashed line) and the cumulative probability that all N waves exerted a stress less than breaking stress (the corresponding point on the cumulative stress curve). (b) The breaking strength distribution corrected from information gained by having survived N waves. The mean is shifted to a higher stress and the curve is no longer normal. (c) The survivorship of the coral *A. cervicornis* corrected for history. The longer the coral survives, the less its risk of being dislodged by the next wave.

or recruited and the rate at which they die. In fact, the analogy is sufficiently precise so that we can use a mathematical approach employed by demographers to construct a model for a stable tenacity distribution.[3]

We begin at time t with a population of N individuals. For simplicity, we assume that tenacity varies among individuals, but that the tenacity of each individual is constant. We can express the tenacity distribution of this population as a set of discrete classes, each encompassing a tenacity range of $2 \cdot 10^6$ Pa (Fig. 16.9). The number of individuals who have tenacities between $\sigma - 2 \cdot 10^6$ and σ at t is $n_{\sigma,t}$, and the sum of all $n_{\sigma,t}$ is N. We examine the population at discrete time intervals; however, the length of the interval is not important.

At time $t + 1$ (one time interval after time t) the number of individuals in tenacity class σ is

$$n_{\sigma,t+1} = P_{s,\sigma} n_{\sigma,t} + P_{s,\sigma} m_\sigma b, \quad (16.35)$$

where $P_{s,\sigma}$ is the probability that individuals with tenacity σ will survive an interval (eq. 16.24). m_σ is the proportion of recruits and immigrants that fall into tenacity class σ, and b is the total number of recruits and immigrants that arrive during an interval. For simplicity we assume that m_σ and b are constant.

Given this formulation, we can easily express the tenacity distribution at time $t + 2$:

$$n_{\sigma,t+2} = P_{s,\sigma} n_{\sigma,t+1} + P_{s,\sigma} m_\sigma b. \quad (16.36)$$

Or, in terms of the distribution at time t,

$$n_{\sigma,t+2} = P_{s,\sigma}^2 n_{\sigma,t} + P_{s,\sigma}^2 m_\sigma b + P_{s,\sigma} m_\sigma b. \quad (16.37)$$

By carrying this process on to time $t + 3$, $t + 4$, etc., we see that the tenacity distribution at any arbitrary time $t + z$ is

$$n_{\sigma,t+z} = P_{s,\sigma}^z n_{\sigma,t} + \sum_{k=1}^{z} P_{s,\sigma}^k m_\sigma b. \quad (16.38)$$

[3] This was brought to my attention by Steven Gaines, for which I am in his debt.

Now $P_{s,\sigma} < 1$, so that as z goes to infinity, $P_{s,\sigma}^z n_{\sigma,t}$ goes to zero. Thus at time $t + z$ the number of individuals in any tenacity class σ is

$$n_{\sigma,t+z} = \sum_{k=1}^{z} P_{s,\sigma}^k m_\sigma b = \left. \frac{m_\sigma b}{1 - P_{s,\sigma}} \right|_{z \to \infty}. \quad (16.39)$$

This number can be expressed as a proportion of the total population size:

$$\frac{\left\{ \dfrac{m_\sigma}{1 - P_{s,\sigma}} \right\}}{\sum\limits_{\sigma=0}^{\infty} \dfrac{m_\sigma}{1 - P_{s,\sigma}}}. \quad \text{as } z \to \infty \quad (16.40)$$

This is the stable tenacity distribution. Note that this distribution is independent of both the initial tenacity distribution and of b, the rate of recruitment and immigration. It depends only on the tenacity distribution of recruits and immigrants. Figure 16.9c shows the stable tenacity distribution for a population whose initial distribution is shown in Figure 16.9a, and whose immigrants and recruits have a truncated Gaussian tenacity distribution as shown in Figure 16.9b. This stable distribution indeed bears little resemblance to the initial distribution. It has a higher mean than that of the recruits and immigrants and is skewed toward high tenacities.

This model can provide a simple means for interpreting tenacity distributions observed in nature. If these distributions are indeed stable over the course of a few years, the tenacity distribution of immigrants and recruits can be inferred.

This simple model has assumed that tenacity varies among individuals in a population but is constant for each individual. This may not be the case, and it is easy to imagine mechanisms by which resident individuals could move among tenacity classes. For instance, adhesives may fatigue as a result of the constant pounding by waves, and individuals in high tenacity classes may therefore slide to lower classes as time passes. Alternatively, an organism

Figure 16.9. Tenacity distributions. (a) The initial, observed distribution. (b) The tenacity distribution of immigrants and recruits, a Gaussian curve truncated at 0 and $2.6 \cdot 10^7$ Pa. (c) The stable tenacity distribution (see text).

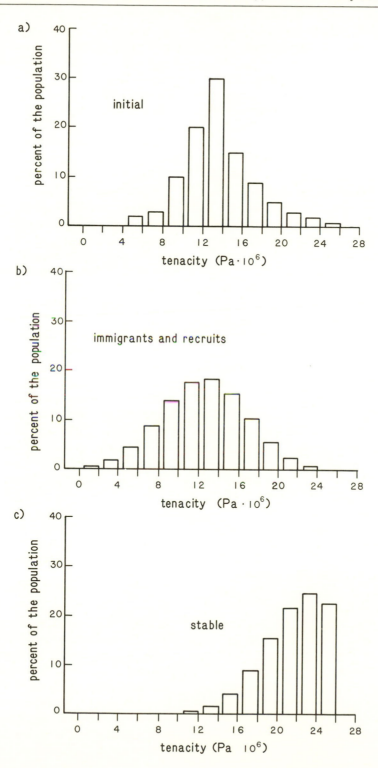

could respond to wave pounding by increasing its tenacity. For example, Price (1980) has shown that the mussel *Mytilus edulis* increases its tenacity in the stormy winter season by laying down more byssal threads. If this kind of shuffling of individuals among tenacity classes is substantial, the model outlined here must be refined. If the shuffling occurs on a short time scale (i.e., the tenacity of an individual changes between the imposition of wave forces), the shuffling itself may be the dominant factor, and the tenacity distribution may be stable over longer time scales due simply to the random variation in individual tenacity. This is probably the case for foraging limpets, for instance, where the tenacity of an individual is likely to change dramatically and quickly as it encounters different microtopographies on the substratum.

This model for a stable tenacity distribution is subject to many of the same problems mentioned in conjunction with the calculations of survivorship. For instance, it assumes that the rms wave height and the wave period remain constant. Although the approaches outlined here provide tools whereby we can begin to examine the environmental control of survivorship, we are a long way from being able to measure all the input parameters required to give an accurate estimate of the survivorship of a real wave-swept organism.

Finally, there is a potential problem in all these calculations associated with our neglect of the acceleration reaction in estimating the stress imposed on wave-swept organisms. Even though the acceleration reaction forms only a small portion of the overall force, this fraction scales with size in a different fashion from the lift and drag forces. The result is that the acceleration reaction can have substantial impact on the survivorship of animals that grow through time. This "problem" is interesting enough to bear closer scrutiny, as done in Chapter 17.

Bottom Slope, Energy Dissipation, and Exposure

The results obtained so far depend on a knowledge of the root mean square wave height. For subtidal organisms, these rms values can be calculated from the height of waves offshore by taking the wave shoaling into account (Chapter 7). For intertidal organisms on steeply sloping shores, the root mean square of the breaking wave heights is the parameter used. In this case we assume that the wave breaks directly on the shore, and that the crest velocity calculated from the breaking height is imposed on the organisms. However, if the broken wave (now a periodic bore) must travel a substantial distance to reach an organsim, it loses height as its energy is dissipated by turbulence, and we can no longer use eq. 16.3 to calculate u_{max}.

This situation would arise on a shore with a gently sloping bed. How should we calculate u_{max} for organisms in the surf zone on such a shore? It has often been proposed that within the surf zone the height of bores is a linear function of water depth: $H_{rms} = \gamma d$ (Thornton and Guza 1983). Given this relationship, we can specify $u_{max,rms}$ for organisms within the surf zone. The celerity of the bore is calculated by inserting γd for H_{rms} in eq. 7.13:

$$\therefore C_{rms} = \sqrt{gd(1 + \gamma)}. \qquad (16.41)$$

The water velocity at the bottom under the crest of the bore is approximately $0.3C$ (Chapter 5), and

$$u_{max,rms} \simeq 0.3\sqrt{gd(1 + \gamma)}. \qquad (16.42)$$

As d decreases, $u_{max,rms}$ decreases and so does the exposure index. Thus, on a gently sloping shore, organisms farther inshore are less exposed (in a structural sense) than organisms farther away from the shoreline. On such shores the exposure index for a particular organism varies greatly as the local water depth changes with the tides.

It seems reasonable to use this approach to provide an educated guess for the local u_{max} for gently sloping bottoms, although the presence of boulders, rock outcroppings, and so forth that are common on rocky shores will undoubtedly lead to some error. Such large-scale relief on an otherwise gently sloping shore should increase the rate of energy dissipation, leading to a smaller value of γ than that measured on sandy beaches ($\gamma \simeq 0.4$). Because the local H_{rms} is independent of the offshore wave height, on a gently sloping shore one might be led to conclude that the exposure of an organism would be entirely independent of the offshore wave conditions. But the size of offshore waves can affect inshore exposure by increasing the local setup, in effect increasing the local depth d and by the periodic changes in sea level due to surf beat (see Chapter 7). On very gently sloping shores, the water velocities associated with surf beat may themselves be a factor in structural wave exposure. Velocities due to surf beat in excess of 2 m/s have been measured on sandy beaches in Oregon (R. Beach, pers. comm.).

All of these conclusions regarding wave exposure on gently sloping rocky shores are extrapolated from findings determined either in laboratory wave tanks or on gently sloping sandy beaches. A great deal of field-work must be done before they can be applied with any assurance to sites on rocky shores.

Effects of Height on the Shore

In the preceding analysis we have assumed that the number of waves to which an individual is subjected is a function simply of time and the wave frequency. This is a good approximation for subtidal organisms, but organisms living in the intertidal zone by definition spend some of their time above the still-water level. When the tide is sufficiently low, these plants and animals are not subjected to wave forces, and the number of waves they experience in a given period is less than that of a subtidal organism. The fewer waves an organism experiences, the less its chance of being dislodged.

We begin an analysis of this problem by considering the shape of a typical wave as it breaks on a steeply sloping shore (Fig. 16.10). Near the time of breaking, the leading surface of the wave is more or less vertical, and the water at this edge is moving with a velocity near that of the crest. Behind and below this leading face the water is moving more slowly. When waves break, the crest of the waves, and often much of the leading surface, is formed of foam. Although this foam moves at the same velocity as the front of the wave, its density is likely to be much less than that of water. As a result, the hydrodynamic forces imposed by this foam, all of which are directly proportional to the fluid density, will be less than those caused by the water farther down the wave face. Taking this qualitative description into account, we can draw a very simplified view of a breaking wave. We define a "danger zone," the vertical distance on the shore that lies between the toe of the wave and the lower limit of the foam. A plant or animal on the shore at the level of this danger zone will be subjected to the maximum force that the wave can impose. Organisms lower than the danger zone are subjected to decreased water velocities, and organisms higher on the shore are initially subjected to foam flow and experience the bulk flow of unfoamed water only after the wave has run up the shore. During this run-up the broken wave (a bore) loses height, and the water velocity is less than what we would calculate on the basis of the wave height at breaking. This notion of a danger zone is intentionally fuzzy and not intended to imply that we could take a snapshot of a breaking wave and accurately draw on it the limits of the danger zone.

As the tide rises and falls, the danger zones of the waves also rise and fall. For a

Figure 16.10. Definitions for calculating the effect of height on the shore on the probability of encountering a large imposed force (see text). (Redrawn from Denny 1985 by permission of the American Society of Limnology and Oceanography.

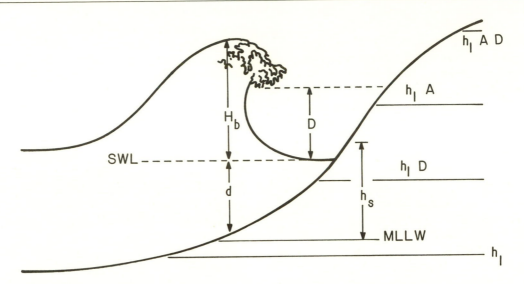

stationary organism present at one height on the shore, the result is a certain time spent in the danger zone. For example, a barnacle living in the midintertidal zone will be above the danger zone when the tide is very low, in the danger zone for some period during midtide, and below the danger zone at high tide. If we know the height of the danger zone, and the pattern of the tidal fluctuations, we can calculate the time for which a given organism occupies the danger zone. From this we can calculate the number of forceful waves to which the organism is subjected on a given tide and thereby gain entry to the survivorship calculations we have derived.

Here we assume that the time course of the tidal fluctuation is sinusoidal between a low tide and the succeeding high, or between high tide and the succeeding low. Given this we can write an expression that describes the height of the still-water level at any time:

$$h(t) = \frac{1}{2}A \left\{ \cos \left\{ \frac{\pi t}{T_t} \right\} + 1 \right\} + h_l \quad (16.43)$$

where A is the difference in height between high and low tides, and T_t is the time be-

tween high and low tides (usually about 6.15 hours). h_l is the height of the still water level at low tide. If the heights of succeeding high tides or low tides differ (as they do on the west coast of North America), this function must be evaluated for each half-tide—that is, from one high to the next low, or for one low to the next high.

To calculate the time during which an organism occupies the danger zone, consider the situation shown in Figure 16.10. A stationary organism is stuck to the rock at height h_s, measured for convenience from mean lower low water (MLLW). The height of the danger zone is D. The danger zone just reaches the organism when the still water level reaches $h_s - D$. As the tide rises the danger zone moves upward, until, when the still water level equals h_s, the organism is left behind. Thus we need to know the time between when the still water level reaches $h_s - D$ and h_s. Solving eq. 16.43 for time rather than $h(t)$, we see that the time when the tide reaches $h_s - D$ is

$$t = \frac{T_t}{\pi} \arccos \left\{ \frac{2[h_s - D - h_l]}{A} - 1 \right\} \quad (16.44)$$

and the time when the still water level reaches h_s is

$$t = \frac{T_t}{\pi} \arccos\left\{\frac{2[h_s - h_l]}{A} - 1\right\}. \quad (16.45)$$

Subtracting eq. 16.44 from 16.45 (and taking the absolute value to avoid problems with the sign), we see that the time an organism is in the danger zone Δt is

$$\Delta t = \frac{T_t}{\pi} \left| \arccos\left\{\frac{2[h_s - D - h_l]}{A} - 1\right\} \right.$$
$$\left. - \arccos\left\{\frac{2[h_s - D - h_l]}{A} - 1\right\} \right|.$$
$$(16.46)$$

For a constant danger-zone height, we can thus calculate the time spent in the danger zone. If we further assume a constant wave period, we can convert this time into the number of forceful waves encountered during the half-tide:

$$N = \frac{\Delta t}{T}, \quad (16.47)$$

where T is the wave period. Knowing the number of waves encountered, we can then calculate the probability of encountering a wave imposing a certain stress. For example, from eq. 16.7 we can estimate that a wave must be 3 m high at breaking to exert a force of 7.8 N on a *Lottia gigantea*. If the rms wave height is 1 m, the probability of encountering a wave of this height (or greater) is (from eq. 6.4)

$$P_{(H > 3)} = \exp -\left\{\frac{3}{1}\right\}^2 = 1.23 \cdot 10^{-4}.$$

Following the logic used several times so far, we can calculate that the probability of encountering a wave of at least this height in a random sample of N waves is

$$P = 1 - (1 - [1.23 \cdot 10^{-4}])^N.$$

The value of N used in this equation is that calculated from eq. 16.47. This calculation can be carried out as a function of h_s and D.

Graphs of this function are shown in Figure 16.11. For each individual tidal fluctuation, the curve of probability vs. h_s can look a bit weird. The dip or plateau at midtide level is due to the relatively slow rate of change of SWL near high and low tide and to the assumption of a distinct danger zone (being *near* the danger zone does not count). If the probabilities at each h_s are averaged for the tidal fluctuation occurring during a typical month, these odd peaks and dips tend to disappear, and the average probability of encountering waves of a given frequency of occurrence is maximal near midtide level. A twofold variation in D (from 1 to 2 m) does not change the position of the average probability peak. Given this insensitivity to the exact value of D, it does not seem worthwhile to extend this model to account for the stochastic variability of wave heights and therefore of D.

The values used in these calculations have been chosen to be applicable to Tatoosh Island in the state of Washington, a site where the maximum forces exerted on small spheres were measured at different h_s (Denny 1985a; Table 16.1). The mean maximum force is greatest at the middle site and is less at the higher and lower sites. These measurements were made over the course of seven tides. When this small sample size is taken into account, the differences in the means between sites

Table 16.1
Maximum Wave Forces on a Limpet As a Function of Height on the Shore (from Denny 1985a)

Height above MLLW (h_s) (m)	Mean Maximum Force (N)	Standard Error
1.65	0.864	0.193, $n = 8$
2.75	1.149	0.332, $n = 12$
3.50	0.767	0.171, $n = 10$

MLLW = mean lower low water.

Figure 16.11. The probability, P, of encountering a wave of a certain frequency of occurrence during one tidal excursion (low tide to high tide or vice versa). The dashed lines are for tidal excursions with $h_l = -0.3$ m and $A = 4.0$ m. The dotted lines are tidal excursions with $h_l = 1.4$ m and $A = 1.2$ m. The solid lines are the average for the 119 tidal excursions in January 1980. The upper graphs are calculated for waves imposing a force of magnitude seen on average only once in 1,000 waves, the lower graphs for one wave in 10,000. (a) Calculated for a danger zone 2 m high. (b) Calculated for a danger zone 1 m high (from Denny 1985, by permission of the American Society of Limnology and Oceanography).

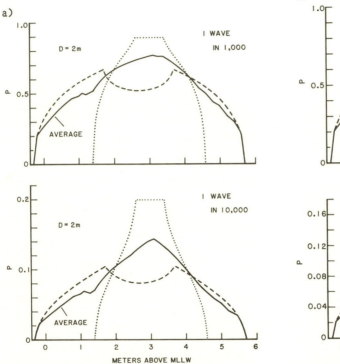

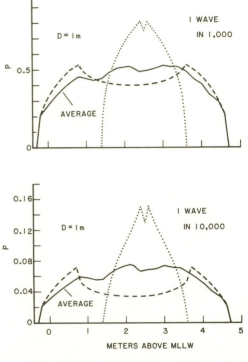

is not significant, and any conclusions made on the basis of these measurements should be taken with a grain of salt. However, they do suggest that the simple model outlined here may be on the right track. It is also interesting to note that the h_s level where the probability is predicted to be the greatest corresponds to the h_s level at which mussel beds are found on Tatoosh. Inasmuch as mussel beds form a self-sheltering environment, they should be a near ideal tactic for coping with wave forces. The spatial correspondence between the predicted area of highest probability of wave force and the presence of the organism best suited to cope with that force is tantalizing.

This model is intended to apply only on shores where the bottom slope is steep enough to render the situation depicted in Figure 16.10 reasonably accurate. On a gently sloping shore the situation is different. When the water is deep enough so that waves are not breaking, the velocity experienced by an organism near the bottom is the one calculated in eq. 16.2. For a wave of a given height, this velocity increases gradually as the depth decreases. The waves break when the water depth is roughly equal to the wave height. Except with the smallest waves, this breaking occurs well away from the shoreline, and thus the plunging or collapsing wave front does not land directly on the organisms attached to the substratum—a considerable cushion of water is present between the two. At breaking, the wave is transformed into a turbulent bore, where the water velocity (at the sea bed) is approximately

$$u \simeq 0.3\sqrt{g(H + d)}. \qquad (16.42)$$

Thus, after the wave breaks, the water velocity experienced by an organism decreases the shallower the water. Given this general picture, the maximum water velocity experienced by an organism should occur when the organism is under the breakers. As the tide moves up and down, the breaker zone moves up and down the shore in a manner directly analogous to the danger zone described already, and the time spent in the breaker zone could be calculated in the same manner. The results would be similar to those for organisms on steep shores but the maximum probability of encountering a large water velocity would be associated with a position on the shore shifted downward roughly twice the wave height from that shown in Fig. 16.11.

Functional Wave Exposure

In this chapter we have confined ourselves to an examination of the effects of wave forces on the structural integrity of an individual, but there are many other ways in which wave forces can affect organisms. For instance, on shores exposed to rapid water velocities, predatory organisms may be prohibited from moving about lest they be dislodged. The effect has been recorded for the predatory sea star *Pisaster ochraceus* (Quinn 1979) and is very likely to be true for predatory snails. In such a case the wave-swept environment may have a direct effect on the intensity of predation, and thereby on many aspects of community structure. Unfortunately, to examine this kind of "functional wave exposure" we need to know more than we do about the foraging

behavior of the animals in question. How much force can a crawling sea star resist? How often does a sea star move at high tide relative to movement at low tide? Is there an additional risk associated with using tube feet for feeding rather than adhering to the substratum? All of these questions need to be answered before we can begin to assess the effect wave forces have on foraging behavior.

Wave exposure is likely to have substantial effects on other biological processes as well. For instance, barnacles reproduce using internal fertilization, and to do so each barnacle must maneuver its penis to one of its neighbors. Performing this feat at low tide may expose the barnacle to a large risk of predation or desiccation. If this is so, the barnacle may need to copulate while it is washed by waves. The ability to copulate successfully would then depend on the amount of time available between the imposition of hydrodynamic forces too large to be resisted by the penis. If no sufficient period were available, the barnacle could not successfully copulate. This kind of quantized biological process—requiring a set amount of time between forces—is open to analysis by the same sort of logic followed in this chapter. Unfortunately, we do not yet know enough about the biology of these processes to be able to apply these methods productively.

Summary

In this chapter we have examined one particular aspect of the wave environment—the relationship between the structural capabilities of individual organisms and the stresses placed on them by waves. We have seen that, at least in theory, it is possible to account for the mechanical survivorship of organisms in a probabilistic sense. The calculations have often been based on incomplete data or pure, educated guesswork, but the results obtained bear a heartening resemblance to what one would expect from observations of the real world. It is probable that this kind of mechanistic approach will soon be used to predict the survivorship of actual wave-swept organisms in the absence of disturbing factors such as predation and herbivory. This approach is therefore a potentially useful tool for examining the relative importance of physical and biological processes in the structuring of wave-swept communities.

~ ~ ~ ~ ~ ~ ~ ~ ~ ~ ~

Mechanical Determinants
of Size and Shape

In the last chapter we briefly examined the role that wave forces play in determining the survivorship of an organism of a given size and shape. In this chapter we extend these ideas by exploring the possibility that the wave-force environment may have been an important selection pressure in determining the shape of these organisms and the size to which they grow.

The basic premise behind this discussion is familiar to any biologist—we measure the success of an organism by how many young it produces. The more surviving progeny an organism produces, the larger its relative contribution to the next generation, and the more "fit" it is. As simple and appealing as the idea of fitness is, it is very difficult to measure directly the fitness of most marine organisms. Because most wave-swept organisms have planktonic larval stages, it is well nigh impossible to keep track of the progeny produced by any one individual. If the young cannot be traced, it is impossible to measure fitness directly. This problem is usually circumvented by making a simplifying assumption—the more young an organism produces, the greater its chance of contributing to the next generation. Thus, the number of larvae or spores produced can be taken as an approximate measure of the fitness of an organism. This method of assessing fitness is far from foolproof. For instance, it tacitly assumes that there are no important selective processes occurring during the planktonic portion of an organism's life. If there were, an organism producing a few very robust young could

end up with more progeny surviving to settle than an organism producing vast numbers of maladapted offspring. Any estimation of fitness through measurement of reproductive output is at best a rough approximation.

Three major factors govern reproductive output: generation time, survivorship to reproductive age, and size at first reproduction. The shorter the generation time, the greater the survivorship, and the larger the adult, the more young produced in a period of time. Of these three factors, survivorship and size may be directly affected by the mechanics of the wave-swept environment.

There are many reasons for large size being advantageous: the larger the organism, the larger the gonad it can contain, with direct advantages for reproductive output. Furthermore, large size often increases survivorship. For example, large mussels (*Mytilus californianus*) are too big to be eaten by the starfish *Pisaster ochraceus* and are therefore immune to starfish predation (Paine 1976). Large organisms may be less susceptible to being overgrown (Sebens 1980; Jackson 1977), and because they have a smaller surface area in proportion to their volume they may be better able to cope with problems of desiccation and heat stress than are smaller organisms.

In this chapter we explore the mechanisms by which the forces of the wave-swept environment can oppose these advantages of large size. Any successful organism must be able to cope structurally with the environment. If in attempting to get unduly large prior to reproduction an organism encounters sufficient stress to dislodge it, its reproductive output (and thereby its fitness) is nil.

The mechanical problems associated with large size are often closely tied to questions of shape. A shape that may be mechanically feasible at small size may be impractical at a larger size. Therefore, in order to grow, many plants and animals must change their shape in response to environmentally imposed forces. If they are incapable of sufficiently changing their shape, they may be limited in the size they can attain.

Size

Sea Anemones

Sebens (1979, 1982) has provided a careful analysis of how the availability of food (in this case, a function of wave force) effects the optimal size of the great green anemone, *Anthopleura xanthogrammica*. The mechanical structure of this animal has been discussed in Chapter 13, where we noted that the organism is shaped so as to keep its oral surface pointing upward. By this means the animal presents the greatest area with which to intercept prey that fall down from above. Many low intertidal pools and surge channels on the Pacific Northwest coast are carpeted with these creatures.

The limits to size in these animals are due to the manner in which their energy budget changes as they grow. Here we use the mass of the anemone as a measure of its size. Other dimensions are then derived from the mass by assuming that the animal stays the same shape as it grows, i.e., it grows isometrically. This is a valid assumption for *A. xanthogrammica*. As with any geometrically similar structure, a linear dimension is proportional to $(mass)^{1/3}$ and surface area is proportional to $(mass)^{2/3}$. Thus we can specify the catching

area of an anemone (the oral disk plus tentacles) as

$$\text{catching area} = k_1 m^{2/3}, \quad (17.1)$$

where k_1 is the appropriate constant of proportionality. If we know the average rate, Q, at which food items fall on an area of anemones (calories/m²/s), we can specify the rate at which food is ingested as a function of an anemone's mass:

$$\text{feeding rate} = Q k_1 m^{2/3}. \quad \text{(calories/s)}$$
$$(17.2)$$

To calculate the energy balance, we must be able to relate tissue mass to the rate at which energy is used. Anemone tissue (or any other tissue) consumes oxygen at a rate that is a function of the mass of the organism. In general, the larger the organism, the smaller the oxygen consumption on a per mass basis; the rule of thumb is that oxygen consumption/gram scales as $(mass)^{-1/4}$ (Schmidt-Nielsen 1984). Thus, to calculate the overall energy consumption of an anemone, we write

$$\frac{\text{energy consumption}}{\text{mass}} = k_2 m^{-1/4}$$

$$\text{total energy consumption} = k_2 m^{-1/4} m^1$$
$$= k_2 m^{3/4}. \quad (17.3)$$

The relationships between energy input and energy consumption are graphed in Figure 17.1a.

The rate of energy consumption increases faster than the rate of energy input. At small sizes, more energy is taken in than is consumed by the resting metabolism of the organism, and the excess energy is available for growth. However, as the organism increases in size, a point is reached where the energy intake is only just sufficient to provide the maintenance energy of the anemone's body. At this point further growth becomes impossible, setting an upper limit to the size of the organism. We can calculate this size by setting eq. 17.2 equal to eq. 17.3:

Figure 17.1. Energy input and energy consumption in organisms where food capture is proportional to feeding area. (a) A typical organism (see text). (b) The sea anemone *Anthopleura xanthogrammica*.

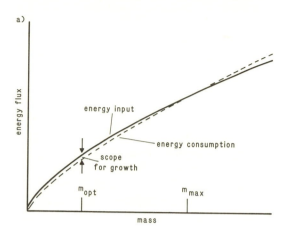

a)

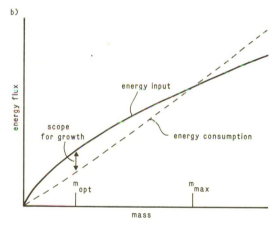

b)

$$Qk_1 m^{2/3} = k_2 m^{3/4}$$

$$Q\frac{k_1}{k_2} = \frac{m^{3/4}}{m^{2/3}} = m^{1/12}$$

$$m_{max} = \left\{ Q\frac{k_1}{k_2} \right\}^{12}. \qquad (17.4)$$

Thus the maximum size an anemone can reach depends on the ratio of the food availability ($Q \cdot k_1$) to metabolic rate (k_2). As long as the water temperature does not change drastically, the metabolic rate of the organism should be constant through time, and the main determining factor of maximum size is the availability of prey. In

the case of *A. xanthogrammica*, the primary prey items (78% of the diet by weight) are mussels that have been dislodged from the substratum. The usual assumption is that these mussels are dislodged by wave forces. Thus the rate at which waves dislodge mussels is hypothesized to be a direct limiting factor in the maximum size to which anemones can grow.

Sebens points out, however, that it would not be advantageous to the anemone to grow to its maximum size. At the maximum size there is only enough energy available to keep up with maintenance metabolic demands—there is none left over to grow a gonad. The reproductive output (and thereby the fitness) of an individual of maximum size would be zero. Instead, the rate at which progeny can be produced is maximized when the surplus of energy taken in is at its greatest. The difference between energy intake and maintenance energy consumption is termed *scope for growth*, and from eqs. 17.2 and 17.3 is

$$\text{scope for growth} = Qk_1 m^{2/3} - k_2 m^{3/4}. \qquad (17.5)$$

We can calculate the size at which the scope for growth is maximized by taking the derivative of eq. 17.5 and setting it equal to zero:

$$(\tfrac{2}{3})Qk_1 m^{-1/3} - (\tfrac{3}{4})k_2 m^{-1/4} = 0 \qquad (17.6)$$

Solving for m,

$$m_{opt} = \left\{ \frac{8}{9} Q\frac{k_1}{k_2} \right\}^{12} = 0.24 \left\{ Q\frac{k_1}{k_2} \right\}^{12}. \qquad (17.7)$$

The scope for growth is maximized at 24% of the maximum possible mass. We assume that come time for reproduction, the anemone takes all of this excess energy and diverts it to gonad production. Thus, given sufficient space within the anemone in which to house the gonad, the maximum reproductive output occurs at this optimum size. As before, this optimal size is dependent on the rate at which mussels are made available to the anemone. The greater the availability of mussels, the larger Q is, and the larger the optimum size. This prediction is borne out by field measurements (Sebens 1982).

In theory, we could predict from a calculation of structural exposure (Chapter 16) the rate of mussel dislodgement near a given patch of anemones, and thereby calculate Q based solely on the root mean square wave height and the tenacity distribution of mussels. We could then trace a direct causal line between the wave force environment and the fitness of *A. xanthogrammica*.

Note that the values of the various exponents used in this analysis have been "typical" values for animals in general. The precise values for any particular organism are likely to be different. For example, Sebens (1981, 1982) found that metabolic rate was nearly independent of anemone size (proportional to $m^{1.08}$) and used this accurate value in his model. The calculations then predict that optimum size is approximately 30% of maximum size (Fig. 17.1b).

Sebens also discusses the possible confounding factor of prey size. For instance, an anemone 10 cm in diameter cannot fit a 15 cm mussel in its coelenteron, while a 20 cm anemone can. If mussels too large to fit in a 10 cm anemone form a large fraction of the mussels available, the effective availability of prey is smaller for a 10 cm anemone than for a larger anemone. In this case the constant of prey availability, Q, is itself size-dependent, and this factor must be taken into account in an accurate calculation of the optimum mass.

Barnacles

The same sort of calculations can be made for other organisms for which food intake is proportional to the area of a feeding structure (e.g., tunicates, mussels [see Vahl 1973; Griffiths and King 1979], and sabbellid polychaetes). For instance, acorn barnacles deploy a feeding area by which they make their living. Unlike anemones, which capture the fairly large prey dislodged by wave action, barnacles use a sievelike structure, the cirral net (Fig. 17.2), to filter small particles from the surrounding fluid. When the water is still, the cirral net can be waved about to create the animal's own feeding current. However, when sufficient water motion is present, barnacles of the genera *Chthamalus*, *Balanus*, and *Tetraclita* merely extend the cirral net into flow, allowing the ambient water motion to bring food to them. Although no direct observations have been made of acorn barnacles feeding in the surf, it seems safe to assume that they use the same strategy as that of the goose-neck barnacle, *Pollicipes polymerus* (Barnes and Reese 1958)—they keep their cirral net furled inside the shell while the very rapid water flow of an upsurge passes by, and then they extend the net during the less stressful backwash.

The use of a feeding area by a growing barnacle would seem to present the same possibility for size limitation as the oral disk area does for an anemone, but in this case there is a strange twist. For an isometrically growing barnacle the size of the feeding net increases as $(mass)^{2/3}$. However, Barnes and Barnes (1959) have shown that the metabolic rates (ml O_2/hr/gram) of a variety of barnacles scale as $(mass)^{-1/3}$. Thus the carbon demand of the animal (mass × mass-specific metabolic rate) scales as $(mass)^{2/3}$, exactly the same as food input. Given this state of affairs, barnacles would not be limited in size by their capacity to capture food.

However, this argument assumes that barnacles can grow isometrically, and this seems not to be the case. No study has been made comparing feeding area as a function of size within a single species, but we can get some idea of the possible limitations to isometric growth by examining individuals from different species. In Monterey Bay the size of common barnacles ranges from 2–3 mm basal diameter (*Chthamalus dalli* and *fissus*) to 10–20 cm for a few

Figure 17.2. The feeding structure of an acorn barnacle—the cirral net. (a), (b) The entire net. (c) A single cirrus showing the setae.

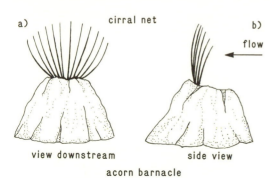

acorn barnacle

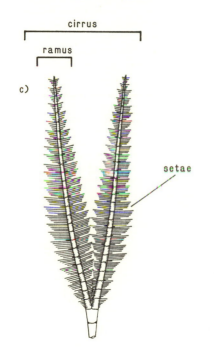

exceptionally large individual *Balanus nubilis*—a range in size of nearly two orders of magnitude. Across this size range there is an obvious change in the size of cirral nets relative to body mass: *Chthamalus* have enormous cirral nets (relative to their diminutive dimensions) while the large *Balanus nubilis* have rather embarrassingly small nets. A preliminary measurement of this variability (Fisher, unpublished) suggests that cirral-

net area scales as (body mass)$^{0.19}$, a far cry from the relationship expected for isometric scaling. It is probable that something about the mechanics of cirral nets does not allow them to be made arbitrarily large. Unfortunately, we do not have enough information about the functional morphology of cirral nets to explore their limitations. We can, however, examine the results of the process by which cirral nets scale.

Given the relationships cited above for net area and metabolic demand, we can carry out the same sort of calculation used for anemones to determine maximum and optimal size. Using Q, k_1, and k_2 in the same sense as before,

$$m_{max} = \left\{ Q \frac{k_1}{k_2} \right\}^{2.08}$$

$$m_{opt} = 0.07 \left\{ Q \frac{k_1}{k_2} \right\}^{2.08}. \qquad (17.8)$$

Here the optimal mass is a considerably smaller fraction of the maximum mass (7%) than for a "typical" organism (24%) or an anemone (30%).

These calculations have been made assuming that food intake is directly proportional to cirral-net area. This assumes that all barnacles see the same flow, and that the same amount of water/area goes through nets of all sizes. The first of these assumptions is likely to be valid. All the barnacles described here live intertidally, many in very close proximity to each other. We would be hard pressed to say that one size sees a much different flow than another. However, the assumption that flow per area is independent of net size is probably not valid. A quick examination of cirral nets of various sizes shows that the spacing between filtering elements (ramal setae; Fig. 17.2c) is a function of the size of each ramus. In general, small cirri have small spacings between setae, and large cirri have large spacings. It seems likely that less water per area goes through the fine-meshed small cirral net than through the coarse-meshed large net when the two are exposed to the same flow. A rough calculation based on information about the flow resistance of screens (Idel'chik 1966) suggests that if the spacing between setae in the cirral net scales isometrically with mass [i.e., $(mass)^{1/3}$], volume filtered per area scales as $(mass)^{0.15}$. If food intake is proportional to the volume of water filtered, food intake scales as

$$\frac{\text{volume filtered}}{\text{area}} \cdot \text{area} = m^{0.15}m^{0.19} = m^{0.34}.$$

(17.9)

This revised figure can be used to calculate m_{max} and m_{opt}:

$$m_{\text{max}} \simeq \left\{ Q\frac{k_1}{k_2} \right\}^3$$

$$m_{\text{opt}} \simeq 0.12 \left\{ Q\frac{k_1}{k_2} \right\}^3. \quad (17.10)$$

In this revised estimate, the optimum size is 12% of the maximum size, still a much smaller fraction than that found for anemones. These calculations have been based on very scant data, and should not be taken too seriously. It is tempting, however, to speculate that the size of barnacles is limited by their ability to support a large cirral net.

The Role of the Acceleration Reaction

In calculating exposure indices (Chapter 16), we always began by making some assumptions that simplify the calculations regarding the operative forces imposed on a particular organism. In each case we dismissed the acceleration reaction as being a minor component of the total force, and therefore a quantity we could neglect. This is a valid approach as long as we are only concerned with calculating the exposure index for a given size of organism. However, the acceleration reaction can be important in determining how the stress on an organism scales as the organism changes size.

To see how this works, consider a hypo-

thetical cubic organism (Fig. 17.3a). In this case we use the length of one side of the cube, L, as a measure of size. The force in the direction of flow experienced by the cube is the vector sum of drag and the acceleration reaction:

$$\text{force} = \tfrac{1}{2}\rho u^2 S_p C_d + \rho C_m V a. \quad (11.5)$$

For this cube the areas projected in the direction of flow (S_p) is L^2. The volume of fluid displaced by each cube is L^3. Inserting these values into eq. 11.5,

$$\text{force} = \tfrac{1}{2}\rho u^2 L^2 C_d + \rho C_m L^3 a. \quad (17.11)$$

Figure 17.3. (a) Flow forces on a cube (see text). (b) Flow forces on a cone (see text).

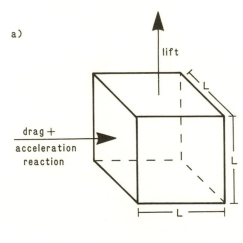

a)

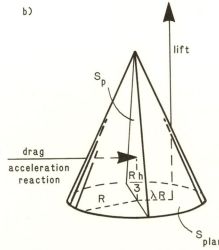

b)

What stresses does this force place on the object? For the moment we consider only the shear force. The average shear stress is simply the total shear force (drag + acceleration reaction) divided by the attachment area (the cube's base, L^2). Thus the average shear stress is

$$\bar{\tau} = \tfrac{1}{2}\rho u^2 C_d + \rho C_m L a. \qquad (17.12)$$

For given values of C_d, u, C_m, and du/dt, the larger the size of the cube (L), the greater the shear stress imposed on its base. Thus, all other things being equal, the larger the organism, the larger the shear stress with which it must cope. As we saw in Chapter 16, the exposure index depends on the ratio of imposed stress to tenacity. Unless larger organisms increase their tenacity, they have a higher risk of dislodgement than smaller individuals of the same species. If this increased risk is sufficiently great, an increase in size ceases to have any advantage in terms of reproductive output, and the risk of dislodgement can serve as a mechanical limit to the size of wave-swept organisms.

The logic involved in this argument is simple, but implementing it in a testable form becomes somewhat involved. We start with a conical shape, one that can be used as a model for limpets and barnacles (Fig. 17.3b). The forces imposed on this shape by moving water are the same three that we have dealt with previously: drag, lift, and the acceleration reaction. As we have already seen, drag and the acceleration reaction impose a shear force:

$$\text{shear force} = \tfrac{1}{2}\rho u^2 S_p C_d + \rho C_m V a.$$

In order to examine the relationship between size and the shear force, we need to express the projected area (S_p) and the volume (V) in terms of a linear dimension of the cone. We define the shape of the cone by two numbers, the radius of the base, R, and the ratio, h, between height and the radius. Thus the area of the cone projected in the direction of flow

is $R^2 h$ and the volume of the cone is $(\tfrac{1}{3})R^3 h$.

Inserting these values into the above equation, we see that

$$\text{shear force} = \tfrac{1}{2}\rho u^2 R^2 h C_d + (\tfrac{1}{3})a\rho C_m R^3 h. \qquad (17.13)$$

This shear force is applied over the basal area (πR^2), and the average shear stress (force/area) is

$$\bar{\tau} = \left\{\frac{1}{2\pi}\right\}\rho u^2 C_d h + \left\{\frac{1}{3\pi}\right\}a\rho C_m R h. \qquad (17.14)$$

The result is similar to that obtained for the cube—the larger the radius, the greater the shear stress.[1]

The shear stress is not the only stress acting on the cone. The combined force of drag and acceleration may be assumed to act at the center of the projected area—for the cone, a distance $(\tfrac{1}{3})Rh$ up from the base. The applied force acting over this distance imposes a bending moment. This particular moment tends to place the downstream side of the cone in compression and the upstream side in tension.

The bending moment due to the combined action of drag and acceleration reaction is not the only moment present. The lift force acts in a direction perpendicular to the flow and applies a force:

$$\text{lift force} = \tfrac{1}{2}\rho u^2 \pi R^2 C_l. \qquad (17.15)$$

This force need not be evenly distributed, and in practice the *center of lift* is located a fraction λ of the radius downstream of the cone's apex.[2] The center of lift is the

---

[1] The maximum shear stress may be somewhat larger than the average shear stress, because (as noted in Chapter 13) the shear stress is not usually constant across a cantilever. However, the shear stress calculated here will eventually be compared to the organism's tenacity. For practical reasons, tenacity in shear is measured as the average shear strength. Thus, for this particular comparison, the results are not compromised by failing to correct for shear distribution.

[2] The precise location of the center of lift depends on the shape of the cone, a factor we will deal with in greater detail later in this chapter. λ varies between 0.1 and 0.4.

point where all the lift force could be applied and have the same mechanical effect as the distributed force. Thus the lift force acts over a moment arm of λR (Fig. 17.3b) and imposes a bending moment with the opposite sense of that imposed by the drag and acceleration reaction. The *net* bending moment imposed by the flow is the difference between these two moments:

$$\text{net moment} = \left| \left\{ \frac{Rh}{3} \right\} [\tfrac{1}{2}\rho u^2 R^2 h C_d + \rho C_m R^3 ha] \right.$$
$$\left. - \lambda R[\tfrac{1}{2}\rho u^2 \pi R^2 C_l] \right|. \quad (17.16)$$

If the moment due to lift is less than that due to drag and the acceleration reaction, the leading edge of the cone is in tension; otherwise the trailing edge is in tension. The net moment is small unless h is large (in other words, unless the cone is tall and skinny). In Chapter 13 we derived a formula for calculating the stresses resulting from a bending moment:

$$\sigma = \frac{Mz}{I}, \quad (13.9)$$

where M is the applied moment, y is the distance from the neutral surface, and I is the second moment of area. We can apply this formula here by assuming that the neutral surface runs through the center of the base. In this case the maximum stress occurs at the upstream and downstream edges of the cone, a distance R from the neutral surface, and

$$\sigma = \frac{MR}{I}. \quad (17.17)$$

The lift force tends to put the entire base of the cone in tension, and this tends to shift the neutral axis (the surface where neither compression nor tension are applied) downstream somewhat, but the effect is minor and we will not take it into account here. The second moment of area for a circular cross section such as that of the cone is

$$I = \frac{\pi R^4}{4}. \quad \text{(Table 13.1)}$$

Inserting these values into eq. 17.17, we arrive at an expression for the maximum tensile and compressive stresses imposed by the net bending moment:

$$\sigma = \left| \frac{4h^2}{3\pi} [\tfrac{1}{2}\rho u^2 C_d + \rho C_m Ra] - 2\lambda \rho u^2 C_l \right|. \quad (17.18)$$

All the terms contributing to the bending stress are independent of size (R) except for the term due to the acceleration reaction, which increases directly with the radius of the cone. Thus the larger the cone, the larger the bending stress, all other factors staying constant.

Most rigid wave-swept organisms such as barnacles and limpets are likely to be much stronger in compression than in tension, and it seems reasonable to assume that if a bending stress is placed on such an organism it will give way first on the side placed in tension. The tensile stress at the leading or trailing edge is directly augmented by the lift force. If the cone is rigid, the stress produced in the basal adhesive by the lift force is evenly distributed across the base even though the lift force itself is not evenly distributed. Thus the tensile stress caused by the net bending moment is increased by an amount equal to the average tensile stress applied by the lift force (life force/basal area). Taking this into account, we arrive at our final expression for the maximum tensile force acting on the cone:

$$\sigma = \left| \frac{4h}{3\pi} [\tfrac{1}{2}\rho u^2 h C_d + \rho C_m Rha] - 4\lambda[\tfrac{1}{2}\rho u^2 C_l] \right|$$
$$+ \tfrac{1}{2}\rho u^2 C_l. \quad (17.19)$$

We now have expressions relating the water flow (u, a) to the tensile and shear stress as a function of the size and shape of the cone. To apply these expressions, we use values for the lift, drag, and added mass coefficients and for λ as given in Table 17.1.

Table 17.1
Force Coefficients for Cones

H/D	C_d	C_i	C_m	λ
0.2	0.360	0.246	1.68	0.242
0.5	0.405	0.223	1.68	0.303
0.8	0.450	0.200	1.68	0.364

Notes: Inertia coefficient is that of a cone-shaped impet (Denny et al. 1985). In all likelihood C_m varies directly with H/D. All remaining data from Ullensvang, unpublished.

Figure 17.4 shows tensile and shear stresses as a function of radius for a standard water velocity of 10 m/s and an acceleration of 100 m/s². Values are given parametrically for different ratios of height to radius ($H/R = h$). The tensile stresses are larger than the shear stresses, irrespective of h. For cones with a low value of h, tensile stress is high because of a high value of C_l and a low value of C_d, and the tensile stress is imposed primarily by the lift stress. The stress imposed on tall, skinny cones is dominated by the bending stress due to the long moment arm for drag and acceleration reaction.

Figure 17.4. Shear stresses (eq. 17.14) and tensile stresses (eq. 17.19) as a function of cone radius for cones of different ratios of height to diameter (h). Calculated assuming $u = 10$ m/s and $a = 100$ m/s². Force coefficients taken from Table 17.1.

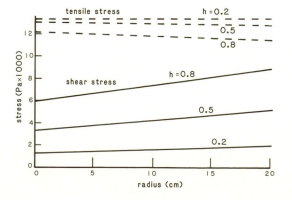

Tensile stress decreases with increasing h as the moment due to drag and acceleration increases and approaches the value of the moment due to lift. For $H/R = 0.8$ the tensile stress decreases as the radius increases, due to the same effect. At larger sizes or faster accelerations, the moment due to drag and acceleration reaction would exceed the moment due to lift, and tensile stress would increase with increasing radius.

The fact that the tensile stresses are larger than the shear stresses does not necessarily mean that a conical organism would fail in tension. We noted in Chapter 15 that many adhesive systems are much weaker in shear than in tension. Thus the mode of failure caused by the stresses analyzed here must be determined for each particular organism.

If we know the strength of the basal adhesive for our cone, we can calculate a rough estimate of the maximum size to which the shape can grow without being dislodged by water moving with a certain velocity and acceleration. Consider an example. A typical shear strength for crawling gastropods is on the order of 10^4 Pa (Miller 1974a), and typical values for C_d, C_m, and h are 0.48, 1.7, and 1, respectively. Inserting these values into eq. 17.14, we have

$$10^4 \text{ Pa} = \frac{1}{2\pi}(1025 \text{ kg/m}^3)u^2(1.0)(0.48)$$

$$+ \frac{1}{3\pi}(1025 \text{ kg/m}^3)(1.7)R(1.0)\frac{du}{dt}.$$

(17.20)

Using values of 10 m/s for u and 100 m/s² for du/dt yields $R = 14$ cm. This then is a rough estimate of the maximum size to which a conical gastropod can grow with this particular strength of basal adhesive before it would be dislodged in shear. Higher values for u and du/dt would result in a lower estimated maximum size.

The tensile strength of crawling gastropods is typically $1-5 \cdot 10^4$ Pa (Miller 1974a). Using a value of $2.5 \cdot 10^4$ Pa and the same values as above for velocity and acceleration, and inserting the appropriate force coefficients into eq. 17.19 ($C_m = 1.7$, $C_d = 0.48$, $C_l = 0.17$ for $h = 1$), we calculate that a conical gastropod would be limited to $R < 21$ cm if they were not to be dislodged in tension. Again, higher values for u and du/dt would lead to a smaller estimated maximum size.

The values calculated here for maximum attainable size are in the right ballpark for the organisms found in the wave-swept environment. Purely on the basis of mechanical factors, we can provide an argument as to why limpets the size of cattle do not exist, and why predatory snails are smaller than lions.

This sort of mechanical argument is fine as far as it goes, but we have reached the end of the line for the argument in its present state. To predict a specific maximum size to which an organism can grow, we must be able to specify exactly the maximum velocity and acceleration the organism encounters and the strength of its basal adhesive. As shown in Chapter 16, these sorts of exact specifications are not possible in the real world. How then can we apply the theory outlined here?

We first deal with the variability in tenacity and follow the same sort of reasoning used in calculating exposure. Although we cannot a priori specify the tenacity of a given organism, we can gain information about an organism's tenacity in a probabilistic sense. By dislodging a number of randomly chosen members of a species, we estimate the mean tenacity for the population, and the standard deviation about that mean. Assuming that tenacity is normally distributed, we can then calculate the probability that an organism chosen at random has a certain tenacity. The probability that an organism will be dislodged by a stress σ is then

$$P_d = \int_0^\sigma \left\{ \frac{1}{\mathscr{S}\sqrt{2\pi}} \right\} \exp - \left\{ \frac{[\sigma' - \bar{\sigma}_b]^2}{2\mathscr{S}^2} \right\} d\sigma'.$$

(17.21a)

Or, for shear forces

$$P_d = \int_0^\tau \left\{ \frac{1}{\mathscr{S}\sqrt{2\pi}} \right\} \exp - \left\{ \frac{[\tau' - \bar{\tau}_b]^2}{2\mathscr{S}^2} \right\} d\tau',$$

(17.21b)

where σ and τ are functions of u, du/dt, and size as determined by eq. 17.14 or 17.19. For given values of u and du/dt, this risk (P_d) can be calculated as a function of size (Fig. 17.5). The slope of this curve at any point is the change in the probability of dislodgement with a change in size, a handy measurement of the risk of growing larger.

It is not necessary to construct this sort of graph to measure the effects of growth; the slope dP_d/dR can be calculated directly. First, we take advantage of a fundamental rule of differentiation and note that

$$\frac{dP_d}{dR} = \frac{dP_d}{d\sigma} \cdot \frac{d\sigma}{dR}.$$

(17.22)

We already have expressions for P_d (eqs. 17.21) and can take their derivative with respect to stress:

$$\frac{dP_d}{d\sigma} = \left\{ \frac{1}{\mathscr{S}\sqrt{2\pi}} \right\} \exp - \left\{ \frac{[\sigma - \bar{\sigma}_b]^2}{2\mathscr{S}^2} \right\}$$

(17.23a)

$$\frac{dP_d}{d\tau} = \left\{ \frac{1}{\mathscr{S}\sqrt{2\pi}} \right\} \exp - \left\{ \frac{[\tau - \bar{\tau}_b]^2}{2\mathscr{S}^2} \right\}.$$

(17.23b)

Similarly, we already have an expression for the applied stress (eq. 17.14 or 17.19), and we can take its derivative with respect to R:

$$\frac{d\sigma}{dR} = \frac{4}{3\pi} \rho C_m h^2 \frac{du}{dt}$$

(17.24a)

$$\frac{d\tau}{dR} = \frac{1}{3\pi} \rho C_m h \frac{du}{dt}$$

(17.24b)

Multiplying eqs. 17.23 by eqs. 17.24 yields the desired equations for the change in the probability of dislodgement as a function of change in size:

$$\frac{dP_d}{dR} = \frac{4}{3\pi}\,\rho C_m h^2 \frac{du}{dt}\int_0^\sigma \left\{\frac{1}{\mathscr{S}\sqrt{2\pi}}\right\}$$

$$\cdot \exp - \left\{\frac{[\sigma' - \bar{\sigma}_b]^2}{2\mathscr{S}^2}\right\} d\sigma' \quad (17.25a)$$

$$\frac{dP_d}{dR} = \frac{1}{3\pi}\,\rho C_m h \frac{du}{dt}\int_0^\tau \left\{\frac{1}{\mathscr{S}\sqrt{2\pi}}\right\}$$

$$\cdot \exp - \left\{\frac{[\tau' - \bar{\tau}_b]^2}{2\mathscr{S}^2}\right\} d\tau'. \quad (17.25b)$$

These expressions give dP_d/dR values for an absolute change in size. Of more interest in the present context is how much the probability of dislodgement changes for a relative change in length. For example, it is more useful to think about how much greater risk an organism incurs by doubling its present length rather that how much its risk is increased by growing a fixed amount. Thus we want to know $dP_d/(dR/R)$, the change in probability per change in relative length. Now, $dP_d/(dR/R) = R \cdot dP_d/dR$, so to arrive at this value we multiply eqs. 17.25 by R:

$$R\frac{dP_d}{dR} = \frac{4R}{3\pi}\,\rho C_m h^2 \frac{du}{dt}\int_0^\sigma \left\{\frac{1}{\mathscr{S}\sqrt{2\pi}}\right\}$$

$$\cdot \exp - \left\{\frac{[\sigma' - \bar{\sigma}_b]^2}{2\mathscr{S}^2}\right\} d\sigma' \quad (17.26a)$$

$$R\frac{dP_d}{dR} = \frac{R}{3\pi}\,\rho C_m h \frac{du}{dt}\int_0^\tau \left\{\frac{1}{\mathscr{S}\sqrt{2\pi}}\right\}$$

$$\cdot \exp - \left\{\frac{[\tau' - \bar{\tau}_b]^2}{2\mathscr{S}^2}\right\} d\tau' \quad (17.26b)$$

This is the *size-specific increment in risk*.

Denny et al. (1985) carried out this sort of analysis for a variety of wave-swept organisms, and we examine the case of a sea urchin in an attempt to make some biological sense of the values we have calculated.

The purple sea urchins of the Pacific Northwest (*Strongylocentotus purpuratus*) live on both exposed and protected shores. On the most exposed shores the urchins usually confine themselves to cuplike burrows that they wear into the rock. In these situations the urchins are not ex-

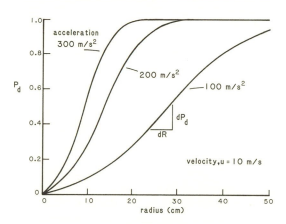

Figure 17.5. The probability of dislodgement as a function of radius for three different water accelerations (eq. 17.21). A tenacity of 10^4 Pa ($\mathscr{S} = 2 \cdot 10^3$ Pa) is assumed. $h = 0.8$. Force coefficients taken from Table 17.1.

posed to the sort of flow patterns that have been assumed in this analysis. In semiexposed areas, such as on gently sloping intertidal shelves, the wave-induced water motion can be substantial, but the animals do not confine themselves to burrows. In this situation, the calculations made here can be applied.

Values for P_d as a function of water velocity and acceleration, and $R \cdot dP_d/dR$ as a function of R, are shown in Figure 17.6. The first thing to note is the variation in P_d with acceleration. The probability of dislodgement increases substantially as the acceleration increases. This curve was calculated assuming R equal to the maximum value observed among those urchins sampled. Smaller values of R would have led to a smaller dependence on acceleration. This size dependence is better shown in Figure 17.6b, where $R \cdot dP_d/dR$ is graphed as a function of size. The proportional change in risk increases with a change in size up to sizes well above those observed in nature. The bigger the animal is, the greater is the increase in risk it incurs by getting even larger. At very large sizes the size-specific increment in risk reaches a peak and declines, but this effect is more apparent than real. At these large sizes, it

Figure 17.6. Probability of dislodgement, P_d, in the purple sea urchin *Strongylocentrotus purpuratus*. (a) P_d as a function of acceleration and velocity for an urchin of the size found in nature (test diameter = 7 cm). (b) The size-specific increment in risk for *S. purpuratus* as a function of size. (Redrawn from Denny et al. 1985 by permission of the Ecological Society of America.)

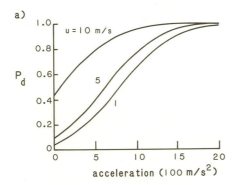

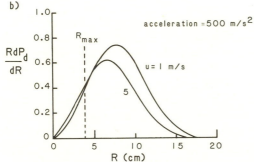

is a near certainty that the organism will be dislodged. If the animal is almost sure to be dislodged at its present size, it cannot increase this probability much by growing, so $R \cdot dP_d/dR$ must decrease. However, if the animal is almost certain to be dislodged, it has little biological viability. Thus, for all practical purposes, the size-specific increment in risk is a monotonically increasing function of size.

These calculations provide values of P_d and $R \cdot dP_d/dR$ for an urchin encountering one wave with specific values of u and du/dt. These are interesting mechanical data but have little direct biological significance. Are these values for $R \cdot dP_d/dR$ large enough to limit effectively the size of urchins? Unfortunately, this is not an easy

question to answer definitely. In evolutionary terms, we are really asking whether the increase in risk of dislodgement with an increase in size is sufficient to lower the organism's fitness. To answer this question, we must relate the size-specific increment in risk ($R \cdot dP_d/dR$) to the organism's reproductive output. To do this in any precise sense requires more demographic data than are available, but given a few simplifying assumptions we can come to some reasonable conclusions.

First, we assume that reproductive output is directly proportional to an organism's volume at reproduction. Doubling the size of an organism (measured as a linear dimension) thus results in an eightfold increase in reproductive output. We next assume that the organism produces young only at discrete intervals. In other words, to reproduce at all the organism must survive a specified period of time. For the sake of simplicity, in the examples that follow we assume that this period of time is one year. This assumption is a fair approximation of the actual reproductive cycle of many species—barnacles, urchins, etc.—but it is far from being an accurate analogy for other species that reproduce more or less continuously. Given these simplifying assumptions, we need only calculate the survivorships for organisms of different sizes and compare these to the organism's calculated reproductive output.

We calculate survivorship in much the same fashion as we did in Chapter 16, but in this case the effects of the acceleration reaction are taken into account when calculating P_d (eq. 17.21). To do this in any precise way would require a knowledge of the distributions of accelerations that occur at a wave-swept site and a correlation between the water velocity and acceleration at the time when a maximum force is being exerted. Unfortunately, not enough empirical measurements of wave forces have been made to allow a search for such a correlation. Rather than guess at a correlation, we assume that a certain value of acceleration is reached in each and every wave. This assumption allows us to study the effects of size on survivorship, but it is clearly not an accurate estimate of what happens in the real wave-swept environment. As a result, the survivorship curve we calculate will be correct only in a relative sense, and should not be taken as an accurate indication of the survivorship of organisms in the real world.

Survivorship curves calculated in this fashion depend on the size of the organism, and we want to know whether the size we observe in nature is optimal. In other words, we want to know whether sizes above or below the maximum observed in nature have a lower reproductive output than the size we actually see. For instance, an organism twice the maximum size we observe can produce eight times as many eggs. If this larger organism has better than one-eighth the probability of surviving, its effective reproductive output is greater than that of the organism we observe, and the observed size is less than optimal. Conversely, an organism half the maximum size we observe should have less than an eightfold greater survivorship.

Denny et al. (1985) calculated these relationships for a variety of wave-swept organisms. Two examples are shown in Figure 17.7. At the end of a year an urchin of the maximum size observed is predicted to have a 41-fold greater survivorship than an urchin twice its size (Fig. 17.7a). The realized reproductive output (volume · probability of survival) is thus five times greater at the maximum size observed than at twice this size. Clearly it is not advantageous for the urchin to get larger in this situation. An urchin half the maximum size observed has a realized reproductive output only about 50% of that for the observed size, and it would be advantageous for an urchin of this smaller size to grow. Thus the optimal size (the size with the greatest realized reproductive output) is somewhere near the size observed in nature.

Figure 17.7. (a) Calculated relative survivorship for the purple sea urchin *S. purpuratus*. An urchin half the maximum size found in nature ($R/2$) has a 3.8-fold better chance of surviving to reproduce than an urchin of the maximum size found in nature (R). An urchin twice the size found in nature ($2R$) has a 41-fold smaller chance of surviving to reproduce. (b) Calculated relative survivorships for the acorn barnacle *Semibalanus cariosus*. Even a barnacle twice the maximum size found in nature ($2R$) has a better than 90% chance of surviving to reproduce. R = maximum size found in nature. (Redrawn from Denny et al. 1985 by permission of the Ecological Society of America.)

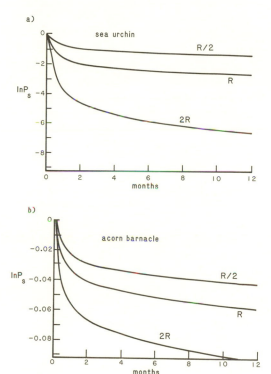

The barnacle *Semibalanus cariosus* is in clear contrast to the urchins (Figure 17.7b). Barnacles twice the size of those observed at an exposed site are predicted to have better than 90% survivorship at the end of a year. Obviously their survivorship cannot be eightfold less than that of smaller individuals, and as a result their realized reproductive output would be much larger than that calculated for the maximum observed size. The high survivorship predicted for these barnacles is due both to their high average tenacity ($3.3 \cdot 10^5$ Pa) and their small size relative to organisms such as urchins. In this case it is reasonably clear that the effects of the acceleration reaction do not act as a limit to size. Perhaps the size of barnacles is governed by the energy-balance problems discussed earlier.

The results for other organisms are less clear-cut. Solitary mussels and some limpets are likely to have their size limited by the effects of the acceleration reaction. A blade-shaped fire coral was found to grow to a size quite a bit larger than the optimum predicted by theory, but in this case there is a possibility that a broken coral can re-attach and grow somewhere else on the reef. Because breakage may serve as a dispersal mechanism for this species, the evolutionary advantages of dispersal would have to be weighed against the disadvantages of breakage before any firm conclusion could be drawn regarding the mechanism of size limitation.

All of these conclusions have been drawn on the basis of relative survivorship. It would be much better to have conclusions based on more accurate calculations, but it will be some time before these calculations are possible. For instance, accurate calculations can be made only once the relationship is found between wave height and the water acceleration occurring in the intertidal zone. The conclusions drawn here also assume that the rms wave height and the size of the organism remain constant throughout the year. Clearly this is a gross oversimplification. And, finally, we need to account for demography. As already mentioned, it may be advantageous for an organism to reproduce at a size less than the optimum predicted here if it saves sufficient time in doing so. A few progeny produced early and maturing in a month or two may produce more young in a year than a single large adult that requires an entire year to mature.

Mechanical Determinants of Shape

Corals

In Chapter 13 we briefly examined the optimal shape for beams under various sorts of loads. By assuming that a beamlike organism grows in a manner that minimizes the outlay of structural materials, we were able to predict what shape would be the best for each type of loading—point loading and loading from drag and the acceleration reaction. However, in the wave-swept environment it would be very unusual to find one or another of these loading regimes acting in the absence of others. For example, a coral blade is continually subjected to both the water's velocity and acceleration. What is the optimal shape of the beam in this case? It is also likely that the blade is periodically subjected to point loads from pieces of material suspended in the water. For a coral beam subjected to all three types of loads, what is the optimal shape?

These questions can be answered using the kind of computer program described in Chapter 13. First, we decide on the cross-sectional shape of the coral blade. We then gradually grow the blade by adding small blocks at its free end. Any combination of point load, drag force, and accelerational force can be applied to the beam. As new blocks are added, the dimensions of each older block are adjusted to maintain a constant maximum stress in the beam. When the beam has grown to some arbitrary length, its shape is described by the allometric law

$$z = \text{constant} \cdot x^\beta, \qquad (17.27)$$

where z is the depth of the beam parallel to the applied forces, and x is the distance from the free end. For a beam with a circular cross section, z is the radius at any point x. In Chapter 13 we concluded that for beams with circular or elliptical cross section, β is $\frac{1}{3}$ for point loads applied at the free end. For drag forces $\beta = 1$, and for accelerational loads $\beta = 2$.

Figure 17.8. The shape coefficient β for optimally shaped beams. (a) In steady flow. For small point loads or fast flows, the shape coefficient is approximately 1. Only in slow flows with large point loads does the shape coefficient approach that predicted for a point load alone (0.33; see text). (b) In flow accelerating at 500 m/s². Even with this rapid acceleration there is little effect of the acceleration reaction on the optimal shape (see text).

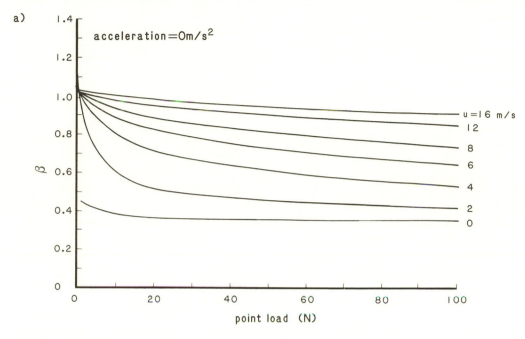

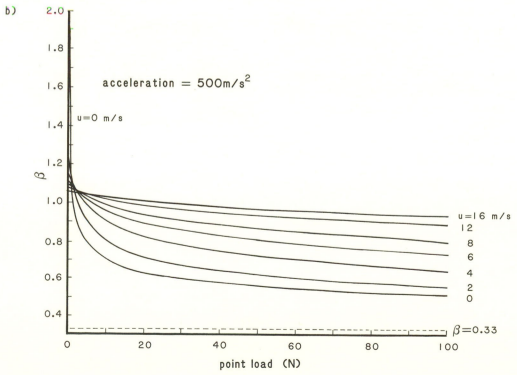

Figure 17.8 shows how the shape factor β varies as a function of the type of load applied. These calculations were made for a beam with a circular cross section and a breaking strength of 10^7 Pa. When no accelerational load is applied (Fig. 17.8a), β varies between 0.33 and 1, as might be expected. For small point loads and rapid water velocities, the shape approaches that predicted for drag forces alone. At low water velocities and large point loads, the shape approximates that predicted for point loads alone.

Surprisingly, when acceleration is included, the variation in β is not drastically changed. Figure 17.8b was calculated using a water acceleration of 500 m/s^2, a value that is probably very high for the subtidal environment in which corals are found. Only in still water and at minuscule point loads (< 1 N) does β approach 2. Otherwise the only effect of the acceleration is to shift β to higher values than obtained in the absence of acceleration, and the shift is usually less than 0.2.

These findings reinforce the general conclusion reached in Chapter 13. Point loads and drag forces are much more potent than accelerational loads in determining optimal shape for biological cantilevers. Given the likelihood that any coral blade will be subjected to both drag and point loads, we would expect to find most corals growing with a β between 1/3 and 1. This prediction seems to be borne out in nature.

The Shape of Limpets

We now return to a subject we briefly touched upon in Chapter 11—the optimum shape of wave-swept organisms. At that time we tentatively concluded that a shallow cone or section of a sphere with a height to diameter ratio of about 0.2 had the least drag per enclosed body volume. However, before we can decide on an optimal shape we need to take into account not only the drag, but the lift and the acceleration reaction as well. The methods

outlined in this chapter provide us with a mechanism for doing this. Unfortunately, we only have sufficient information to make these calculations for one shape—a cone. But this shape has some biological relevance because it approximates the shape of many limpets. By searching for the optimal cone shape, we may gain information about why limpets have the shapes they do.

Figure 17.9 gives values of C_d, C_l and the location of the center of pressure (λ) for cones of different height-to-diameter ratios (Ullenszvang, unpublished). C_m has not been determined for these shapes, so we must make our usual assumption that the acceleration reaction is small compared to the lift and drag. We can readily see that there is a trade-off between lift and drag.

Figure 17.9. Force coefficients (C_d and C_l) and the location of the center of pressure (λ) as a function of the ratio of height to diameter (H/D) for cones. Solid lines are regressions. Values measured at Re = $7.6 \cdot 10^4$. (Data from Ullensvang, unpublished.)

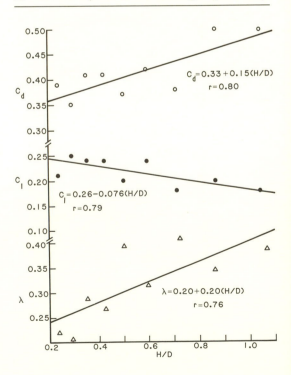

Figure 17.10. Tensile stress per volume and shear stress per volume for different ratios of cone height to diameter (H/D). There is a distinct minimum in tensile stress per volume at $H/D = 0.7$. Calculations based on the coefficients shown in Figure 17.10. (Data from Ullensvang, unpublished.)

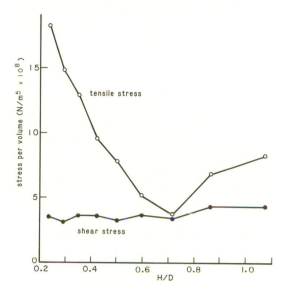

Tall cones have high drag coefficients but low lift coefficients, and for low cones it is just the reverse. We are immediately led to suspect that a cone of some intermediate peakedness experiences the least force per volume. To be precise, the measures used here in determining the optimal shape are tensile stress per body volume and shear stress per body volume.

Tensile stress is calculated by inserting the values for C_d, C_l, and λ (Fig. 17.9) into eq. 17.19. Stress is then divided by the volume of each cone (Fig. 17.10). There is a definite minimum in tensile stress per volume for cones with a height/diameter of about 0.7. Average shear stress is calculated by inserting the values for C_d (Fig. 17.9) into eq. 17.14. Shear stress is then divided by the volume of each cone (Fig. 17.10). The shear stress is much lower than the tensile stress, and varies less with peakedness.

On the basis of this analysis we can tentatively conclude that a cone with a height-to-diameter ratio of 0.6–0.7 is optimal in that it exhibits the lowest stress for the volume it contains. In biological terms, an organism of this shape would have the maximum ratio of gonad volume to risk of dislodgement, and we predict that this shape will be favored in nature.

The shapes of actual limpets are reasonably consistent with this prediction. Because most limpet shells do not have a circular aperture, they have two measurements analogous to the height-to-diameter ratio of a cone—height to length (H/L) and height to width (H/W). On the west coast of North America, limpets range in shape from very low-peaked (e.g., *Notoacmea scutum*, $H/D = 0.2$, $H/W = 0.3$) to very high peaked (e.g., *Acmaea mitra*, $H/D = 0.7$, $H/W = 0.8$). For most limpets, $H/L = 0.4–0.5$ and $H/W = 0.5–0.6$. These values are slightly lower than we predict, but not excessively so. Perhaps when the appropriate lift and drag measurements are made on actual limpet shells, the prediction of the optimum shape will be closer to that actually observed. In any case, the calculations made here help explain why all limpets are not shaped like very shallow cones.

CHAPTER 18

~ ~ ~ ~ ~ ~ ~ ~ ~ ~ ~

Whither Hence?

Our exploration now draws to a close. The previous chapters have presented a particular perspective on the study of wave-swept organisms, providing an introduction to the tools through which this perspective can be explored. Though we have not yet explored all the possible questions and answers regarding the mechanics of the wave-swept environment, we have run out of data on which to base useful speculation. It is tempting to extend this exploration out of the realm of science and into the world of "fun with numbers," but we should resist this temptation. Before the examination of mechanics in the wave-swept environment can fruitfully proceed, we need more basic information about the biology of the organisms that live there.

This lack of information applies virtually to all aspects of biology in the wave-swept environment, but three areas of research cry out for particular attention.

Animal Behavior

We have been dealing almost entirely with the structural consequences of wave-induced water motion, and we have largely ignored the possibility that an organism can respond to the forces imposed on it and act to save itself. Although this assumption is valid when applied to plants and some sessile animals (at least on a time scale of single waves), it is an oversimplification when applied to mobile animals. Even the lowliest invertebrate is capable of sensing the nature of its environment and responding appropriately. Therefore, if we are to account fully for the effects of wave-induced water motion on organisms, we must examine the behavioral response of animals.

These behavioral responses are likely to have wide-reaching effects on community structure. Time and again ecological studies have documented the importance of grazing and predation on the species composition and diversity of communities. If grazing and predation are themselves affected by water motion, it may well be water motion that is the ultimate controlling factor in community structure. This idea is certainly not new—it is a basic reason for classifying wave-swept communities by their "wave exposure." Despite its venerability, the idea that water motion affects wave-swept community structure by affecting foraging behavior has been the subject of very little research.

These aspects of behavior have not been studied for some obvious reasons. It is difficult to observe the behavior of organisms in the swash zone or the shallow subtidal zone, and these observations are most difficult just when they are most interesting—when the wave forces are largest. However, these difficulties are not insurmountable. If the experimental site is appropriately chosen, the behavior of intertidal organisms can be observed through a telescope. Alternatively, a television camera in a waterproof housing can be installed to observe subtidal behavior. Such observations, coupled with the measurement of water velocities, hydrodynamic force coefficients, and adhesive tenacities as outlined here may shed important light on the role of wave-induced forces in controlling behavior, and thereby community structure.

Large-Scale Mixing

In Chapter 10 we examined the role of turbulent mixing in two aspects of biology—larval settlement and external fertilization. In both cases we were concerned with mixing on a fine spatial scale, centimeters to meters. The effects of mixing clearly extend to larger scales. A prime example is the transport of planktonic larvae during dispersal. Several orderly processes may account for the offshore and subsequent onshore movement of dispersing larvae (e.g., transport of water by onshore winds, mass transport by surface waves, transport accompanying internal waves; see Shanks 1983, 1986). However, this transport is likely to be strongly affected by turbulent mixing where the scale of eddies is measured in meters or tens of meters. For instance, a larva will be carried inshore by the wave-induced surface mass transport only if it can maintain its position near the surface. This may be difficult or impossible if the vertical turbulent mixing is intense. These mixing processes have not been well studied. It is exciting to anticipate studies that tie together the fine-scale mixing of the surf zone and nearshore with the very large-scale mixing of oceanic currents and gyres.

Material Properties

Our knowledge of the materials from which wave-swept organisms are constructed is rudimentary at best. We know a great deal about the structure and properties of mollusc shell (for a review, see Currey 1980).

We know a reasonable amount about the chemistry and properties of barnacle glue (Walker 1981), mussel byssal threads (Waite 1983), limpet pedal mucus (Grenon and Walker 1980), and sea-anemone mesoglea (Gosline 1971; Koehl 1977c). Beyond this our information is generally confined to measurements of the modulus and breaking strength of materials.

What is the macromolecular architecture of barnacle glue, and how does it allow this material to act as such a powerful, long-lasting adhesive? What materials do kelps use to adhere to the substratum, and how are the materials constructed? How does the body wall of sea stars manage to be so flexible at certain times and so rigid at others? The tubefeet of *Pisaster brevispinus* can extend by nearly tenfold. How does the collagen of their connective tissue accommodate this extension and still elastically recoil? A wealth of interesting material is found in the wave-swept environment and undoubtedly much practical information can be gained from its study.

This short list of areas requiring further research is by no means complete. Certainly much remains to be done in measuring the life-history parameters of wave-swept organisms, quantifying the behavior of larvae as they settle in turbulent flow, explaining the feeding mechanisms of sessile animals, and examining the role of longshore currents in the dispersal of larvae. Go down to the shore with an open mind, and who knows what interesting ideas will demand your attention.

CHAPTER 19

~ ~ ~ ~ ~ ~ ~ ~ ~ ~

Techniques of Measurement

In the past eighteen chapters we have discussed in detail the nature of the wave-swept environment and its biological consequences. In many places we have assumed that a certain kind of measurement can be accurately made. If we want to know the water velocity in a certain habitat, we grab our trusty, all-purpose velocity meter and go out and measure it. If we want to know the drag coefficient of a certain animal, we stick the creature in our drag meter, and voilà! This cavalier attitude toward the measurement of flows and forces, though it may be handy for discussing theory, can be dangerous in practice. The accurate measurement of flows and forces is an exacting business and demands our attention.

Field Measurements

As we have seen (Chapters 4–11), it is not easy to describe flow patterns and the forces they cause. Different aspects of flow are important to different aspects of the wave-swept environment. For instance, if we were to examine longshore currents, the average flow parallel to the shore would be of interest, and we would ignore the oscillations accompanying waves. In contrast, if we were to examine turbulent mixing, we might well want to ignore the average flow and deal only with the fluctuations in velocity. Our choice of an appropriate technique depends on which aspects of flow we want to measure.

One approach to this sort of measurement problem is to choose the technique that provides the most complete and accurate measurement available, and, after the measurements are made, to throw out

any surplus information. For example, if we had a velocity meter that accurately measured the instantaneous flow velocity, we could easily compute the average velocity. Why not use this meter in all situations? The primary reason is the cost. The more information we want, the more it costs to measure it, and the slope of the cost versus information curve is steep enough to discourage overkill in measurement technique. For example, it costs less than $100 to build a device capable of measuring the average flow in one direction. To measure velocity fluctuations in two dimensions with a temporal resolution of around 50 ms costs at least $5,000. To measure velocity fluctuations in three dimensions with a temporal resolution of less than 1 ms costs well in excess of $20,000. Few scientists can spend $20,000 on a general-purpose device when they plan to throw away most of the information obtained.

Thus we are left with the problem of making the most accurate measurements at the least possible cost. We will start with the simplest and least expensive methods and will work our way up the ladder of technology and price.

Measurement of Net Flow

In some situations we may want to know only how much water has passed a given point in a certain period of time irrespective of the direction of flow, the maximum velocity, or the fluctuation in velocity. A simple and inexpensive technique is available for such measurements. As we have seen (Chapter 10), the rate of exchange between a solid surface and the surrounding fluid is largely a function of turbulent mixing. To a certain extent, the faster the average main-stream velocity, the larger u_*, and the more effective the mixing. Thus, the rate of exchange at a surface is a rough measure of the average mainstream flow. Muus (1968) used this idea as the basis for a method of measuring net flux. He noted that plaster of Paris is slightly soluble in seawater; the rate at which plaster dissolves should thus depend on the rate at which water is moving over the plaster surface. This idea is transformed into a practical measurement technique by molding plaster spheres in an ice-cube tray. The head of a nail is embedded in each sphere to serve as a means of anchoring the sphere in the environment. Spheres are weighed before they are placed in the environment, and are dried and re-weighed after they are recovered a day or two later. The loss in weight is a measure of the net flux of water past the sphere. By placing spheres in known water velocities in the laboratory, we can determine the rate of plaster dissolution as a function of velocity and calibrate the measurements.

This method is elegantly simple, and Muus used it with some success in measuring fluxes on a sand flat. There are, however, practical problems. Unless the plaster is cast as a radially symmetrical shape (e.g., a sphere or cylinder), the shape of the cast affects the rate of dissolution. Casting plaster in a spherical or cylindrical mold may sound easy, but it takes a certain knack that is only acquired after many trials and much frustration, and it may be best to work with precast, commercially available objects such as carpenter's chalk (M. LaBarbera, pers. comm.) or gypsum disks (Santschi et al. 1983). Finally, the rate of dissolution is temperature-dependent. If the water temperature in the environment varies during the course of a measurement, a correction factor must be applied to the observed weight loss, and this correction factor must be determined from calibrations made in the laboratory.

The need for multiple calibrations makes this method very time consuming, and at best we cannot count on an accuracy better than $\pm$ 10%–20%. The elegance of the basic idea loses some of its glamor in the light of practical matters. However, for situations in which gross differences in net flux are suspected (so that a 10%–20% error in measurement would be negligible), this method may be the best choice.

An innovative variation on this measurement scheme has been used by Drs. M.A.R. Koehl and T. Daniel (pers. comm.) to get an estimate of the relative rates of mixing at various points on a kelp frond. They sewed peppermint Lifesaver candies in a grid on the frond, and measured the weight loss of the candies after immersion. The rate of dissolution was calibrated in a flow tank with a known mean velocity and Reynolds shear stress. A similar technique utilizing plaster buttons has been used by Gerard (1982).

Measurement of Maximum Water Velocity

When we examine structural wave exposure, we are much more interested in the maximum water velocity occurring in an environment than in the average water velocity. Several schemes have been devised to measure maximum water velocity.

The simplest of these was introduced by Jones and Demetropoulos (1968), who modified a simple spring scale to record the maximum force exerted on a small "drogue disk" (Fig. 19.1). The faster the water velocity, the more force felt by the disk, and the farther the spring scale is stretched. The maximum excursion of the scale's indicator is recorded by a lever. The spring scale itself is enclosed in a short length of plastic pipe and attached to the substratum by a bolt and swivel fitting. The apparatus can be left in the field for extended periods of time (days to weeks) before it is read, and resetting the apparatus is simple and quick. If the drag coefficient of the drogue disk is known, the force measurements obtained can be

Figure 19.1. A recording spring scale for measuring maximum velocity. (Redrawn from Jones and Demetropoulos 1968 by permission of Elsevier Biomedical Press.)

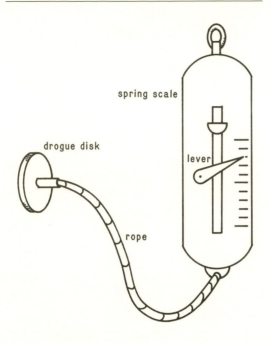

used to estimate maximum water velocity. As discussed in Chapter 11, the drag coefficient of a thin disk is about 1.2 and is independent of Reynolds number over the range encountered in the wave-swept environment. Jones and Demetropoulos used this device to measure maximum water velocities on the exposed coast of Wales, and they correlated their findings with the distribution of macroalgae.

This device is highly satisfactory if its limitations are kept in mind. The primary limitation is one of response time. In order for the spring scale to respond to the force on the drogue disk, the entire apparatus must be aligned with the flow. If the device has just aligned itself with the offshore ebb of a wave, it must swivel 180° to measure the force accompanying the next shoreward upsurge. This reorientation requires time. For instance, if the total length of the device (from bolt to drogue disk) is 0.5 m and the mainstream

velocity in the upsurge is 10 m/s, the device requires at least 50 ms to reorient. Slower flows would result in a longer reorientation period. The response time could be minimized by placing the drogue disk on a short tether, but the drogue should not be too close to the spring scale lest the force it feels be confounded by the altered flow in the scale's wake. In addition to the response time due to reorientation, there is a response time associated with the stretching of the spring in the scale. As a result, the overall response time of the device is likely to be on the order of 0.1 s. This is far too slow to record transient maxima in flow velocity (see Fig. 7.7), and this apparatus probably underestimates the maximum velocity. However, this is a problem only if we are interested in the maximum velocity as experienced by an organism that is "stiffer" than this device. A very stiff organism such as a barnacle or coral "feels" the instantaneous force exerted by rapidly fluctuating velocities (see Chapter 14), and this device would be inappropriate for measuring these velocities. However, the device is a reasonable approximation of the sturcture of a macroalga, and the forces it records would thus be a good approximation of the "effective" maximum velocities that such an alga would experience.

Denny (1983a) has constructed a device with a faster response time than that of Jones and Demetropoulos (Fig. 19.2). This apparatus has the added advantage of recording the direction from which the maximum force is exerted. A Teflon slider is sandwiched between two parallel plastic plates in a circular housing. The top plate has a hole drilled in its center and a small sphere or cylinder is bolted to the slider through this hole. Any fluid-dynamic shear force exerted on the sphere or cylinder is transmitted to the slider. The slider is held centered by three rubber bands; the distance the slider moves in response to an applied force is proportional to that force and inversely proportional to the stiffness

Figure 19.2. (a), (b) A device for measuring the maximum force and direction exerted on an intertidal organism (see text). (a) Top view with top plate removed to show slider and rubber bands. Dashed circle shows the position of the hole in the lower plate through which the scriber extends. (b) Side view of a section made through *a-a* in (a). (c) Typical results gained using the device. Solid lines measured at a site 2.75 m above MLLW; dotted lines for a site 3.5 m above MLLW. (From Denny 1983, by permission of the American Society of Limnology and Oceanography.)

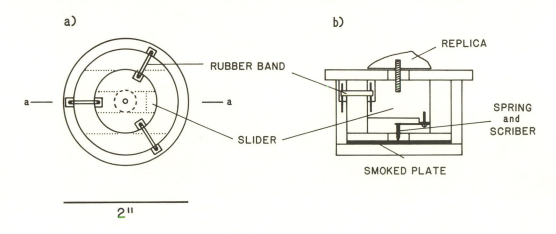

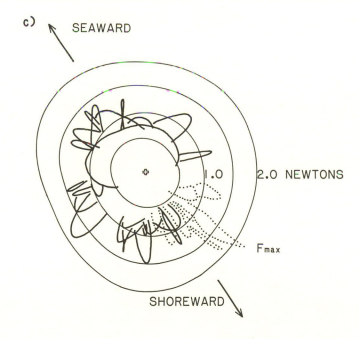

of the rubber bands. In practice, the stiffness of the rubber bands is adjusted so that the slider moves a measurable amount under the influence of forces applied by the environment, but it does not move far enough for the bolt to bump into the sides of the hole in the top plate. The slider is free to move in any direction between the parallel plates, so that the excursion of the slider is a measure not only of the magnitude of the applied force, but of its direction as well.

A small, spring-loaded scriber is attached to the base of the slider. The scriber extends through a hole in the base plate and writes on a smoked glass slide held below the plate (Fig. 19.2b). For use of the device in the field, an emplacement is chiseled into the substratum, and a plastic sleeve is cemented in place. The recording device fits snugly into the sleeve with the top plate flush with the surface of the substratum. A smoked glass slide is carefully installed in the device, and the device is bolted into its emplacement. The device is recovered a day or two later, and the record is photographed.

Calibration of the apparatus is straightforward. The device is placed on its side at the edge of a table, and a known weight is hung from a bolt attached to the slider. The device is then rolled along the table, resulting in a bull's-eye pattern corresponding to the directional excursion of the slider in response to the applied force (Fig. 19.2c) If the drag coefficient is known for the object attached to the device, the recorded force can be used to calculate the maximum velocity.

The response time of the device is controlled by the ratio of the mass of the moving parts (the Teflon slider and the attached sphere or cylinder) to the stiffness of the rubber bands. The response time of the device used by Denny was about 25 ms.

There are two problems with this apparatus. In order for the device to function properly, the slider must move freely between the parallel plates. If the fit is too tight, the friction between slider and plates causes inaccuracies in measurement. If the fit is too loose, the moment exerted on the slider by the attached object causes the slider to bind between the plates much like a drawer binds in its slides if it is not pulled straight out. This problem is avoided by careful construction. The top and bottom plates must be accurately parallel, and each slider must be ground by hand to fit precisely. Teflon and acrylic plastic expand and contract with variations in temperature, so the slider should be ground to fit at the temperature at which the device will be used. This is not difficult, but it does require some patience and a good machine shop. The second minor problem with the device concerns its sensitivity to temperature. The elastic modulus of rubber increases with temperature—about 0.3% per °C. The device should be calibrated at the temperature that will be encountered in the field.

Construction details and other considerations can be found in Denny (1983a).

Continuous Water Velocity Measurement

There are many occasions when a simple measurement of mean or maximum flow will not suffice. All turbulence measurements require information about the temporal fluctuations in velocity, and many applications regarding the distribution of flow forces, feeding effectiveness, and locomotion require that flow be measured continuously. Several techniques have been devised for the continuous measurement of flow; the technique to use depends on the precision required and the amount of money available.

Propeller Flow Meters. A propeller or rotor placed in moving fluid rotates with an angular velocity proportional to the speed of the fluid, and this simple fact has been used as the basis for many commercial flow meters. These meters range in size from those with propellers only 5–10 mm in diameter to large oceanographic meters with propellers 20–40 cm in diameter. Many schemes have been devised for recording the angular velocity of the propeller. In some meters the propeller drives a small generator, the voltage output being a measure of angular velocity. In others, the rotation of the propeller periodically brings a magnet into proximity with a reed switch or Hall-effect transistor. Each passage of the magnet thus completes a circuit, and the number of switch closures per time is a measure of angular velocity. Similarly, the blades of the propeller can be used to block the light falling on a sensor, generating an appropriate impulse in the sensing circuitry.

The choice among propeller meters depends largely on two factors: the spatial scale of the measurements to be made, and the response time. A propeller meter essentially averages the flow across its diameter. Thus, if you want to measure small scale flows, you need a small-diameter propeller. The response time of a propeller is largely a function of its rotational moment of inertia. The more massive the propeller and the larger the fraction of mass located a substantial distance away from the axis of rotation, the more sluggish the response. Thus small propellers made from a material with a low density have a fast response time. A massive propeller is slow to respond to flow and can effectively average out temporal fluctuations in velocity. Which of these alternatives is appropriate depends on which aspects of flow one wishes to measure. Measurement of rapidly fluctuating flows on a small spatial scale are generally better accomplished by other measurement schemes, and small-propeller flow meters have generally not proven useful in the wave-swept environment. Large-propeller flow meters have been a standard research tool for studying ocean currents for many years and have proven quite successful in

this capacity. Many commercial models are available.

An inexpensive propeller flow meter can be constructed from the type of paddle-wheel knot meter commonly used on sailboats (Fig. 19.3a). Typically each paddle of the wheel contains a magnet, and as the paddle passes the base of the meter it induces a current in a coil of wire. A simple circuit can be used to detect this signal and transduce it to a form suitable to be recorded on a tape recorder (Fig. 19.3b). Each time a paddle passes the coil, a "click" (actually a 1 ms duration square wave) is recorded. The number of clicks per time is a record of angular velocity and thereby of water velocity. A stereo tape recorder can record two channels of data simultaneously. This is useful when measuring two directional components of velocity in one site, or one directional component from two different sites. The taped records can be returned to the laboratory and played onto a chart recorder for analysis. This device is particularly convenient for measuring velocities in the surf zone in remote areas. It is simple to build, and inexpensive.

The primary drawback to this particular propeller velocity meter is its slow response time. Because each paddle contains a sizable magnet, the wheel has a large rotational moment of inertia. It takes the better part of a second to get the wheel up to speed, and a similar time for the wheel to slow down once removed from the flow. The device is thus useless for measuring rapidly fluctuating flows, but can be quite useful for measuring time-averaged flows. Note also that there is a minimum flow velocity below which the propeller will not reliably turn. For the meter shown here this minimum detectable velocity is about 20 cm/s.

Electromagnetic Flow Meters. While experimenting with a galvanometer in 1831, Michael Faraday noticed that he could induce an electric current (and hence a voltage) in a coil by moving the coil past a

Figure 19.3. (a) A schematic drawing of a propeller knot meter. (b) A simple circuit used to record the signal from a knot meter. IC1 (a 741 op amp) amplifies the signal from the knot meter. IC2 (a 741 op amp) acts as a comparator. When the output from IC1 goes above 9 V, IC2 drops the voltage at its output (pin 6), triggering the timer (IC3, a 555 timer). All resistors ($\frac{1}{4}$) W. Capacitance values in microfarads. (From Denny 1985b by permission of Cambridge University Press.)

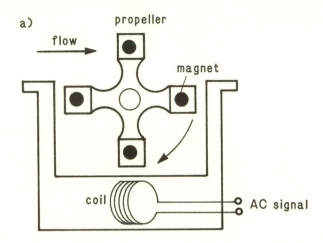

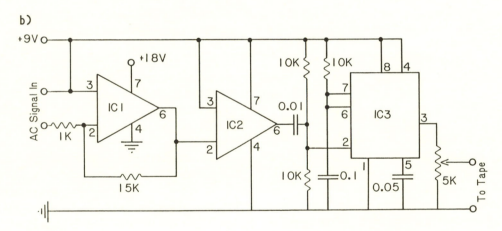

magnet. This discovery led to what is now known as Faraday's law of inductance, the principle behind electromagnetic flow meters. Seawater is a fairly good conductor of electricity, and if we allow it to flow through a magnetic field, we produce a voltage (Fig. 19.4a). The faster the water moves, the higher the voltage produced. If we use a magnet whose field lines are oriented parallel to the z-axis and water flows parallel to the x-axis, the electrical potential is induced parallel to the y-axis. If we were to place two electrodes into the water, separated by

some distance in the y-direction, we would measure a voltage difference proportional to the velocity of the water.

This principle has been utilized by a number of commercially available flow meters, the most widely used of these being the meter manufactured by the Marsh-McBirney company (Fig. 19.4b). A small wire coil is housed in the spherical portion of the sensor. When electric current flows through this coil, a magnetic field is established. Around the equator of the sphere the magnetic field lines are oriented parallel to the

Figure 19.4. (a) A conductive fluid moving through a magnetic field induces a voltage proportional to velocity. (b) A schematic drawing of a typical electromagnetic flowmeter (redrawn from Denny 1985b by permission of Cambridge University Press).

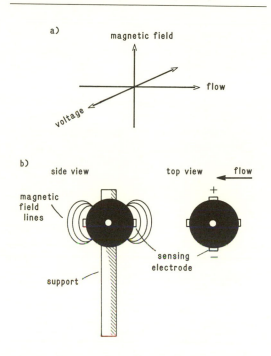

axis of the steel support post. Two pairs of electrodes placed on the equator sense the voltage induced by flow past the sphere. If the sensor is oriented with the support post vertical, these two sets of electrodes measure orthogonal components of horizontal flow. From these measurements one can calculate both the magnitude and the direction of horizontal velocity.

In theory, the response time of such a flow meter should be limited only by the speed of the electronics in sensing the change in voltage. Because electronic circuits are very fast, this would pose no practical limitation. In practice, however, the response time of the device is limited by other considerations. The voltages induced by the flow are very small. In order to detect these voltages in the midst of the random fluctuations in voltage that are always pre-

sent in the environment, the designers of the meter use a pulsed, alternating magnetic field. By pulsing the magnetic field at a known rate, and only measuring those voltages that vary at the same rate, the "true" (i.e., velocity-induced) voltages can be detected. The Marsh-McBirney meter is pulsed thirty times per second, and this pulse rate sets the response time of the meter. In essence, the meter samples the flow environment thirty times per second, and fluctuations in velocity between samples are not measured. Thus the response time is at best 1/30 s (33 ms), and in practice is closer to 50 ms. Moreover, unless the meter is specifically ordered to minimize response time, the output signal is filtered with a 1 s time constant.

There are two practical problems with this sort of electromagnetic flow meter. The meter measures the flow in a torus whose major diameter is roughly three times that of the sensor's sphere. For the standard Marsh-McBirney sensor this is an effective diameter of about 12 cm. Thus the meter cannot be used to measure the fine-scale structure of flow. Given the long response time of the meter (which itself puts practical limits on spatial resolution) this should seldom cause any hardship. A more serious problem emerges when one attempts to use this sort of meter in an intertidal environment. Every time the sensor is immersed or emersed, a large transient voltage is produced. Unless some scheme is devised to take these transients into account, the meter is useless in measuring flows in the swash zone. Finally, because the sensor must extend into the flow to measure velocity, it may be fouled by drift algae and subsequently be bent or broken.

Drag Sphere Velocity Meters. The fact that water velocity exerts a force on objects can be used as a means of measuring velocity. We have already seen how this can be used to estimate maximum water velocity from records of maximum force. The same idea can be applied to the continuous measurement of water velocity.

A simple velocity meter of this sort is described by Denny (1982). This scheme uses three separate force transducers to measure the drag exerted on objects in flow. Each transducer measures force exerted along one axis—two of the transducers are used to measure shear forces and the third to measure normal force.

The basic idea behind these transducers is quite simple. The object on which force is exerted is attached to two plastic beams held in parallel (Fig. 19.5a,b) and any force imposed on the object is transmitted to the beams. This arrangement of the beams ensures that the apparatus deflects appreciably only when forces are applied along one particular axis. As we have seen (Chapter 13), the strain imposed on a bent beam is proportional to the applied bending force. The simplest, most reliable method for sensing strain in this situation is the foil strain gauge, a thin piece of resistance wire arranged in a zigzag pattern on an epoxy backing (Fig. 19.5g). This gauge is glued to the face of one of the beams in the transducer. When no force is applied, the beams are not strained and the strain gauge is not stretched. In its unstretched condition the gauge has a certain electrical resistance (usually 120 or 350 ohms). When a force is applied, the beam deforms and the gauge is stretched or compressed. In stretching, the wire of the gauge gets longer and thinner, and as a result its resistance is increased. If the gauge is compressed, the wire gets shorter and fatter, and its resistance goes down. Thus the force placed on the beam can be transduced to a change in resistance of the strain gauge. By wiring the gauge into a circuit configuration known as a Wheatstone bridge (see Horowitz and Hill 1980), this change in resistance can be converted to a change in voltage. This change in voltage (proportional to the force exerted on the transducer) can be amplified and recorded by standard techniques (an instrumentation tape-recorder or chart recorder, for instance).

The beams of this particular force transducer are housed in a short section of plastic pipe so that the beams themselves are protected from the flow, but the object mounted on the transducers is exposed to the prevailing fluid-dynamic forces. The transducer housing is mounted in the rock in an emplacement similar to that described for the maximum force recorder. Electrical cables from the three transducers are tacked to the substratum appropriately and led to the recording apparatus. Denny (1982) describes a method for transmitting the flow data as an FM radio signal, but this method is somewhat finicky and is worth using only in areas where a direct cable connection between transducer and recorder is not feasible.

The response time of the transducer is (as always) a function of the ratio of moving mass (in this case the beams themselves and the object experiencing the force) to the stiffness of the beams. The apparatus described by Denny (1982) has a response time of about 8 ms.

Two problems are associated with this type of flow meter. First, it measures the three components of flow at three separate points in space. Because the volume of water sampled by each of the three meters is not the same, it is not valid to combine these measurements to calculate the instantaneous flow vector. This problem has been overcome by a design proposed by Donelan and Motycka (1978), but their 3-D transducer is difficult to build and has not been widely used. Second, in order to work backward from the applied force to the velocity and acceleration which caused that force, one must make several assumptions and have careful measurements of the necessary force coefficients. As a practical matter, we must assume that the three-dimensional version of the Morison equation (eq. 11.8) holds exactly. Although this may be approximately true in most cases, it has not been demonstrated for the situation we are dealing with here—a small object near a solid boundary. So, although

Figure 19.5. (a–f) Force transducers used to measure the hydrodynamic forces exerted on intertidal organisms. BP = bottom plate, C = shielded cable, DB = double beam, G = strain gauges, MA = milled area for attaching cable to gauges, MH = mounting hole for object on which force is to be measured, O = object, SR = silicon rubber damping and sealant for beam (from Denny 1981, by permission of the American Society of Limnology and Oceanography). (g) A schematic drawing of a foil strain gauge.

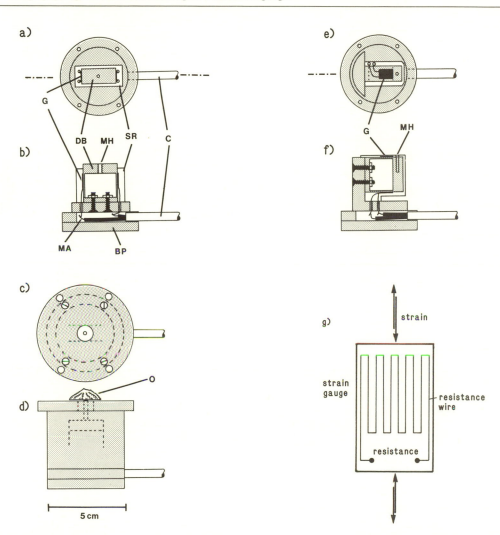

we have no practical alternative to using the Morison equation, we use it more as a matter of faith than as soundly based fact. Even if we assume the Morison equation to be correct, it is not a simple matter to work backward from a time series of forces to the velocities and accelerations that caused these forces (see Denny 1985 and Denny et al. 1985 for a discussion).

If one is willing to accept a certain degree of inaccuracy in the results, these mathematical complications can be avoided. For instance, we have seen (Chapter 11) that the acceleration reaction accounts for only a small percentage of the overall force exerted on an object unless the object is large and the water is very rapidly accelerated. By using a small object to sense the flow, we can ensure that the acceleration reaction is small compared to the combined forces of drag and lift. If the acceleration reaction is sufficiently small, we can ignore it altogether, and solve for velocity simply in terms of lift and drag, for example:

$$u = \sqrt{\frac{2f(t)}{\rho C_d S_p}}, \qquad (19.1)$$

where $f(t)$ is the instantaneous force at time t. These calculations will never be exact because some force is actually due to the acceleration reaction. If the water is accelerating in the direction of velocity, this equation overestimates the water velocity; if the water is accelerating opposite the direction of flow, it underestimates the water velocity.

Regardless of the calculations used to estimate water velocity from force, the estimate will only be as good as the measurement of the force coefficients used in the calculation, and this can be problematical. For example, say we decide to use a sphere to measure the fluid-dynamic force. Spheres are handy because they present the same projected area to flows from all directions. However, the drag coefficient for a sphere is highly dependent on the Reynolds number (Chapter 11). To know what C_d to use in calculating velocity, we need to know the Reynolds number, but to know the Reynolds number we need to know the velocity. There are several ways out of this dilemma. The simplest is to use a sphere small enough so that at the velocities encountered in the field, the sphere lies in the range of Reynolds numbers where its C_d is virtually constant ($10^2 < \mathrm{Re} < 10^5$). Alternatively, one could use some spherically symmetrical shape whose C_d is less variable—a pin-cushion shape resembling a spherical sea urchin might do. Similar problems are associated with accurately determining the lift coefficient and the added mass coefficient.

Heated Probe Flow Meters. Heated probe flow meters work on the same basic prin-

ciple as the plaster-cast flow meters already discussed—the faster the flow, the greater the mixing near an object—but in this case heat is being transferred instead of mass. Heated probe flow meters come in four general varieties: hot wire, metal-clad hot wire, hot film, and thermistor. Of these four, the one best suited for use in the wave-swept environment is the metal-clad hot probe, and it is this system that will be described here.

Figure 19.6a shows a typical metal-clad hot-wire probe. A small coil of wire is enclosed in a rugged steel tube. This wire forms a resistive element in an electrical circuit. By passing current through the wire, the tube can be heated, in exactly the same manner as the heating caused by passing current through the filament of a light bulb. The tube is typically heated to about 60°C above the ambient water temperature. The faster the water moves by the probe, the faster the heat is conducted away. The circuitry of the system is designed so that it maintains the tube at a constant temperature regardless of the water's motion. To do this in the face of changing flows, the electrical current passing through the wire must be constantly adjusted. Thus the amount of electrical current required to heat the probe is a measure of the flow velocity and can be detected, amplified, and recorded. The probe is sensitive to flow perpendicular to its axis, and insensitive to flow along the tube. Because the steel tube has a small thermal mass, it can respond quickly to changes in water velocity, usually requiring less than 0.1 s to adjust to a change in flow, a response time similar to that of an electromagnetic flow meter. Use of these probes is described in Gust (1982).

Another form of heated probe, the hot film, has a much smaller thermal mass and therefore a shorter response time (< 0.001 s). This response time is far better than any other method we have yet examined, and this type of flow meter has long been a standard tool for examining

Figure 19.6. (a) A schematic drawing of a metal-clad heated flow probe. The wire coil acts as a resistor and is heated to a constant temperature above ambient. The amperage required to maintain this temperature is a measure of water velocity. (b) A schematic drawing of a hot-film flow probe.

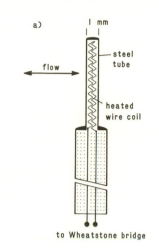

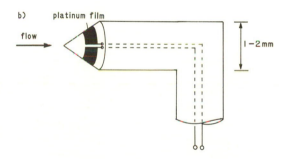

axes (the split film probes); but these specialized probes are very fragile and quite expensive (> $700 per probe), and therefore are probably not well suited for use outside of the laboratory.

All heated probes have a problem with bubbles. If a small bubble attaches itself to the probe, it serves as an effective insulator, radically changing the rate at which heat is conducted to the surrounding water and profoundly affecting the calibration of the device. Bubbles can attach to the probe if the water itself contains bubbles (a common occurrence in the surf zone), or they can be formed *in situ* as the water next to the probe is heated. As long as the water velocity is sufficiently high, bubbles are blown off the probe as soon as they are formed and do not unduly affect the record. However, at velocities below about 0.5 m/s, bubbles may remain attached for considerable periods.

The two other forms of heated-probe flow meters (hot wire and thermistor) have proven very useful in certain situations but are not well suited for the wave-swept environment. Hot-wire probes use a thin platinum wire (on the order of 1 μm diameter) as a resistive element instead of a coil or film. The wire is several millimeters long and is suspended between supports. The response time of this sort of probe is very short (< 0.1 ms) and this type of probe is the standard tool for examining turbulence in air. However, the wire is exceedingly fragile, and in the wave-swept environment it would be broken by the first suspended larva or sand grain to impact the device. Thermistor probes use a small, glass-coated thermistor as the resistive element but otherwise work on the same principle as a hot-film probe. The problem here is that at velocities above about 0.5 m/s, the water carries heat away from the thermistor as fast as it can, and the entire temperature gradient is formed in the glass coating of the thermistor. Above this velocity the meter is entirely insensitive to changes in flow ve-

rapidly fluctuating turbulent flows in laboratory situations. There are, however, practical problems associated with using these probes in the wave-swept environment. Most hot-film probes are designed to measure flow in only one direction. For instance, the conical probe shown in Figure 19.6b is designed to measure flows directed at the apex of the cone. Flows coming from the opposite direction place the hot film in the wake of the probe support, and the readings are therefore biased. There are hot film probes suitable for measuring flows in both directions along an axis, or even along two perpendicular

locity, an undesirable attribute for a flow
meter. Because most flows of interest in
the wave-swept environment exceed
0.5 m/s, this type of probe is generally inappropriate. However, in those situations
where the flow is sufficiently slow (at sufficient depths, or in sheltered microhabi-
tats) thermistor flow meters may be useful.
LaBarbera and Vogel (1976) have described a simple, inexpensive thermistor
flow meter which has gained much popu-
larity among biologists examining slow
flows. A correction to the circuit shown in
LaBarbera and Vogel (1976) is given by
Vogel (1981).

Acoustic and Optical Flowmeters. All of
the flow-measurement techniques de-
cribed so far require that some object be
inserted into the flow. This is less than
ideal because there is always the chance
that the mere presence of the object substantially changes the flow pattern. Three
methods have recently been devised that

Figure 19.7. A scheme for an acoustic flow meter
(see text). *T* = transmitter of sound, *R* = receiver.
(Redrawn from Denny 1985b by permission of
Cambridge University Press.)

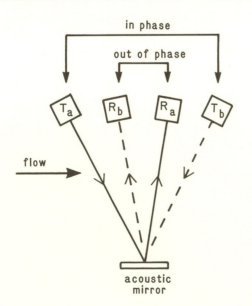

can potentially avoid this problem. At present none of these methods is particularly
well suited for use in the wave-swept environment, but it is probably just a matter
of time before these methods are adapted
for field use, and it is therefore worth describing them here.

Transit Time Flowmeters. Figure 19.7
diagrams an idea for an acoustic transit-
time flowmeter. The device consists of two
sets of sound transmitters and receivers
and an acoustic mirror (a smooth, hard
surface). A pulse of high-frequency sound
produced by transmitter A is directed at
the mirror and the reflected sound (the
echo) is detected some time later by receiver A. If flow is from the left to the
right in this diagram, receiver A is downstream of transmitter A. In the time it
takes the sound to travel from the transmitter to the receiver, the water in which
the sound is traveling has moved toward
the receiver. As a result, the sound reaches
the receiver faster than it would if the
water were stationary. Conversely, sound
produced at transmitter B moves upstream and takes longer to reach receiver
B than if the water were still. If sound
impulses are produced at both transmitters
at the same time, the difference in time
between the arrival of the echoes at the
two receivers is a direct measurement of
the water velocity along the axis connecting the transmitters and receivers. In
practice, sound pulses are produced many
times a second, and the difference in transit time is detected electronically. By arranging sets of transmitters and receivers
along perpendicular axes, this basic scheme
can be used to measure two or more
components of flow.

At present this form of acoustic flow
meter has been developed only in a configuration suitable for measuring midwater
flows (for instance, the meter manufactured by Neill-Brown Instruments). However, it should be possible to construct
such a meter so that the transmitters/
receivers and the acoustic mirror could be

installed across a surge channel or in a tide pool.

Doppler Flow Meters. Consider the apparatus shown in Figure 19.8. A laser produces a beam of monochromatic, coherent light (i.e., all the light waves have the same frequency and are in phase). This beam is split in two, one-half of the beam passing straight through a volume of water, the second half focusing on one small portion of the water from a slight angle. The beam passing straight through the water is known as the reference beam and it falls on a light detector (usually a photomultiplier tube). Because the second beam arrives at some angle, most of it does not fall on the light detector. From this beam only light that is scattered by particles in the water is reemitted in the direction of the detector. If the water is not moving, the scattered light has the same frequency as the light in the reference beam. However, if the water is moving, the light scattered by particles has a slightly different frequency from the incident light: it has undergone a Doppler shift. The faster the water moves, the greater the shift in frequency. This frequency shift is familiar for sound waves as the change in the frequency of the whistle of a passing train. The slight difference in frequency between the reference beam

and the scattered light causes a periodic change in phase between the reference and scattered beams. When the two are in phase at the detector, the light intensity is augmented; when the two are out of phase, the light intensity is decreased. The frequency at which the light intensity is modulated (the beat frequency) is equal to the difference in frequency between the reference and scattered beams. Thus the beat frequency is a direct measure of the Doppler shift, and thereby a measure of the water velocity.

By suitably arranging the optics of the system, this method can be used to measure the velocity of water in an extremely small volume (easily < 1 mm^3). By splitting the light beam more than once and directing it onto the test volume at different angles, all three components of velocity can be measured simultaneously for the same fluid volume. The response time of the apparatus is limited only by the electronics and thus may be measured in microseconds. In these respects a laser Doppler velocity meter is the ideal, general-purpose flow meter.

There are, however, practical problems involved in using this type of device in the field. First, the scheme shown here requires that the light detector be on the opposite side of the test volume from the

Figure 19.8. A scheme for a laser doppler flow meter (see text). (Redrawn from Denny 1985b by permission of Cambridge University Press.)

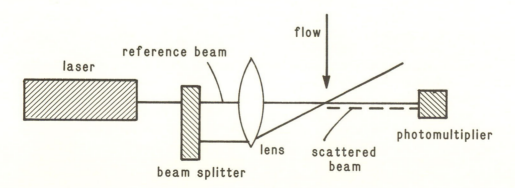

light source. This would be inconvenient in most field situations. It is possible to arrange the device so that the light detector is on the same side of the test volume as the laser, and the light sensor detects the backscattered light from particles in the fluid. The intensity of this backscattered light is much less than that of forward-scattered light, and as a consequence the light detector must be more sensitive, or the light source more intense. Regardless of the precise arrangement of the optical system, this type of flow meter requires that the optics be precisely aligned and very stable. Any wobble in a lens or beam splitter can result in an intensity modulation at the light detector and thereby may be falsely interpreted as a fluid velocity. This sort of stability would be difficult to achieve in the wave-swept environment. An interesting approach is that used by the DISA company in a recent commercial instrument. The optics of a backscatter system are firmly held in a small probe. The probe is attached to the laser and the light detector by a glass fiber-optic cable. It is likely that within a few years an instrument constructed along these lines will be devised for use in the wave-swept environment.

Even when the problem of suitably arranging the optical system has been solved, there are other problems with field use of a laser Doppler flow meter. For example, the system requires the presence of small particles in the flow to scatter the light, but the presence of too many particles or a few big particles can block the light detector's view of the test volume. This has been a common problem in laboratory systems where bubbles cause "signal dropout." In practice, the water must be made very clean to avoid the problem, or the periods during which the signal has dropped out must be taken into account in the computations resulting from the measurements.

The same principal used in the laser Doppler flow meter can be applied using high-frequency sound (5–20 MHz) instead of light. These sorts of acoustic Doppler meters, relying on the backscattered signal from particles in flow, have been used for several years by biologists measuring flow in arteries, and may be useful in measuring boundary-layer flows.

Measurement of Wave Height

There are several standard methods for measuring wave height, none of which is entirely satisfactory, but all of which have their use.

Bottom Pressure Gauges. As we noted in Chapter 5, the passage of a wave crest in shallow and intermediate water depths results in an increase in pressure at the bottom. If we can measure this pressure, we can measure the wave height. The pressure transducer typically used for this purpose is quite similar to the force transducers already described. A thin diaphragm is exposed to the hydrostatic pressure of the water on one side and a known constant pressure (usually very close to 0 Pa) on the other side (Fig. 19.9a). The difference in pressure between the two sides causes the diaphragm to bend, and this bending is detected by a strain gauge. As the water pressure varies during a wave's passage, the deflection of the diaphragm varies, and the voltage output from the transducer varies analogously.

In some pressure transducers the diaphragm is made of stainless steel, and foil strain gauges are used. In more recent designs, the diaphragm is a thin area of silicon etched into one part of an integrated electronic circuit. In this case a semiconductor type of strain gauge is used, and the electronics for detecting and amplifying the voltage signal from the gauge are built onto the same piece of silicon.

In most applications it is best if the pressure transducer itself is not exposed to seawater. Typically the transducer is

Figure 19.9. (a) A schematic diagram of a pressure transducer. (b) A schematic diagram of a mechanical tide gauge.

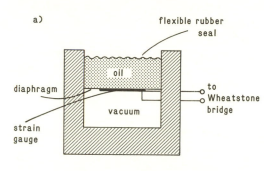

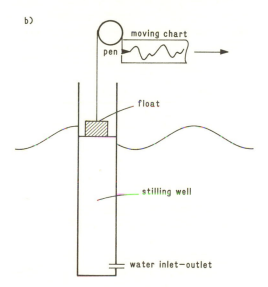

housed in a rigid container filled with oil. A flexible rubber diaphragm separates the oil from the surrounding water and allows pressure to be transmitted to the transducer.

The technology of pressure transducer wave gauges is quite advanced. Commercially available transducers are very sensitive (some can measure the difference in air pressure between the top of your desk and the floor) and very stable. Many are equipped with digital output so that the pressure signal can be fed straight into a computer. The problem in using these transducers lies not with the transducer it-

self, but in the interpretation of the pressure measurements. The pressure under a wave in intermediate depths is not hydrostatic (Chapter 5). Due to the movement of the water, the pressure measured by a bottom-mounted transducer is less than would be measured were an equivalent static head of water imposed, and in working backward from pressure to wave height one relies on the predictions of wave theory. Thus the calculated wave heights will be only as good as the theory used. In practice, linear wave theory is generally used to calculate wave heights, and calculated and actual wave heights agree within about 10%. There is an additional problem because wave-induced changes in pressure are attenuated by depth. The rate of attenuation depends on the wave length (Chapter 5)—the shorter the wave length, the greater the attenuation. Thus a pressure transducer must be located at a sufficiently shallow depth to detect the pressure signal from the shortest wave-length (shortest period) waves of interest. For instance, at the bottom in a depth of 30 m, the pressure signal for a 12 s wave would be 73% of the equivalent pressure signal at the surface, but the signal from a 5 s wave would be only 1.6%. The signal from a bottom-mounted pressure transducer situated at an intermediate depth must be corrected for each wave's period, and in a random sea it is often difficult to determine the period for each individual wave. Bottom pressure transducers can be used in shallow water, where the measured pressures are assumed to be hydrostatic.

Surface-Piercing Wave Staffs. The problems associated with bottom-pressure transducers can be avoided by directly measuring the water's surface elevation. A number of schemes for doing this have been devised; the two most commonly used are resistance and capacitance wave staffs.

A resistance wave staff consists of

two fine wires held parallel and vertical through the water's surface, and the resistance between wires is measured electronically. If the wires were entirely out of the water, the resistance would be nearly infinite. However, seawater is a good electrical conductor, and the greater the length of wire submerged, the less the overall resistance between wires. Thus, as the water rises and falls, the resistance of the wires changes analogously. A circuit diagram for building such a meter is given by Flick et al. (1979).

Capacitance wave staffs work on the same basic idea, except that the change in capacitance rather than the change in resistance, is measured as the water rises and falls. An appropriate circuit is given by Anderson et al. (1972).

This sort of wave-height meter presents two practical problems. The largest drawback is that the meters must be supported at the surface. This is no problem if there is a convenient piling to which the staff may be attached, but most wave-swept shores do not come preequipped with pilings in the proper places. Without a piling, one is left with the sizable task of improvising an appropriate support. Surface-piercing wave staffs are also prone to breakage as they become fouled by drift algae.

The second problem with surface-piercing wave staffs becomes apparent when they are used in the surf zone. Here much of the height of the wave consists of foam. Whereas a bottom-pressure transducer ignores this low-density foam, both resistance and capacitance wave staffs include a certain portion of the foam in their measurement. Thus the method that gives a more accurate measurement depends largely on the definition of wave height. If foam is to be included in calculations, a surface-piercing wave staff may be the instrument of choice. If one wants to exclude foam, a pressure transducer may be better.

Other Methods. There is a host of other methods for measuring wave height. Most of these are not well suited for use in the wave-swept nearshore, but they deserve to be mentioned if only for the ideas they may stimulate.

In the deep ocean, bottom-pressure transducers cannot work because the pressure signal is totally attenuated before it reaches the bottom. It is possible to moor a pressure transducer on a subsurface buoy at an appropriate depth, but if we are going to go to all the trouble of rigging a mooring a more direct method is available. A buoy floating on the surface obviously rises and falls with the passage of each wave. By equipping the buoy with an appropriately sensitive accelerometer, we can measure how fast it is accelerating vertically, and by twice integrating this signal with respect to time we can calculate how far the buoy moves up and down. Such "wave-rider" buoys are a standard method for measuring offshore wave heights. The signal from each buoy is sent via radio to a shore-based recording station.

Finally, three other ideas bear mentioning: (1) It should be possible to use a sonic depth sounder (a "fish finder") mounted on the bottom and directed upward to measure wave height. (2) Alternatively, if there is a convenient piling or pier on which to mount the apparatus, the sonic rangefinder from a self-focusing camera could be used to measure the distance between a fixed point on the piling and the water's surface. (3) Various stereoscopic techniques have been used to measure wave heights (see Kinsman 1965), but these do not seem well adapted to continuous measurement.

Measuring the Tides

Any of the methods described here for measuring wave heights can be used to measure the tidal fluctuation in sea-surface elevation. In this case, only long-period (approximately 12 hours) fluctuations are measured, and the short-period fluctuations from waves are somehow filtered out. This filtering can be accomplished either electronically or computationally.

Alternatively, the tides can be measured by the simple device shown in Figure 19.9b. A large-bore pipe (a "stilling well") is held vertical by a piling. Water enters and leaves the pipe through a small hole in the bottom. If the hole is small enough, it takes a considerable time for the water level to change in the pipe, effectively filtering out fluctuations due to waves, but the water level can track the tidal fluctuation. A float in the stilling well is attached (via a pulley) to a pen. As the water level rises and falls, the fluctuation is traced on a moving chart.

Creating Water Flows

Once the flow has been measured in the natural environment, it is often useful to be able to reproduce aspects of that flow in a controlled fashion in the laboratory. There are three standard techniques for producing flows in the lab: unidirectional flow tanks (or flumes), oscillating flow tanks, and wave tanks.

Unidirectional Flow Tanks

This is the simplest type of flow tank, and many varieties have been used over the years. Those described here have the greatest potential use in the study of the wave-swept environment.

Vogel and LaBarbera (1978) have described a simple, basic design for a low-speed flow tank (Fig. 19.10a). The working section of the tank is rectangular in cross section and constructed from acrylic plastic. Water flows through the working section and is then pumped around a closed circuit (made from PVC

or ABS plastic pipe) to reenter the working section. Allowing the water to recirculate keeps the kinetic energy of the fluid roughly constant, and only a small electric motor is required to provide a reasonable flow. Pumping is accomplished by a series of propellers situated in the vertical leg of the return pipe, and inserting the propeller shaft through the free surface of the water avoids the necessity for a shaft seal. The flow is made uniform across the working section through the use of a soda-straw "flow straightener" at the upstream end. Details of the construction can be found in either Vogel and LaBarbera (1978) or Vogel (1981).

This type of flow tank has proven to be extremely useful when experimenting with low-speed flows (< 1 m/s). However, three difficulties are inherent in using this design in any attempt to achieve the sorts of rapid flows often found in the wave-swept environment: (1) At rapid speeds surface waves form in the working section. (2) Practical constraints on the length of the propeller shaft require that the propeller be quite near the water's surface. At high speeds the vortex created as water leaves the working section can suck air into the propeller, resulting in cavitation and limiting the speed at which the tank can be operated. (3) The small diameter of the return piping causes large losses to viscous processes.

These problems can be avoided by using a design similar to that shown in Figure 19.10b. Here the working section has a smaller cross section than the return pipe. Water moves relatively slowly through the return pipe (minimizing viscous losses) and is rapidly speeded up to pass through the working section. This "choking down" of the flow as it enters the working section has the effect of minimizing the variation in flow across the working section, and flow straighteners (another source of viscous losses) are generally not required. After passing through the work-

Figure 19.10. Flow tanks. (a) A low-speed flow tank (redrawn from S. Vogel, Life in Moving Fluids, 1981, Wadsworth, Inc., by permission of Brooks/Cole Publishing Co., Monterey, Calif.). (b) A high-speed flow tank.

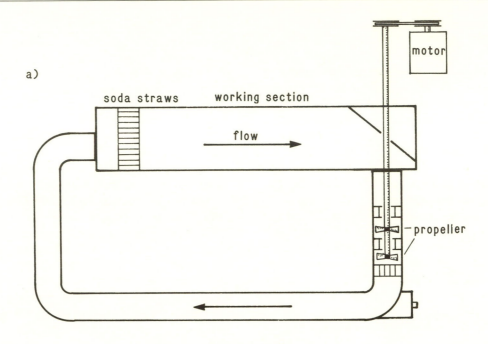

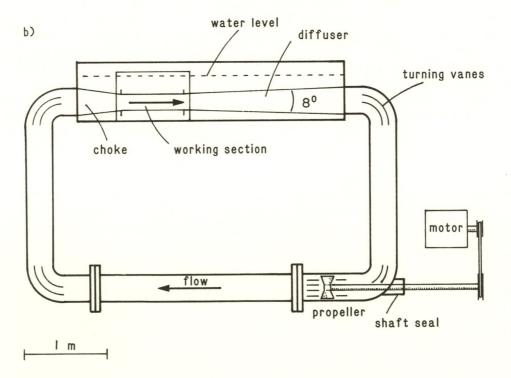

ing section, the flow cross section is gradually expanded back to that of the return pipe. This gradual expansion is necessary lest the flow separates from the walls of the tank, resulting in turbulence and large viscous losses. It has been experimentally determined that an angle of 8° is optimal for this "diffuser" (Tritton 1977). Much of the viscous loss of energy in the flow through the return pipe is associated with the right-angle bends—each right angle has the same resistance to flow as about thirty pipe diameters of straight pipe. This factor can be reduced considerably (to about one-third) by using turning vanes in the elbows. These vanes ensure that water that is in the middle of the pipe before it enters the elbow remains in the middle of the pipe as it turns and leaves the elbow. To avoid the need for sealing the working section, the choke, working section, and diffuser are submerged in a larger tank of water. The propeller is situated in the bottom leg of the return pipe, and the propeller shaft exists through a shaft seal.

The dimensions given here are for a tank in my laboratory. A five-horsepower electric motor powers the device and is not at all strained in achieving water velocities in excess of 5 m/s. This tank design works very well, but it is much more difficult and expensive to build than that of Vogel and LaBarbera.

A third type of flow tank has recently been constructed by Peter Jumars and Arthur Nowell at the University of Washington's Friday Harbor Laboratory. Nicknamed "the race track," it is a large, horizontal oval. The working section is approximately 1 m deep by 1.5 m wide and about 10 m long. Water is moved by the conveyor-belt equivalent of a paddle wheel located in the back stretch of the race track. The tank can attain speeds of 1–2 m/s and is well suited to experimentation with depth-limited boundary layers. Another design serving the same purpose is described by Muschenheim et al. (1986).

Design Considerations

In general, the role of a flow tank is to reproduce a particular flow that has been measured in the field. Various aspects of this flow may be important, including the mean velocity, the turbulence-intensity and eddy-diffusion coefficients, and the boundary-layer thickness. Which of these aspects will be important design considerations depends on the type of measurement to be conducted. For example:

1. When measuring the drag or lift coefficient of an organism it is necessary to ensure that the velocity gradient in the tank is similar to that in the field. As we have seen (Chapter 9), boundary layers are generally quite thin in the wave-swept environment, and many organisms are exposed to mainstream flows. It would be useless to reproduce the mainstream velocity if in our flow tank the boundary layer is unreasonably thick and the test object is buried in the velocity gradient. The boundary layer thickness can be adjusted by varying the position of the test object relative to the nearest leading edge.

2. If we make measurements of the boundary layer itself, we must take special care. For instance, if we want to measure the characteristics of an equilibrium boundary layer, the measurements must be made at least $50\,\delta$ downstream from a leading edge or change in roughness. This can require a very long working section. We should also remember that the side walls of the working section have a boundary layer. The working section must be wide enough for an object in the middle not to be affected by the velocity gradient extending from the sides. Additional considerations for designing a flow tank for work with boundary layers are found in Nowell and Jumars (1984), Muschenheim et al. (1986), and the literature cited therein.

3. Measurement of drag and lift can be affected by the consequences of the work-

ing section's solid walls. For example, consider an object large enough so that its projected area (measured along the direction of flow) is a substantial portion of the cross-sectional area of the working section. The presence of the object reduces the area through which water can flow, and we know from the concept of continuity (Chapter 3) that the water in the rigid working section must speed up as it flows through this constriction. This increase in speed must be taken into account when calculating C_d and C_l. The speed in the working section is also increased as the boundary layer thickens downstream (leading to an effect known as "horizontal buoyancy," which augments drag) and by the effects of the test object's wake. As long as the projected area of the test object is less than 10% of the cross-sectional area of the working section, these effects are negligible, and it is best to design a working section large enough to avoid these problems. For a more complete discussion of these phenomena, consult Appendix 1 of Webb (1975).

4. The turbulence structure in a flow tank can effect many measurements. For example, Patterson (1984) found a distinct pattern in the feeding efficiency of polyps on an octocoral when the colony was immersed in laminar flow. When the flow was manipulated to better match the turbulence regime found in the field, the pattern disappeared. Unfortunately, it is very difficult to duplicate the turbulence structure of field flows in a laboratory setting. For example, the width and height of the working section of a flow tank place an upper limit on the size of turbulent eddies. Thus, unless the working section has the same height and width as the field site, the results are liable to be biased. This may not be a problem for research concerning flows in small surge channels or tidepools, but in other cases there may be substantial effects. To a certain degree, the intensity and size structure of turbulent eddies can be controlled by placing a mesh

screen or a series of cylinders upstream of the working section. The type of turbulence created in the wake of a screen has been well characterized (see any standard text on turbulent flow), and the manipulation of an upstream screen is perhaps the best tool for "designing" the turbulence in a flow tank.

Oscillating Flow Tanks

Two types of oscillating flow tanks have been used in the majority of studies on oscillating flows. The two are quite similar, differing primarily in the manner by which the period of oscillation is controlled.

The simpler of these two consists solely of an upright U formed of pipe with the working section in the base of the U (Fig. 19.11a). By pumping water up into one arm of the U and then releasing it, the water can be made to oscillate back and forth through the working section. The period of oscillation is determined by the mass of water in the tank and by the restoring force, which in this case is due to the acceleration of gravity. For example, a vertical pipe with a cross-sectional area S feels a restoring force of $g\rho S$ for every meter that water is displaced from its equilibrium level. When the level in one arm of a U-tube rises, the level in the other arm falls, so that the total restoring force is $2g\rho S$ per meter. This is the effective "stiffness" of the system, k, for insertion into eq. 14.18. The appropriate mass is the total mass of the water in the U-tube—ρSL, where L is the overall length of the water column (Fig. 19.11a). The natural period of this tank is thus

$$T_n = 2\pi \sqrt{\frac{L}{2g}}. \qquad (19.2)$$

For instance, a tank with a total length of 10 m has a natural period of 4.5 s.

Although the period of the oscillation is determined by eq. 19.2, the amplitude of the oscillation depends on the initial deflection of the water level and the number

Figure 19.11. Tanks for producing oscillating flow. (a) A tank driven at its resonant frequency. (b) A tank with forced oscillation.

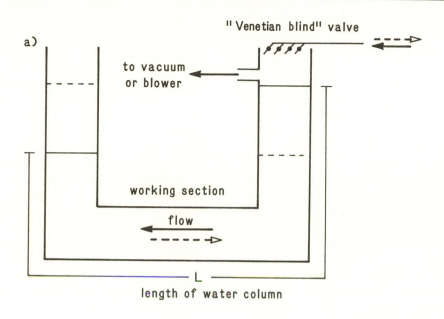

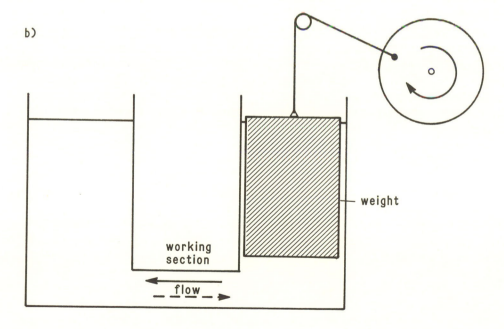

of cycles after the water is released. Each time the water moves through the working section, energy is lost to viscous processes and the oscillation is damped. To maintain the amplitude of oscillation, we must replace the energy lost to viscosity. This can be done by applying a vacuum to one of the upright arms as the water is rising in that arm. The lowered pressure pulls the water higher than it would otherwise go. A very large tank using this basic design has been built by Dr. Tergut Sarpkaya at the Naval Post-graduate School in Monterey, California. A valve constructed along the lines of a Venetian blind controls the timing of the air pressure applied to one arm of the tank, and thereby controls the amplitude of oscillation. The working section of this tank is approximately 1 m by 1 m and the natural period is 5.5 s.

A second type of oscillatory tank does not rely on the natural period of the water oscillation. Because it does not work at its resonant period, it takes more energy to drive, but for a small tank this may not be a serious drawback. The tank consists of two large reservoirs connected by the working section (Fig. 19.11b). A large weight is suspended in one of the reservoirs, the height of the weight being controlled by the rotation of an arm attached to an electric motor. As this weight is lowered into the reservoir, water is displaced, the displaced water flowing through the working section into the other reservoir. The period with which the weight rises and falls can easily be adjusted by varying the speed of the electric motor, and the amplitude of the oscillation can be adjusted by changing the length of the arm. A tank of this sort has been used by Svoboda (1970).

Wave Tanks

The basic design of a simple wave tank has already been discussed (Chapter 4), and only a few practical details need be added

here. A typical laboratory wave tank is 0.5–1.0 m wide, and 20–30 m long, with a water depth of 0.3 to 0.5 m. The length of the tank is determined by two factors. First, waves are often somewhat "messy" in the immediate vicinity of the wave-making paddle. By the time the waves have traveled 5–10 wave lengths, they are much more regular. For a wave with a 1 s period this amounts to 7.5–15 m. Second, if a shallow beach slope is to be used, a long tank is needed. For instance, a 1:30 slope is commonly used when examining breaking wave patterns and run-up. If the water in the tank is 0.5 m high, this requires a beach that extends at least 15 m down the wave tank.

The construction of the wave tank itself is straightforward. Usually most of the effort in design goes into the wavemaker. The simple paddle-and-motor arrangement discussed in Chapter 4 is sufficient for producing monochromatic waves. If it is desirable to simultaneously produce waves of several periods and heights, there are a variety of options. Waves can be produced by a partition that moves on a horizontal slide, the partition being moved either by a motor-driven lead screw or a hydraulic ram. In either case, the partition can be made to move at several frequencies at once, each frequency having its own amplitude. Alternatively, waves can be generated in response to a change in air pressure in a chamber enclosing one end of the wave tank. The change in pressure is provided by a fan pumping air either into or out of the chamber, and the pattern of pressure change is controlled by an appropriate valve. These are but a few of the many schemes used to produce waves traveling in one direction. The interested reader should consult back issues of the *Journal of Fluid Mechanics* or the *Journal of Geophysical Research* for information and ideas on two-dimensional wave tanks, and for the wonderful schemes wave researchers have devised for generating waves.

Motors

All of the flow tanks described here require some sort of electric motor. Electric motors come in a bewildering array of makes and models, and the motor must fit the design. Vogel (1981) presents a useful introduction to variable-speed motors, and the various manufacturers listed at the end of this chapter can be consulted for advice on particular applications.

Techniques for Measuring Material Properties

At its most basic level, the measurement of the mechanical properties of a material consists of obtaining a stress-strain curve at an appropriate strain rate. Three pieces of apparatus are required for these measurements: a mechanism for applying a deformation to a sample, a method for measuring the resulting force, and a method for accurately measuring the applied deformation. Many different apparatuses have been devised to fill these requirements, and we only discuss two of the most basic designs here.

The simplest way to measure a stress-strain curve is to attach the sample to the ceiling, hang weights off its lower end, and measure the deformation with a ruler. For a reasonably large, reasonably elastic sample (such as an algal frond or stipe) this method works wonderfully and is satisfyingly field-portable.

The more typical way of measuring stress-strain curves is by using a tensometer (Fig. 19.12a), a device similar to a medieval rack. The sample is clamped at one end to a stationary force transducer (similar in principle to that described earlier). The other end of the sample is clamped to a crosshead that can be moved toward or away from the force transducer, placing the sample in compression or tension. Usually the crosshead is moved by a lead screw. By attaching a variable-

Figure 19.12. (a) A schematic drawing of a tensometer. (b) A schematic drawing of a linearly variable differential transformer (see text).

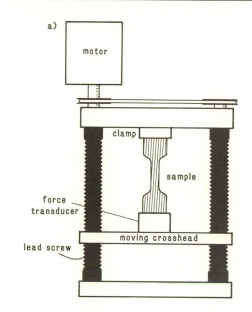

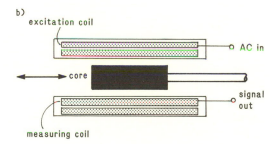

speed electric motor to the lead screw, the tensometer's crosshead can be made to move at a constant, controlled rate. The industry standard tensometer is made by the Instron Corporation.

The deformation of the sample can be measured directly using a device known as a linearly variable differential transformer (LVDT). The device consists of two cylindrical coils of wire (potted in a protective housing) with a sliding iron core extending into the center of the coils (Fig. 19.12b). One of the coils is driven by a high-frequency alternating current, inducing a voltage in the second coil. The voltage induced depends on the position of the core. The core is attached to the sample, and as the sample deforms, the core moves. A change in position of the core thus results in a change in induced voltage and serves as a measure of deformation. The coils are specially wound so that the voltage output of the device varies linearly with the position of the coil. The device is frictionless, and thus avoids any bias of the measured force.

There are many elaborate variations on this basic theme, and the interested reader should consult special texts in materials science before embarking on a research program in this field. Ferry (1980) is especially useful in discussing the many, varied techniques of materials testing.

The largest problem with measuring a stress-strain curve is often encountered when one tries to get a grip on a sample. Many biological materials are soft and slimy and very difficult to clamp reliably. Two pieces of advice may be helpful: (1) An hourglass-shaped sample (as shown in Fig. 19.12a) provides a large end area to be clamped while reducing the cross-sectional area of the central portion of the sample. As long as the sample is gradually tapered toward the center, there are no problems with stress concentrations, and stress is evenly applied to the central portion. This central portion is considered the "test area." Strain measurements should be taken directly from this test area. (2) Cyanoacrylate glues are excellent tools for holding onto slimy samples.

Sources and Suppliers

Two books serve as general reference sources for measurement techniques: Horowitz and Hill (1980) is an invaluable source of information on the electronic aspects of measurement. Beckwith et al. (1982) covers a wide variety of topics concerning the mechanics of measurement.

The following list of suppliers may be of some use. By no means is this an exhaustive list of the companies and products; rather, it is a potpourri of the technology that is available—a selection that can serve as an entry into the commercial world of measurement.

Sailboat Knot Meters, Propeller Flow Meters

Swoffer Marine Instruments, Inc.
  1048 Industry Drive
  Seattle, Washington 98188
Heathkit Company
  Benton Harbor, Michigan 49022
Omega Engineering, Inc.
  One Omega Drive, P.O. Box 4047
  Stamford, Connecticut 06907

Electromagnetic Flow Meters

Marsh-McBirney, Inc.
  8595 Grovemont Circle
  Gaithersburg, Maryland 20760
InterOcean Systems, Inc.
  3540 Aero Court
  San Diego, California 92123
Omega Engineering, Inc. (see above)

Acoustic Flow Meters

EG&G Neill Brown Instruments Systems
  1140 Route 28A, P.O. Box 498
  Cataumet, Massachusetts 02534
Valpey-Fisher, Inc.
  75 South Street
  Hopkinton, Massachusetts 01748
Bioengineering Section
  College of Medicine
  56 M.R.F.
  The University of Iowa
  Iowa City, Iowa 52242
Omega Engineering, Inc. (see above)

Strain Gauges and Accessories

BLH Electronics
  42 Fourth Ave.
  Waltham, Massachusetts 02254
Omega Engineering, Inc. (see above)

Heated Probe Flow Meters, Laser Doppler Flow Meters

TSI, Inc.
   P.O. Box 43394
   St. Paul, Minnesota 55164
DISA Electronics, Inc.
   779 Susquehanna Avenue
   Franklin Lakes, New Jersey 07417

Thermistors

Thermometrics, Inc
   808 U.S. Highway 1
   Edison, New Jersey 08817

Pressure Transducers, Pressure-sensing Wave Height Gauges

Paroscientific, Inc.
   4500 148th Ave. NE
   Redmond, Washington 98052
Sensym, Inc.
   1255 Reamwood Ave.
   Sunnyvale, California 94086
Schaevitz, Inc.
   P.O. Box 505
   Camden, New Jersey 08101
Omega Engineering, Inc. (see above)
InterOcean Systems, Inc. (see above)

Capacitance Wave Gauges

Data Industries, Inc. (Ocean Data Division)
   883 Waterman Ave. East
   East Providence, Rhode Island 02914

Electric Motors

Minarik, Inc.
   224 E. 3d St.
   Los Angeles, California 90013
Bodine Electric Company
   2500 W. Bradley Place
   Chicago, Illinois 60618
W. W. Grainger, Inc.
   Sales offices in most large cities

Tensometers and Accessories

Instron Corporation
   100 Royall St
   Canton, Massachusetts 02021
Tequipment, Inc.
   P.O. Box 1074
   Acton, Massachusetts 01720

Linearly Variable Differential Transformers

Schaevitz, Inc. (see above)

Sealants, Casting Rubber, etc.

Dow Corning Corporation
   Midland, Michigan 48604

Useful Bits and Pieces

Small Parts, Inc.
   P.O. Box 381736
   Miami, Florida 33238
PIC Design Corporation
   P.O. Box 335, Benrus Center
   Ridgefield, Connecticut 06877

APPENDIX

~ ~ ~ ~ ~ ~ ~ ~ ~ ~

List of Symbols

Symbol		Equation Where First Used
a	acceleration	11.3
a_n	normal acceleration	8.3
a_t	tangential acceleration	8.4
A	wave amplitude	Fig. 4.3
	amplitude coefficient	5.52
	amplitude coefficient	14.15
A_{rms}	rms wave amplitude	6.3
$\overline{A^2}$	mean square amplitude	6.2
A_s	significant wave amplitude	6.5
B	function of L and d	5.52
	fertilization coefficient	10.37
	bulk modulus	11.19
	elliptical integral	13.10
	amplitude coefficient	14.15
B_a	bulk modulus of air	11.19
B_b	breaking parameter	7.11
B_w	bulk modulus of water	11.19
c	concentration	10.1
	speed of sound	11.18
	damping coefficient	14.29
$\bar{c}$	mean concentration	10.2
c'	fluctuation from mean concentration	10.2
c_c	critical damping coefficient	14.36
C	wave celerity	4.1
C_a	added mass coefficient	11.3
C_d	drag coefficient	11.1
$C_{d,friction}$	friction drag coefficient	11.12
C_g	group velocity	5.46
C_l	lift coefficient	11.6
C_m	inertia coefficient	11.4
C_O	deep-water wave celerity	7.5
d	depth below still water	4.10

Symbol		Equation Where First Used
d_b	water depth at breaking	7.10
d_0	zero plane displacement	9.24
D	height of roughness elements	9.20
	distance to neighbor	
	height of danger zone	16.44
D_m	molecular diffusion coefficient	10.6
e	volume fraction of air	11.19
E	wave energy	7.1
	Young's modulus	12.5
E_O	deep-water wave energy	7.5
E_t	tangential modulus	
f	frequency	6.13
f_a	acceleration reaction	11.3
f_d	drag	11.1
f_l	lift	11.6
f_n	normal force	15.1
f_0	amplitude of external force	14.57
f_s	shear force	15.1
f_x	force in x-direction	3.1
$\hat{f}$	impulse	14.65
F	fraction of eggs fertilized	10.37
g	acceleration due to gravity	3.4
G	universal gravitational constant	8.1
	shear modulus	12.6
G_t	tangential shear modulus	
h	height from reference	3.4
	ratio of beam width to depth	13.17
	response function	14.71
	height of SWL	16.43
	ratio of height to radius	17.13
h_{crest}	height of crest	

Symbol	Equation Where First Used	Symbol	Equation Where First Used
h_l water level at low tide	16.43	m_{opt} optimum mass	17.7
h_s height of organism on shore	16.44	m' mass per length of beam	14.20
h_{trough} height of trough		m_σ proportion of immigrants and recruits in stress range σ	16.35
H wave height	4.2	M parameter of solitary wave	Table 5.6
elliptical integral	13.10	moment	13.6
H_b wave height at breaking	7.10	M_{total} total moment	13.7
H_i initial wave height	6.1	n shoaling coefficient	7.3
H_O deep-water wave height	7.6	N newtons	
H_{rms} root mean square wave height	7.13	N parameter of solitary wave	Table 5.6
$H_{\sigma'}$ wave height that causes stress σ'	16.19	number of observations	6.2
		number of particles	10.1
		number of waves	16.47
I intensity function	6.16	p pressure	3.6
second moment of area	13.8	p_{crest} pressure at the crest	
i square root of -1	14.45	p_{trough} pressure at the tough	
J number of spectral estimates averaged	6.23	p_i pressure inside	15.2
k wave number	5.23	p_o pressure outside	15.2
modulus of elliptical integral	13.10	P probability	6.3
beam stiffness	14.5	$P_{(A>A')}$ probability that $A > A'$	6.4
k_f coefficient of static friction	15.1	P_d probability of dislogement or breakage	16.17
k_{Re} length coefficient	3.17	$P_{d,cumulative}$ cumulative probability of dislodgement or breakage	16.25
k_1 catching area coefficient	17.1		
k_2 energy-consumption coefficient	17.3	P_s probability of survival	16.22
$K(t)$ time-varying available energy	5.11	$P_{s,cumulative}$ cumulative probability of survival	16.23
K period parameter	11.10	$P_{(\sigma')}$ probability of having $\sigma_b = \sigma'$	16.33
K_R refraction coefficient		$P_{(\sigma>\sigma')}$ probability that $\sigma > \sigma'$	16.18
l_* mixing length	10.16	$P_{(\sigma<\sigma')}$ probability that $\sigma < \sigma'$	16.29
L wave length	4.1	$P_{(\sigma<\sigma'),cumulative}$ cumulative $P_{(\sigma<\sigma')}$	16.32
length of beam	13.11		
L_c critical crack length	15.7	$q \sqrt{\dfrac{3H}{4d}}\dfrac{x-Ct}{d}$	5.60
$L_{characteristic}$ characteristic length	3.19		
L_O deep-water wave length	7.18	$N/2$, half the number of observations	6.21
m mass	3.1	Q wave-energy flux	7.1
m_b mass of the beam	14.25	rate of particle release	10.30
m_e mass of the earth	8.1	food flux	17.2
end mass of the beam	14.26	r distance from release point	
m_m mass of the moon	8.1	R distance between centers of earth and moon	8.1
m_{max} maximum mass	17.4		

Symbol	Equation Where First Used
distance form x-axis	10.30
radius	13.1
R' see Figure 8.2	8.3
R_e radius of the earth	8.2
Re Reynolds number	3.19
Re_x local Reynolds number	9.18
Re_* roughness Reynolds number	9.20
s distance from sea bed	Fig. 5.1, Eq. 5.18
s_0 roughness height	9.24
s^+ dimensionless distance	9.23
S area	3.9
$\mathscr{S}$ standard deviation of σ_b	16.15
S_p area projected in direction of flow	11.1
S_{plan} planform area	11.6
$[S]$ sperm concentration	10.37
St Strouhal number	11.11
SWL still water level	
S_{xx} radiation stress in x-direction	7.22
S_{yy} radiation stress in y-direction	7.23
S_{xy} x-directed radiation stress in y-direction	7.24
t time	3.3
time increment	10.19
wall thickness	13.13
t_{stretch} time to extend algal stipe	16.13
T wave period	4.1
T_d damped period of oscillation	14.54
T_n natural period of oscillation	14.17
T_t time between high and low tide	16.44
u velocity along x-axis	3.1
$\bar{u}$ mean transport velocity	5.57
mean velocity	9.24
u_∞ mainstream velocity	9.1
u_{max} maximum water velocity	16.2
$\bar{u}_{\text{max}}$ mean maximum transport velocity	5.59
$u_{\text{max,rms}}$ maximum rms water velocity	16.42

Symbol	Equation Where First Used
u_0 initial water velocity	7.14
maximum water velocity	9.2
u_* friction velocity	9.17
u' fluctuation from mean velocity	9.11
u_{crest} speed at the crest	
u_{trough} speed at the trough	
U speed, $\sqrt{u^2 + w^2}$	4.3, 5.11
v velocity along the y-axis	5.1
V volume	3.6
w velocity along the z-axis	Fig. 4.5
w_{max} maximum water velocity	
w' fluctuation from mean velocity	9.13
W time to settlement	10.19
beam width	13.19
work of fracture	15.7
$\overline{W}$ mean time to settlement	10.27
x distance in x-direction	3.7
distance from free end of beam	14.20
x_{max} maximum excursion along a beach	7.16
y distance along y-axis	Fig. 3.1
length of crest	7.8
deflection of end of beam	13.11
y_0 maximum beam deflection	14.1
z distance from SWL	5.1
z_{max} run-up height	7.17
α cosine amplitude coefficient	6.6
dispersion coefficient	10.12
α_f cosine amplitude coefficient at frequency f	6.14
β sine amplitude coefficient	6.11
dispersion coefficient	10.12
beam shape parameter	17.27
β_f sine amplitude coefficient at frequency f	6.14
γ bore height coefficient	7.13
eddy viscosity	10.7
shear strain	12.4
γ_{lv} surface energy between water and air	15.3
Γ wave power spectrum	6.19
δ boundary-layer thickness	9.1, 9.6, 9.8
logarithmic decrement	14.56

Symbol	Equation Where First Used
Δ distance	10.19
ϵ engineer's strain	12.2
time	14.66
ϵ_t true strain	12.3
ζ damping ratio	14.38
η surface elevation	Fig. 4.3, Eq. 5.11
θ $kx - \omega t$	5.52
beach slope	5.65
angle; see Figure 8.2	8.3
angle of deflection	13.10
κ von Karmann's constant	9.24
K_x eddy diffusivity in x-direction	10.11
K_y eddy diffusivity in y-direction	10.11
K_z eddy diffusivity in z-direction	10.9
λ extension ratio	12.1
distance to center of pressure as a fraction of radius	17.18
μ dynamic viscosity	3.8
ν kinematic viscosity, μ/ρ	3.20
Poisson's ratio	12.8
ν_{yx}, ν_{zx} directional Poisson's ratio	12.7
ξ wave-height similarity index	7.21
curvature	13.3
Π_d exact probability of dislodgement	16.16
ρ density, usually of seawater	3.9
ρ_c density of coral skeleton	13.21
ρ_w density of seawater	13.21

Symbol	Equation Where First Used
σ^2 variance in distance to center of gravity of the group	10.10
σ standard deviation in distance	10.11
stress	12.5
σ_b breaking stress	13.16
$\bar{\sigma}_b$ mean breaking stress	16.15
σ_c critical buckling stress	13.13
σ_n nominal stress	13.2
$\bar{\sigma}$ mean stress	
τ shear stress	3.8
$\bar{\tau}_b$ mean breaking shear stress	17.21
τ_R Reynolds shear stress	9.16
$\tau_{R,bottom}$ Reynolds shear stress acting at the bottom	9.17
ϕ phase	6.8, 14.51
angle	8.3
argument in elliptical integral	13.10
Φ velocity potential function	5.5
χ structural exposure index	16.1
Ψ stream function	5.63
ω wave frequency	5.24
circular frequency	14.1
ω_n natural frequency	14.12
ω_d damped natural frequency	14.50
Ω alternate structural exposure index	16.27

LITERATURE CITED

~ ~ ~ ~ ~ ~ ~ ~ ~ ~ ~

Abbott, I. A., and G. J. Hollenberg. 1976. *Marine Algae of California.* Stanford, Calif.: Stanford University Press.

Abramowitz, M., and Irene Stegun. 1965. *Handbook of Mathematical Functions.* New York: Dover.

Airy, G. B., 1845. Tides and waves. *Encyc. Metrop. Art.* 192: 241–396.

Alexander, R. M. 1968. *Animal Mechanics.* Seattle: University of Washington Press.

———. 1983. *Animal Mechanics,* 2d ed. Oxford: Blackwell.

Anderson, A. L., D. J. Shirley, and L. H. Wilkins. 1972. An improved capacitive wave-staff for water surface wave measurements. *Proc. IEEE Oceans* 72: 483–486.

Antonia, R. A., and R. E. Luxton. 1971. The response of a turbulent boundary layer to a step change in surface roughness. Part 1: Smooth-to-rough. *J. Fluid Mech.* 48: 721–761.

———. 1972. The response of a turbulent boundary layer to a step change in surface roughness. Part 2: Rough-to-smooth. *J. Fluid Mech.* 53: 737–757.

Arnold, G. P., and D. Weihs. 1978. The hydrodynamics of rheotaxis in the plaice (*Pleuronectes platessa* L.). *J. Exp. Biol.* 75: 147–170.

Baier, R. E., E. G. Shafrin, and W. A. Zisman. 1968. Adhesion: Mechanisms that assist or impede it. *Science* 162: 1360–1368.

Barnes, H., and M. Barnes. 1959. Studies on the metabolism of cirripedes: The relation between body weight, oxygen uptake and species habitat. *Veroff. Inst. Meeresforsch. Bremerhaven* 6: 515–523.

Barnes, H., and E. S. Reese. 1958. Feeding in the pedunculate cirripede *Pollicipes polymerus* J. B. Sowerby. *Proc. Zool. Soc. London* 132(4): 569–585.

Basco, D. R. 1982. Surf zone currents. Vol. I, State of knowledge. U.S. Army Corps of Engineers Misc. Report, MR 82–7(I).

Bascom, W. 1980. *Waves and Beaches.* Garden City, N.Y.: Anchor Press/Doubleday.

Batchelor, G. K. 1967. *An Introduction to Fluid Mechanics.* London: Cambridge University Press.

Battjes, J. A. 1974. A computation of set-up, longshore currents, run-up and overtopping due to wind-generated waves. Ph.D. dissertation, Delft University of Technology, The Netherlands.

Battjes, J. A., and T. Sakai. 1981. Velocity field in a steady breaker. *J. Fluid Mech.* 111: 421–437.

Beckwith, T. G., N. L. Buck, and R. D. Marangoni. 1982. *Mechanical Measurement,* 3d ed. Reading, Mass: Addison-Wesley.

Bendat, J. S., and A. G. Picrsol. 1971. *Random Data: Analysis and Measurement Procedures,* 2d ed. New York: Wiley.

Berg, H. C. 1983. *Random Walks in Biology.* Princeton, N.J.: Princeton University Press.

Bird, R. B., W. E. Stewart, and E. N. Lightfoot. 1960. *Transport Phenomena.* New York: Wiley.

Blackmore, P. A., and P. J. Hewson. 1984. Experiments on full-scale wave pressures. *Coastal Engineering* 8: 331–346.

Blevins, R. D. 1977. *Flow-Induced Vibration.* New York: Van Nostrand-Reinhold.

Box, G. P., and G. M. Jenkins. 1970. *Time Series Analysis—Forecasting and Control.* San Francisco: Holden-Day.

Branch, G. M., and A. C. Marsh. 1978. Tenacity and shell shape in six *Patella* species: Adaptive features. *J. Exp. Mar. Biol. Ecol.* 34: 111–130.

Broadwell, J. E., and R. E. Breidenthal. 1982. A simple model of mixing and chemical reaction in a turbulent shear layer. *J. Fluid Mech.* 125: 397–410.

Cantwell, B. J. 1981. Organized motion in turbulent flow. *Ann. Rev. Fluid. Mech.* 13: 457–515.

Carefoot, T. 1977. *Pacific Seashores.* Seattle: University of Washington Press.

Carstens. T. 1968. Wave forces on boundaries and submerged bodies. *Sarsia* 34: 37–60.

Charters, A. C., M. Neushul, and C. Barilotti.

1969. The functional morphology of *Eisenia arborea*. Proc. Intl. Seaweed Symp. 6: 89–105.

Chatfield, C. 1984. *Introduction to Time Series Analysis*. New York: Chapman and Hall.

Clancy, E. P. 1968. *The Tides*. Garden City, N.Y.: Doubleday.

Cokelet, E. D. 1977. Breaking waves. *Nature* 267: 769–774.

————. 1979. Breaking waves—the plunging jet and interior flow-field. In T. L. Shaw, *Mechanics of Wave-Induced Forces on Cylinders*, pp. 287–301. San Francisco: Pitman.

Connell, J. H. 1961a. Effects of competition, predation by *Thais lapillus* and other factors on natural populations of the barnacle *Balanus balanoides*. *Ecol. Monogr.* 31: 61–104.

————. 1961b. The influence of interspecific competition and other factors on the distribution of the barncale *Chthamalus stellatus*. *Ecology* 42: 710–723.

Crank, J. 1975. *The Mathematics of Diffusion*, 2d ed. Oxford: Oxford University Press.

Crapper, G. D. 1957. An exact solution for progressive capillary waves of arbitrary amplitude. *J. Fluid Mech.* 2: 532–540.

Crisp, D. J. 1960. Mobility of barnacles. *Nature* 188: 1208–1209.

————. 1974. Factors influencing the settlement of marine invertebrate larvae. In P. T. Grant and A. M. Mackie, eds., *Chemoreception in Marine Organisms*, pp. 177–264. New York: Academic Press.

Csanady, G. S. 1973. *Turbulent Diffusion in the Environment*. Boston: D. Reidel.

Currey, J. D. 1980. Mechanical properties of mollusc shell. *Symp. Soc. Exp. Biol.* 34: 75–78.

Darwin, C. R. 1952. Note on hydrodynamics. *Proc. Cambridge Phil. Soc.* 49: 342–354.

Dean, R. G. 1974. Evaluation and development of water wave theories for engineering application. Special Report No. 1, prepared for U.S. Army Corps of Eng. Coast. Eng. Res. Center. Washington, D.C.: U.S. Government Printing Office.

Defant, A. 1958. *Ebb and Flow*. Ann Arbor: University of Michigan Press.

Delf, E. M. 1932. Experiments with the stipes of *Fucus* and *Laminaria*. *J. Exp. Biol.* 9: 300–313.

Denman, K. L. 1975. Spectral analysis: A summary of the theory and techniques. Canadian Fisheries and Marine Service Tech. Report No. 539. Dartmouth, Nova Scotia.

Denny, M. W. 1981. A quantitative model for the adhesive locomotion of the terrestrial slug, *Ariolimax columbianus*. *J. Exp. Biol.* 91: 195–217.

————. 1982. Forces on intertidal organisms due to breaking ocean waves: Design and application of a telemetry system. *Limol. Oceanogr* 27(1): 178–183.

————. 1983a. A simple device for recording the maximum force exerted on intertidal organisms. *Limnol. Oceanogr.* 28(6): 1269–1274.

————. 1983b. Molecular biomechanics of molluscan mucous secretions. In P. W. Hochachka, ed., *Metabolic Biochemistry and Molecular Biomechanics*, pp. 432–465. Vol. 1 of K. M. Wilbur, ed., *The Mollusca*. New York: Academic Press.

————. 1984. Mechanical properties of pedal mucus and their consequences for gastropod structure and performance. *Amer. Zool.* 24: 23–36.

————. 1985a. Wave forces on intertidal organisms: A case study. *Limnol. Oceanogr.* 30(6): 1171–1187.

————. 1985b. Water motion. In M. Littler and D. Littler, eds., *Handbook of Phycological Methods—Phycology*, pp. 8–32. New York: Academic Press.

Denny, M. W., and J. M. Gosline. 1980. The physical properties of the pedal mucus of the terrestrial slug, *Ariolimax columbianus*. *J. Exp. Biol.* 88: 375–393.

Denny, M. W., T. L. Daniel, and M.A.R. Koehl. 1985. Mechanical limits to size in wave-swept organisms. *Ecol. Monogr.* 55: 69–102.

Donelan, M. A., and J. Motycka. 1978. Miniature drag sphere velocity probe. *Rev. Sci. Instrum.* 49: 298–304.

Eckart, C. 1951. Surface waves on water of variable depth. Ref. No. 51-12. La Jolla, Calif.: Scripps Institute of Oceanography.

————. 1952. The propagation of waves from deep to shallow water. In *Gravity Waves*, Circular No. 521: 165–73, National Bureau of Standards.

Eckman, J. E. 1982. Hydrodynamic effects exerted by animal tubes and marsh grasses and their importance to the ecology of soft-substratum marine benthos. Ph.D. dissertation. University of Washington, Seattle.

———. 1983. Hydrodynamic processes affecting benthic recruitment. *Limnol. Oceanogr.* 28(2): 241–257.

Eckman, J. E., A.R.M. Nowell, and P. A. Jumars. 1981. Sediment destabilization by animal tubes. *J. Mar. Res.* 39: 361–372.

Emson. R. H., and R. J. Faller-Fritsch. 1976. An experimental investigation into the effect of crevice availability on abundance and size structure in a population of *Littorina radis* (Maton). Gastropoda: Prosobranchia. *J. Exp. Mar. Biol. Ecol.* 23: 285–297.

Fenton, J. D. 1979. A high-order cnoidal wave theory. *J. Fluid Mech.* 94: 129–161.

———. 1985. A fifth-order Stokes theory for steady waves. *J. Waterway, Port, Coast, Ocean. Engr.* 111: 216–233.

Ferry. J. D. 1980. *Viscoelastic Properties of Polymers*, 3d ed. New York: Wiley.

Flick, R. E., R. L. Lowe, M. H. Freilich, and J. C. Boylls. 1979. Coastal and laboratory wavestaff system. *IEEE Oceans '79*, p. 63.

Flory, P. J. 1953. *Principles of Polymer Chemistry*. Ithaca, N.Y.: Cornell University Press.

———. 1969. *Statistical Mechanics of Chain Molecules*. New York: Wiley.

Fox, R. W., and A. T. McDonald. 1978. *Introduction to Fluid Mechanics*, 2d ed. New York: Wiley.

Gaines, S., S. Brown, and J. Roughgarden. 1985. Spatial variation in larval concentration as a cause of spatial variation in settlement for the barnacle, *Balanus glandula*. *Oecologia* 67: 267–272.

Galvin, C. J. 1972. Wave breaking in shallow water. In R. E. Meyer, ed., *Waves on Beaches and Resulting Sediment Transport*, pp. 413–455. New York: Academic Press.

Gerard, V. A. 1982. *In situ* water motion and nutrient uptake by the giant kelp *Macrocystis pyrifera*. *Mar. Biol.* 69: 51–54.

Gere, J. M., and S. P. Timoshenko. 1984. *Mechanics of Materials*. Monterey, Calif.: Brooks/Cole Engineering Division.

Godin. G. 1972. *The Analysis of Tides*. Toronto: University of Toronto Press.

Gordon, J. E. 1978. *Structures*. New York: Plenum Publishing company.

Gosline. J. M. 1971. Connective tissue mechanics of *Metridium senile*. Vol. II: Viscoelastic properties and macromolecular model. *J. Exp. Biol.* 55: 775–795.

———. 1980. The elastic properties of rubber-like proteins and highly extensible tissues. *Symp. Soc. Exp. Biol.* 34: 331–358.

Grace. J. 1977. *Plant Response to Wind*. London: Academic Press.

Grant. W. D., and O. S. Madsen. 1979. Combined wave and current interaction with a rough bottom. *J. Geophys. Res.* 84: 1797–1808.

———. 1982. Movable bed roughness in unsteady oscillatory flow. *J. Geophys. Res.* 87: 469–481.

———. 1986. The continental-shelf bottom boundary layer. *Ann. Rev. Fluid Mech.* 18: 265–305.

Grenon. J. F., and G. Walker. 1980. Biomechanical and rheological properties of the pedal mucus of the limpet, *Patella vulgata* L. *Comp. Biochem. Physiol.* 66: 451–458.

Griffiths. C. L., and J. A. King. 1979. Some relationships between size, food availability and energy balance in the ribbed mussel *Aulacomya ater. Mar. Biol.* (Berlin) 51: 141–149

Gust, G. 1982. Tools for oceanic small-scale, high-frequency flows: Metal-clad hot wires. *J. Geophys. Res.* 87: 447–455.

Guza, R. T., and R. E. Davis. 1974. Excitation of edge waves by waves incident on a beach. *J. Geophys. Res.* 79: 1285–1291.

Hoerner, S. F. 1965. *Fluid-Dynamic Drag*. Bricktown, N.J.: Hoerner Fluid Dynamics.

Holman, R. A. 1983. Edge waves and the configuration of the shoreline. In P. D. Komar, ed., *CRC Handbook of Coastal Processes and Erosion*. Boca Raton, Florida: CRC Press.

Horowitz. P., and W. Hill. 1980. *The Art of Electronics*. Cambridge, Eng.: Cambridge University Press.

Idel'chik, I. E. 1966. *Handbook of Hydraulic Resistance*. Israel Program for Scientific Translations, Jerusalem.

Inglis, C. E. 1913. Stresses in a plate due to the presence of cracks and sharp corners. *Trans. Ind. Naval Arch.* 55: 219–30.

Jackson, J.B.C. 1977. Competition on marine hard substrata: The adaptive significance of solitary and colonial strategies. *Am. Nat.* 111: 743–767.

Jenkins, G. M., and D. G. Watts. 1968. *Spectral Analysis and Its Applications*. San Francisco: Holden Day.

Jones, W. E., and A. Demetropoulos. 1968. Exposure to wave action: Measurements of an important ecological parameter on rocky shores on Anglesey. *J. Exp. Mar. Biol. Ecol.* 2: 46–63.

Jonsson, I. G. 1980. A new approach to oscillatory rough turbulent boundary layers. *Ocean. Eng.* 7: 109–152.

Jumars, P. A., and A.R.M. Nowell. 1984. Fluid and sediment dynamic effects on marine benthic community structure. *Am. Zool.* 24: 45–55.

Ker. R. F. 1977. Some structural and mechanical properties of locust and beetle cuticle. Ph.D. thesis, Oxford University.

Kinsman, B. 1965. *Wind Waves.* Englewood Cliffs, N.J.: Prentice-Hall.

Koehl, M.A.R. 1976. Mechanical design in sea anemones. In *Coelenterate Ecology and Behavior*, ed. G. O. Mackie, pp. 23–31. New York: Plenum.

———. 1977a. Effects of sea anemones on the flow forces they encounter. *J. Exp. Biol.* 69: 87–105.

———. 1977b. Mechanical organization of cantilever-like sessile organisms: Sea anemones. *J. Exp. Biiol.* 69: 127–142.

———. 1977c. Mechanical diversity of connective tissue of the body wall of sea anemones. *J. Exp. Biol.* 69: 107–125.

———. 1977d. Water flow and the morphology of zooanthid colonies. Proc. 3d Int. Coral Reef Symp. I: *Biology*: 437–444.

Koehl, M.A.R., and R. S. Alberte. In press. Flow, flapping and photosynthesis: Functional consequences of undulate blade morphology. *Mar. Biol.*

Koehl. M.A.R., and S. A. Wainwright. 1977. Mechanical adaptations of a giant kelp. *Limnol. Oceanog.* 22(6): 1067–1071.

Komar. P. D. 1976. *Beach Processes and Sedimentation.* Englewood Cliffs, N.J.: Prentice-Hall.

Komar, P. D., ed. 1983. *CRC Handbook of Coastal Processes and Erosion.* Boca Raton, Florida: CRC Press.

Korteweg, D. J., and G. deVries. 1895. On the change of form of long waves advancing in a rectangular channel, and on a new type of long stationary waves. *Phil. Mag.*, series 5, 39: 422–443.

LaBarbera, M., and S. Vogel. 1976. An inexpensive thermistor flowmeter for aquatic biology. *Limnol. Oceanog.* 21: 750–756.

Laitone, E. V. 1959. Water Waves IV: *Shallow Water Waves.* Inst. Eng. Res. Tech. Report. No. 82-11, University of California, Berkeley.

Lamb, H. 1945. *Hydrodynamics.* New York: Dover Publications.

Lienhard, J. H. 1966. Synopsis of lift, drag, and vortex frequency data for rigid circular cylinders. Res. Div. Bulletin No. 300, Washington State University College of Engineering.

Longuet-Higgins, M. S. 1952. On the statistical distribution of the heights of sea waves. *J. Mar. Res.* 11: 245–266.

———. 1953. Mass transport in water waves. *Proc. Roy. Soc. Lond.* A245: 535–581.

———. 1983. Wave set-up, percolation and undertow in the surf zone. *Proc. Roy. Soc. Lond.* A309: 283–294.

Longuet-Higgins, M. S., and R. W. Stewart. 1962. Radiation stress and mass transport in gravity waves with applications to "surf-beats." *J. Fluid Mech.* 13: 481–504.

———. 1963. A note on wave set-up. *J. Mar. Res.* 21: 4–10.

———. 1964. Radiation stress in water waves: A physical discussion with applications. *Deep-Sea Res.* 11: 529–563.

Longuet-Higgins, M. S., and J. S. Turner. 1974. An "entraining plume" model of a spilling breaker. *J. Fluid Mech.* 63: 1–20.

Lugt, H. J. 1983. *Vortex Flow in Nature and Technology.* New York: Wiley.

McGowan, J. 1891. On the solitary wave. *Phil. Mag.* series 5, 32: 45–58.

———. 1894. On the highest wave of permanent type. *Phil. Mag.*, series 5, 38: 351–357.

Massey, B. S. 1983. *Mechanics of Fluids*, 5th ed. New York: Van Nostrand-Reinhold.

Middleton, G. V., and J. B. Southard. 1984. *Mechanics of Sediment Movement.* Tulsa, Okla.: Society of Economic Paleontologists and Mineralogists.

Miller, S. L. 1974a. Adaptive design of locomotion and foot form in prosobranch gastropods. *J. Exp. Mar. Biol. Ecol.* 14: 99–156.

———. 1974b. The classification, taxonomic distribution, and evolution of locomotor types among prosobranch gastropods. *Proc Malac. Soc. Lond.* 14: 233–272.

Morison, J. R., M. P. O'Brien, J. W. Johnson, and S. A. Schaaf. 1950. The forces exerted by surface waves on piles. *Petroleum Trans. AIME* 189: 149–157.

Morris, R. H., D. P. Abbott, and E. C. Haderlie.

1980. *Intertidal Invertebrates of California.* Stanford, Calif.: Stanford University Press.

Munk, W. H. 1944. Proposed uniform procedure for observing waves and interpreting instrument records. Wave Project, Scripps Institute of Oceanography, La Jolla, Calif.

———. 1949. The solitary wave theory and its application to surf problems. *Annals N.Y. Acad. Sci.* 51: 376–424.

Muschenheim, D. K., J. Grant, and E. L. Mills. 1986. Flumes for benthis ecologists: Theory, construction and practice. *Mar. Ecol.,* Progr. Ser., 28: 185–196.

Muus, B. J. 1968. A field method for measuring "exposure" by means of plaster balls. *Sarsia* 34: 61–68.

Nachtigall, W. 1974. *Biological Mechanisms of Attachment.* New York: Springer-Verlag.

Nowell, A.R.M., and M. Church. 1979. Turbulent flow in a depth-limited boundary layer. *J. Geophys. Res.* 84: 4816–4824.

Nowell, A.R.M., and P. A. Jumars. 1984. Flow environments of aquatic benthos. *Ann. Rev. Ecol. Syst.* 15: 303–328.

Okubo, A. 1971. Oceanic diffusion diagrams. *Deep Sea Res.* 18: 789–802.

———. 1980. *Diffusion and Ecological Problems: Mathematical Models.* Biomathematics Series, Vol. 10. New York: Springer-Verlag.

Paine, R. T. 1969. A note on trophic complexity and community stability. *Am. Nat.* 103: 91–93.

———. 1976. Size-limited predation: An observational and experimental approach with the *Mytilus-Pisaster* interaction. *Ecology* 57: 858–873.

———. 1977. Controlled manipulations in the marine intertidal zone, and their contributions to ecological theory. In *The Changing Scene in Natural Sciences,* pp. 245–270. Acad. Nat. Sciences (USA), Special Pub. No. 12.

Panofsky, H. A. 1967. A survey of current thought on wind properties relevant for diffusion in the lowest 100 m. In *Symposium on Atmospheric Turbulence and Diffusion,* pp. 47–58. Albuquerque, N. M.: Sandia Laboratories.

Pasquill, F., and F. B. Smith. 1983. *Atmospheric Diffusion,* 3d ed. New York: Wiley.

Patterson, M. R. 1984. Patterns of whole colony prey capture in the octocoral *Alcyonium siderium. Biol. Bull.* 167: 613–629.

Pennington, J. T. 1985. The ecology of fertilization of echinoid eggs: The consequences of sperm dilution, adult aggregation, and synchronous spawning. *Biol. Bull.* 169: 417–430.

Peregrine, D. H. 1983. Breaking waves on beaches. *Ann. Rev. Fluid Mech.* 15: 149–178.

Peregrine, D. H., and I. A. Svendsen. 1978. Spilling breakers, bores and hydraulic jumps. *Proc. 16th Coast. Eng. Conf.:* 540–550.

Pierson, W. J., G. Newmann, and R. W. James. 1955. *Practical Methods for Observing and Forecasting Ocean Waves.* U.S. Navy Hydrographic Office Pub. No. 603.

Price, H. A. 1980. Seasonal variation in the strength of byssal attachment of the common mussel, *Mytilus edulis* L. *J. Mar. Biol. Assoc. U.K.* 60: 1035–1037.

Purcell, E. M. 1977. Life at low Reynolds number. *Am. J. Physics* 45(1): 3–11.

Quinn, J. F. 1979. Disturbance, predation and diversity in the rocky intertidal zone. Ph.D. dissertation, University of Washington, Seattle.

Ricketts, E. F., J. Calvin, and J. W. Hedgpeth. 1968. *Between Pacific Tides,* 4th ed. Stanford, Calif.: Stanford University Press.

Rienecker, N. M., and J. D. Fenton. 1981. A Fourier approximation method for steady water waves. *J. Fluid Mech.* 104: 119–137.

Roark, R. J., and W. C. Young. 1975. *Formulas for Stress and Strain.* New York: McGraw-Hill.

Rouville, M. A., P. Besson, and P. Petry. 1938. Etat actual des études internationales sur efforts dus aux lames. *Ann. Ponts Chauss.* 108: 5–113.

Sabersky, R. H., A. J. Acosta, and E. G. Hauptmann. 1971. *Fluid Flow,* 2d ed. New York: Collier Macmillan.

Santschi, P. H., P. Bower, U. P. Nyffeler, A. Azevedo, and W. S. Broecker. 1983. Estimates of the resistance to chemical transport posed by the deep-sea boundary layer. *Limnol. Oceanogr.* 28: 899–912.

Sarpkaya, T. 1976a. In-line and transverse forces on smooth and sand-roughened cylinders in oscillatory flow at high Reynolds number. Report No. NPS-69SL76062, Naval Postgraduate School, Monterey, Calif.

———. 1976b. Forces on cylinders near a plane boundary in a sinusoidally oscillating fluid. *Trans. A.S.M.E. J. Fluid Eng.* 98: 499–505.

Sarpkaya, T., and M. Storm. 1985. In-line force on a cylinder translating in oscillatory flow. *Applied Ocean Res.* 7: 188–196.

Sarpkaya, T., and M. Isaacson. 1981. *Mechanics of Wave Forces on Offshore Structures.* New York: Van Nostrand-Reinhold Co.

Schlichting, H. 1979. *Boundary Layer Theory,* 7th ed. New York: McGraw-Hill.

Schmidt-Nielsen, K. 1984. *Scaling: Why Is Animal Size So Important?* Cambridge, Eng.: Cambridge University Press.

Sebens, K. P. 1979. The energetics of asexual reproduction and colony formation in benthic marine invertebrates. *Amer. Zool.* 19: 683–697.

———. 1980. The control of asexual reproduction and indeterminate body size in the sea anemone *Anthopleura elegantissima* (Brandt). *Biol. Bull.* 158: 370–382.

———. 1981. The allometry of feeding energetics, and body size in three sea anemone species. *Biol. Bull.* 161: 152–171.

———. 1982. The limits to indeterminate growth: An optimal size model applied to passive suspension feeders. *Ecology* 63(1): 209–222.

Shadwick, R. E., and J. M. Gosline. 1983. Molecular biomechanics of protein rubbers in molluscs. In P. W. Hochachka, ed., *Metabolic Biochemistry and Molecular Biomechanics,* pp. 399–430. Vol. 1 of K. M. Wilbur, ed., *The Mollusca.* New York: Academic Press.

Shanks, A. L. 1983. Surface slicks associated with tidally forced internal waves may transport pelagic larvae of benthic invertebrates and fishes shoreward. *Mar. Ecol. Prog. Ser.* 13: 311–315.

———. 1986. Tidal periodicity in the daily settlement of intertidal barnacle larvae and an hypothesized mechanism for the cross-shelf transport of cyprids. *Biol. Bull.* 170: 429–440.

Shanks, A. L., and W. G. Wright. 1986. Adding teeth to wave action: The destructive effects of wave-borne rocks on intertidal organisms. *Oecologia* 69: 420–428.

Shapiro, A. H. 1961. *Shape and Flow.* Garden City, N.Y.: Doubleday.

Shepard, F. P., and D. L. Inman. 1950. Nearshore circulation related to bottom topography and wave refraction. *Trans. Am. Geophys. Union.* 31, No. 4: 555–565.

Sleath, J.F.A. 1983. *Sea Bed Mechanics.* New York: Wiley.

Sousa, W. P. 1979. Experimental investigations of disturbance and ecological succession in a rocky intertidal algal community. *Ecol. Monogr.* 49: 227–254.

Steinbeck, J. 1951. *The Log from the Sea of Cortez.* New York: Viking Press.

Stephenson, T. A., and A. Stephenson. 1949. The universal features of zonation between tide marks on rocky coasts. *J. Ecol.* 37: 289–305.

Stokes, G. G. 1847. On the theory of oscillatory waves. *Trans. Cambridge Phil. Soc.* 8: 441.

Streeter, V. L., and E. B. Wylie. 1979. *Fluid Mechanics,* 7th ed. New York: McGraw-Hill.

Sutton, O. G. 1953. *Micrometeorology.* New York: McGraw-Hill.

Svoboda, A. 1970. Simulation of oscillatory water movement in the laboratory for cultivation of shallow water sedentary organisms. *Helgolander Wiss. Meeresunters.* 20: 676–684.

Thomson, W. T. 1981. *Theory of Vibration with Applications,* 2d ed. Englewood Cliffs, N.J.: Prentice-Hall.

Thornton, E. B., J. J. Galvin, F. L. Bub, and D. P. Richardson. 1976. Kinematics of breaking waves. *Proc. 15th Coastal Eng. Conf. ASCE,* pp. 461–476. Honolulu, Hawaii.

Thornton, E. B., and R. T. Guza. 1982. Energy saturation and phase speeds measured on a natural beach. *J. Geophys. Res.* 87, No. C12: 9499–9508.

———. 1983. Transformation of wave height distributions. *J. Geophys. Res.* 88, No. C10: 5925–5938.

Tritton, D. J. 1977. *Physical Fluid Dynamics.* New York: Van Nostrand-Reinhold.

Tunnicliffe, V. 1981. Breakage and propagation of the stony coral *Acropora cervicornis. Proc. Natl. Acad. Sci. USA* 78(4): 2427–2431.

Tyler, A., A. Monroy, and C. B. Metz. 1956. Fertilization of fertilized sea urchin eggs. *Biol. Bull.* 110: 184–195.

U.S. Army Corps of Engineers. 1984. *Shore Protection Manual,* 4th ed. Washington, D.C.: U.S. Government Printing Office.

U.S. Army Corps of Engineers. 1985. Coastal Data Information Program Monthly Report: April 1985. Washington, D.C.: U.S. Government Printing Office.

U.S. Department of Commerce. 1980. *Statis-*

tical Abstract of the United States, 101st ed. Washington, D.C.: U.S. Government Printing Office.

Vahl, O. 1973. Pumping and oxygen consumption of *Mytilus edulis* of different sizes. *Ophelia* 12: 45–52.

Van Dorn, W. G. 1976. Set-up and run-up in shoaling breakers. *Proc. 15th Coastal Eng. Conf. ASCE,* pp. 738–751. Honolulu, Hawaii.

Verley, R.L.P., and G. Moe. 1979. The forces on a cylinder oscillating in a current. River and Harbor Laboratory, The Norwegian Institute of Technology, Report No. 5TF60 A 79061.

Vogel, H., G. Czihak, P. Chang, and W. Wolf. 1982. Fertilization kinetics of sea urchin' eggs. *Math. Biosci.* 58: 189–216.

Vogel, S. 1974. Current-induced flow through the sponge. *Halichondria Biol. Bull.* 147: 443–456.

———. 1977. Current-induced flow through living sponges in nature. *Proc. Natl. Acad. Sci. USA* 74(5): 2069–2071.

———. 1978. Organisms that capture currents. *Sci. Am.* 239: 128–139.

———. 1981. *Life in Moving Fluids.* Boston: Willard Grant Press.

Vogel, S., and M. LaBarbera. 1978. Simple flow tanks for research and teaching. *Bio-Science* 28(10): 638–643.

Vosburgh, F. 1982. *Acropora reticulata*: Structure, mechanics and ecology of a reef coral. *Proc. R. Soc. Lond.* 214: 481–499.

Wainwright, S. A., W. D. Biggs, J. D. Currey, and J. M. Gosline. 1976. *Mechanical Design in Organisms.* London: Edward Arnold.

Waite, J. H. 1983. Adhesion in byssally attached bivalves. *Biol. Rev.* 58(2): 209–231.

Wake, W. C. 1982. *Adhesion and the Formulation of Adhesives,* 2d ed. New York: Applied Science Publishers.

Walker, G. 1981. The adhesion of barnacles. *J. Adhesion* 12: 51–58.

Webb, P. W. 1975. Hydrodynamics and energetics of fish propulsion. Ottawa *Bull. Fish. Res. Board Canada,* No. 190.

Wertheim, A. 1985. *The Intertidal Wilderness.* San Francisco: Sierra Club Books.

Wheeler, W. N. 1980. The effects of boundary layer transport on the fixation of carbon by the giant kelp *Macrocystis pyrifera. Mar. Biol.* 56: 103–110.

Wiegel, R. L. 1964. *Oceanographical Engineering.* London: Prentice-Hall.

Wright, W. G. 1978. Aspects of the ecology and behavior of the owl limpet, *Lottia gigantea,* Sowerby, 1834. *West. Soc. Malac. Ann. Rep.* 11: 7.

Yonge, C. M. 1949. *The Sea Shore.* London: Collins.

INDEX

~ ~ ~ ~ ~ ~ ~ ~ ~ ~ ~ ~

Italicized page numbers indicate a figure or table.